Lecture Notes of the Institute for Computer Sciences, Social Informatics and Telecommunications Engineering 404

More information about this series at https://link.springer.com/bookseries/8197

Qianchuan Zhao · Li Xia (Eds.)

Performance Evaluation Methodologies and Tools

14th EAI International Conference, VALUETOOLS 2021
Virtual Event, October 30–31, 2021
Proceedings

 Springer

Editors
Qianchuan Zhao
Tsinghua University
Beijing, China

Li Xia
Sun Yat-sen University
Guangzhou, China

ISSN 1867-8211 ISSN 1867-822X (electronic)
Lecture Notes of the Institute for Computer Sciences, Social Informatics
and Telecommunications Engineering
ISBN 978-3-030-92510-9 ISBN 978-3-030-92511-6 (eBook)
https://doi.org/10.1007/978-3-030-92511-6

This Springer imprint is published by the registered company Springer Nature Switzerland AG
The registered company address is: Gewerbestrasse 11, 6330 Cham, Switzerland

Preface

We are delighted to introduce the proceedings of the fourteenth edition of the European Alliance for Innovation (EAI) International Conference on Performance Evaluation Methodologies and Tools (VALUETOOLS 2021), which was scheduled to be held during October 30–31, 2021, in Guangzhou, China. Due to the safety concerns and travel restrictions caused by COVID-19, EAI VALUETOOLS 2021 took place online in a live stream. This conference series has brought together researchers, developers, and practitioners around the world who are from different communities including computer science, networks and telecommunications, operations research, optimization, control theory, and manufacturing. The focus of VALUETOOLS 2021 was methodologies and practices in modeling, performance evaluation, and optimization of complex systems.

The technical program of VALUETOOLS 2021 consisted of 16 full papers and 13 abstract papers. Aside from the high-quality technical paper presentations, the technical program also featured three keynote speeches. The three keynote speakers were Xi-Ren Cao from The Hong Kong University of Science and Technology, China, with a talk entitled "Optimization of average-cost nonhomogeneous Markov chains"; Christos G. Cassandras from Boston University, USA, with a talk entitled "Bridging the gap between optimal and real-time safe control: Making autonomous vehicles a reality"; and Eitan Altman from Inria, France, with a talk entitled "Optimal control and game theory applied to epidemics".

Coordination with the steering chairs, Imrich Chlamtac, Xi-Ren Cao, Christos G. Cassandras, and Qianchuan Zhao, was essential for the success of the conference. We sincerely appreciate their constant support and guidance. It was also a great pleasure to work with such an excellent Organizing Committee team, led by Li Xia and Xianping Guo; we are thankful for their hard work in organizing and supporting the conference. In particular, we are grateful to the Technical Program Committee, who completed the peer-review process for technical papers and helped to put together a high-quality technical program. We are also grateful to the Conference Manager, Natasha Onofrei, for her support and all the authors who submitted their papers to the VALUETOOLS 2021 conference.

The VALUETOOLS conference was initiated in 2006 and aims at promoting the research on performance evaluation methodologies and tools for various engineering or social complex systems. We strongly believe that the VALUETOOLS conference provides a good forum for all researchers, developers, and practitioners to discuss all science and technology aspects that are relevant to performance evaluation, analysis, and optimization of complex systems. We also expect that the future conferences will be as successful and stimulating as VALUETOOLS 2021, as indicated by the contributions presented in this volume.

November 2021

Qianchuan Zhao
Li Xia
Qing-Shan Jia

Organization

Steering Committee

Imrich Chlamtac	University of Trento, Italy
Xi-Ren Cao	The Hong Kong University of Science and Technology, Hong Kong, China
Christos G. Cassandras	Boston University, USA
Qianchuan Zhao	Tsinghua University, China

Organizing Committee

General Chairs

Qianchuan Zhao	Tsinghua University, China
Li Xia	Sun Yat-sen University, China

Technical Program Committee

Qing-Shan Jia	Tsinghua University, China

Sponsorship and Exhibit Chair

Zhanbo Xu	Xi'an Jiaotong University, China

Local Chair

Li Xia	Sun Yat-sen University, China
Xianping Guo	Sun Yat-sen University, China

Workshops Chair

Quan-Lin Li	Beijing University of Technology, China

Publicity and Social Media Chair

Yongcai Wang	Renmin University, China

Publications Chair

Xi Chen	Tsinghua University, China

Web Chair

Yilin Mo   Tsinghua University, China

Industrial Chair

Huiying Xu   Huawei Technologies, China

Technical Program Committee

Konstantin Avratchenkov	Inria, France
Marko Boon	Eindhoven University of Technology, The Netherlands
Vivek Borkar	IIT Bombay, India
Anyue Chen	Southern University of Science and Technology, China
Wanyang Dai	Nanjing University, China
Dieter Fiems	Ghent University, Belgium
Jean-Michel Fourneau	University of Versailles, France
Giuliana Franceschini	University of Eastern Piedmont, Italy
Reinhard German	Friedrich-Alexander-Universität Erlangen-Nürnberg, Germany
Pengfei Guo	City University of Hong Kong, Hong Kong, China
Xianping Guo	Sun Yat-sen University, China
Yongjiang Guo	Beijing University of Posts and Communications, China
Yezekael Hayel	Universite de Avignon, France
Qi-Ming He	Waterloo University, Canada
Atsushi Inoie	Kanagawa Institute of Technology, Japan
Alain Jean-Marie	Inria, France
Yoav Kerner	Ben Gurion University of the Negev, Italy
William Knottenbelt	Imperial College London, UK
Pierre L'Ecuyer	University of Montreal, Canada
Lasse Leskela	Aalto University, Finland
Bin Liu	Anhui Jianzhu University, China
Yuanyuan Liu	Central South University, China
Zaiming Liu	Central South University, China
Catalina M. Llado	The University of the Balearic Islands, Spain
Michel Mandjes	Universiteit van Amsterdam, The Netherlands
Andrea Marin	Ca' Foscari University of Venice, Italy
Jose Merseguer	University of Zaragoza, Spain
Jayakrishnan Nair	IIT Bombay, India
Francesco de Pellegrini	University of Avignon, France

Contents

A TTL-based Approach for Content Placement in Edge Networks

Nitish K. Panigrahy[1(✉)], Jian Li[2], Faheem Zafari[3], Don Towsley[1], and Paul Yu[4]

[1] University of Massachusetts, Amherst, MA 01003, USA
{nitish,towsley}@cs.umass.edu
[2] Binghamton University, SUNY, Binghamton, NY 13902, USA
lij@binghamton.edu
[3] Imperial College London, London SW72BT, UK
faheem16@imperial.ac.uk
[4] U.S. Army Research Laboratory, Adelphi, MD 20783, USA
paul.l.yu.civ@mail.mil

Abstract. Edge networks are promising to provide better services to users by provisioning computing and storage resources at the edge of networks. However, due to the uncertainty and diversity of user interests, content popularity, distributed network structure, cache sizes, it is challenging to decide where to place the content, and how long it should be cached. In this paper, we study the utility optimization of content placement at edge networks through timer-based (TTL) policies. We propose provably optimal distributed algorithms that operate at each network cache to maximize the overall network utility. Our TTL-based optimization model provides theoretical answers to how long each content must be cached, and where it should be placed in the edge network. Extensive evaluations show that our algorithm outperforms path replication with conventional caching algorithms over some network topologies.

Keywords: TTL cache · Utility maximization · Edge network

1 Introduction

Content distribution has become a dominant application in today's Internet. Much of these contents are delivered by Content Distribution Networks (CDNs), which are provided by Akamai, Amazon, etc. [18]. There usually exists a stringent requirement on the latency between the service provider and end users for these applications. CDNs use a large network of caches to deliver content from a location close to the end users. This aligns with the trend of edge networks,

---

N. K. Panigrahy and J. Li—Authors with equal contribution.

© ICST Institute for Computer Sciences, Social Informatics and Telecommunications Engineering 2021
Published by Springer Nature Switzerland AG 2021. All Rights Reserved
Q. Zhao and L. Xia (Eds.): VALUETOOLS 2021, LNICST 404, pp. 1–21, 2021.
https://doi.org/10.1007/978-3-030-92511-6_1

where computing and storage resources are provisioned at the edge of networks. If a user's request is served by a nearby edge cache, the user experiences a faster response time than if it was served by the backend server. It also reduces bandwidth requirements at the central content repository.

With the aggressive increase in Internet traffic over past years [6], CDNs need to host content from thousands of regions belonging to web sites of thousands of content providers. Furthermore, each content provider may host a large variety of content, including videos, music, and images. Such an increasing diversity in content services requires CDNs to provide different quality of service to varying content classes and applications with different access characteristics and performance requirements. Significant economic benefits and important technical gains have been observed with the deployment of service differentiation [11]. While a rich literature has studied the design of fair and efficient caching algorithms for content distribution, little work has paid attention to the provision of multi-level services in edge networks.

Managing edge networks requires policies to route end-user requests to the local distributed caches, as well as caching algorithms to ensure availability of requested content at the cache. In general, there are two classes of policies for studying the performance of caching algorithms: *timer-based, i.e., Time-To-Live (TTL)* [5,10,13] and *non-timer-based* caching algorithms, e.g., Least-Recently-Used (LRU), Least-Frequently-Used (LFU), First In First Out (FIFO), and RANDOM [7]. Since the cache size is usually much smaller than the total amount of content, some contents need to be evicted if the requested content is not in the cache. Exact analysis of LRU, LFU, FIFO and RANDOM has proven to be difficult, even under the simple *Independence Reference Model* (IRM) [7], where requests are independent of each other. The strongly coupled nature of these eviction algorithms makes implementation of differential services challenging. In contrast, a TTL cache associates each content with a timer upon request and the content is evicted from the cache on timer expiry, independent of other contents. Analysis of these policies is simple since the eviction of contents are decoupled from each other.

Most studies have focused on the analysis of a single edge cache. When an edge network is considered, independence across different caches is usually assumed [25]. Again, it is hard to analyze most conventional caching algorithms, such as LRU, FIFO and RANDOM, but some accurate results for TTL caches are available [3,13].

In this paper, we consider a TTL-based edge network, where a set of network caches host a library of unique contents, and serve a set of users. Figure 1 illustrates an example of such a network, which is consistent with the YouTube video delivery system [23]. Each user can generate a request for a content, which is forwarded along a fixed *path* from the edge cache towards the server. Forwarding stops upon a cache hit, i.e., the requested content is found in a cache on the path. When such a cache hit occurs, the content is sent over the reverse path to the edge cache initializing the request. This raises the questions: *where to cache the requested content on the reverse path* and *what is the value of its timer?* Answer-

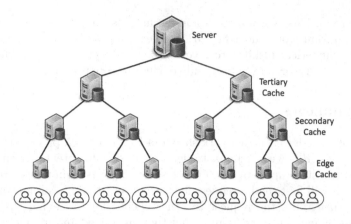

Fig. 1. An edge network with a server holding all contents and three layers of caches. Each edge cache serves a set of users with different requests. The blue line illustrates a unique path between the edge cache and the server. (Color figure online)

ing these questions can provide new insights in edge network design. However, it may also increase the complexity and hardness of the analysis.

Our goal is to provide thorough and rigorous answers to these questions. To that end, we consider moving the content one cache up (towards the user) if there is a cache hit on it and pushing the content one cache down (away from the user) once its timer expires in the cache hierarchy, since the recently evicted content may still be in demand. This policy is known as "Move Copy Down with Push" (MCDP) policy.

We first formulate a utility-driven caching framework for linear edge networks, where each content is associated with a utility and content is managed with a timer whose duration is set to maximize the aggregate utility for all contents over the edge network. Building on MCDP policy, we formulate the optimal TTL policy as a non-convex optimization problem in Sect. 3. One contribution of this paper is to show that this non-convex problem can be transformed into a convex one by change of variables. We further develop distributed algorithms for content management over linear edge networks.

Informed by our results for linear edge networks, we consider a general cache network where each edge cache serves distinct contents, i.e., there are no common contents among edge caches, in Sect. 4.2. We show that this edge network can be treated as a union of different linear edge networks between each edge cache and the central server.

We further consider a more general case where common content is requested among edge caches in Sect. 4.3. This introduces non-convex constraints, resulting in a non-convex utility maximization problem. We show that although the original problem is non-convex, the duality gap is zero. Based on this, we design an iterative primal-dual algorithm for content placement in edge network.

In Sect. 5, through numerical evaluations we show that our algorithm outperforms path replication with traditional caching algorithms over a broad array of network topologies. We provide some discussions on extension of our utility maximization framework under MCDP to a general graph based cache networks in Sect. 5.4. Numerical results are given on how to optimize the performance. We discuss related works in Sect. 6 and conclude the paper in Sect. 7.

2   Preliminaries

We represent the edge cache network shown in Fig. 1 by a graph $G = (V, E)$. We use $\mathcal{D} = \{d_1, \cdots, d_n\}$ with $|\mathcal{D}| = n$ to denote the set of contents. Each network cache $v \in V$ can store up to B_v contents to serve requests from users. We assume that each user will first send a request for the content to its local network cache, which may then route the request to other caches for retrieving the content. Without loss of generality, we assume that there is a fixed and unique path from the local cache towards a terminal cache that is connected to a server that always contains the content.

To be more specific, a request (v, i, p) is determined by the local cache, v, that firstly received the user request, the requested content, i, and the path, p, over which the request is routed. We denote a path p of length $|p| = L$ as a sequence $\{v_{1p}, v_{2p}, \cdots, v_{Lp}\}$ of nodes $v_{lp} \in V$ such that $(v_{lp}, v_{(l+1)p}) \in E$ for $l \in \{1, \cdots, L\}$, where $v_{Lp} = v$. We assume that path p is loop-free and terminal cache v_{1p} is the only cache on path p that accesses the server for content i.

We assume that the request processes for distinct contents are described by independent Poisson processes with arrival rate λ_i for content $i \in \mathcal{D}$. Denote $\Lambda = \sum_{i=1}^{n} \lambda_i$. Then the popularity (request probability) of content i satisfies [2]

$$\rho_i = \frac{\lambda_i}{\Lambda}, \quad i = 1, \cdots, n. \tag{1}$$

We consider TTL caches in this paper. Each content i is associated with a timer T_{ij} at cache j. Suppose content i is requested and routed along path p. There are two cases: (i) content i is not in any cache along path p, in which case content i is fetched from the server and inserted into the terminal cache (denoted by cache 1)[1]. Its timer is set to T_{i1}; (ii) if content i is in cache l along path p, content i is moved to cache $l + 1$ preceding cache l in which i is found, and the timer at cache $l + 1$ is set to $T_{i(l+1)}$. Content i is pushed one cache back to cache $l - 1$ and the timer is set to $T_{i(l-1)}$, once the timer expires. We call it Move Copy Down with Push (MCDP) [24]. Denote the *hit probability* of content i as h_i, then the corresponding *hit rate* is $\lambda_i h_i$.

Denote by $\mathcal{P}$ the set of all requests, and $\mathcal{P}^i$ the set of requests for content i. Suppose a network cache v serves two requests (v_1, i_1, p_1) and (v_2, i_2, p_2), then there are two cases: (i) non-common requested content, i.e., $i_1 \neq i_2$; and (ii) common requested content, i.e., $i_1 = i_2$. In the following, we will focus on how to design optimal TTL policies for content placement in an edge cache network under these two cases.

---

[1] Since we consider path p, for simplicity, we move the dependency on p and v, denote it as nodes $1, \cdots, L$ directly.

While classical cache eviction policies such as LRU provide good performance and are easy to implement, Garetto et al. [14] showed that K-LRU[2] can significant improve LRU even for very small K. Furthermore, Ramadan et al. [23] proposed K-LRU with big cache abstraction (K-LRU(B)) to effectively utilize resources in a hierarchical network of cache servers. Thus, in the rest of the paper, we compare the performance of MCDP to K-LRU and K-LRU(B).

3   Linear Edge Network

We begin with a linear edge network, i.e., there is a single path between the user and the server, composed of $|p| = L$ caches labeled $1, \cdots, L$. A content enters the edge network via cache 1, and is promoted to a higher index cache whenever a cache hit occurs. In the following, we consider the MCDP replication strategy when each cache operates with a TTL policy.

3.1   Stationary Behavior

Requests for content i arrive according to a Poisson process with rate λ_i. Under TTL, content i spends a deterministic time in a cache if it is not requested, independent of all other contents. We denote the timer as T_{il} for content i in cache l on the path p, where $l \in \{1, \cdots, |p|\}$.

Denote by t_k^i the k-th time that content i is either requested or the timer expires. For simplicity, we assume that content is in cache 0 (i.e., server) when it is not in the cache network. We then define a discrete time Markov chain (DTMC) $\{X_k^i\}_{k \geq 0}$ with $|p| + 1$ states, where X_k^i is the index of the cache that content i is in at time t_k^i. The event that the time between two requests for content i exceeds T_{il} occurs with probability $e^{-\lambda_i T_{il}}$; consequently we obtain the transition probability matrix of $\{X_k^i\}_{k \geq 0}$ and compute the stationary distribution. The timer-average probability that content i is in cache $l \in \{1, \cdots, |p|\}$ is

$$h_{i1} = \frac{e^{\lambda_i T_{i1}} - 1}{1 + \sum_{j=1}^{|p|} (e^{\lambda_i T_{i1}} - 1) \cdots (e^{\lambda_i T_{ij}} - 1)}, \tag{2a}$$

$$h_{il} = h_{i(l-1)} (e^{\lambda_i T_{il}} - 1), \quad l = 2, \cdots, |p|, \tag{2b}$$

where h_{il} is also the hit probability for content i at cache l.

Remark 1. The stationary analysis of MCDP is similar to a different caching policy LRU(m) considered in [15]. We relegate its explicit expression to Appendix A.1, and also refer interested readers to [15] for more detail.

---

[2] K-LRU adds $K - 1$ meta-caches ahead of the real cache. Only "popular" contents (requested at least $K - 1$ times) are stored in real cache.

3.2   From Timer to Hit Probability

We consider a TTL cache network where requests for different contents are independent of each other and each content i is associated with a timer T_{il} at each cache $l \in \{1, \cdots, |p|\}$ on the path. Denote $\boldsymbol{T}_i = (T_{i1}, \cdots, T_{i|p|})$ and $\boldsymbol{T} = (\boldsymbol{T}_1, \cdots, \boldsymbol{T}_n)$. From (2), the overall utility on path p is given as

$$\sum_{i \in \mathcal{D}} \sum_{l=1}^{|p|} \psi^{|p|-l} U_i(\lambda_i h_{il}(\boldsymbol{T})), \tag{3}$$

where the utility function $U_i : [0, \infty) \to \mathbb{R}$ is assumed to be increasing, continuously differentiable, and strictly concave of content hit rate, and $0 < \psi \leq 1$ is a discount factor capturing the utility degradation along the request's routing direction. Since each cache is finite in size, we have the capacity constraint

$$\sum_{i \in \mathcal{D}} h_{il}(\boldsymbol{T}) \leq B_l, \quad l \in \{1, \cdots, |p|\}. \tag{4}$$

Therefore, the optimal TTL policy for content placement on path p is the solution of the following optimization problem

$$\begin{aligned}
\max_{\boldsymbol{T}} \quad & \sum_{i \in \mathcal{D}} \sum_{l=1}^{|p|} \psi^{|p|-l} U_i(\lambda_i h_{il}(\boldsymbol{T})) \\
\text{s.t.} \quad & \sum_{i \in \mathcal{D}} h_{il}(\boldsymbol{T}) \leq B_l, \quad l \in \{1, \cdots, |p|\}, \\
& T_{il} \geq 0, \quad \forall i \in \mathcal{D}, \quad l = 1, \cdots, |p|,
\end{aligned} \tag{5}$$

where $h_{il}(\boldsymbol{T})$ is given in (2). However, (5) is a non-convex optimization with a non-linear constraint. Our objective is to characterize the optimal timers for different contents on path p.. To that end, it is helpful to express (5) in terms of hit probabilities. In the following, we discuss how to change the variables from timer to hit probability.

Since $0 \leq T_{il} \leq \infty$, it is easy to check that $0 \leq h_{il} \leq 1$ for $l \in \{1, \cdots, |p|\}$ from (2a) and (2b). Furthermore, it is clear that there exists a mapping between $(h_{i1}, \cdots, h_{i|p|})$ and $(T_{i1}, \cdots, T_{i|p|})$. By simple algebra, we obtain

$$T_{i1} = \frac{1}{\lambda_i} \log\left(1 + \frac{h_{i1}}{1 - (h_{i1} + h_{i2} + \cdots + h_{i|p|})}\right), \tag{6a}$$

$$T_{il} = \frac{1}{\lambda_i} \log\left(1 + \frac{h_{il}}{h_{i(l-1)}}\right), \quad l = 2, \cdots, |p|. \tag{6b}$$

Note that

$$h_{i1} + h_{i2} + \ldots + h_{i|p|} \leq 1, \tag{7}$$

must hold during the operation, which is always true for our caching policies.

3.3   Maximizing Aggregate Utility

With the change of variables discussed above, we can reformulate (5) as follows

$$\max \quad \sum_{i \in \mathcal{D}} \sum_{l=1}^{|p|} \psi^{|p|-l} U_i(\lambda_i h_{il}) \tag{8a}$$

$$\text{s.t.} \quad \sum_{i \in \mathcal{D}} h_{il} \le B_l, \quad l = 1, \cdots, |p|, \tag{8b}$$

$$\sum_{l=1}^{|p|} h_{il} \le 1, \quad \forall i \in \mathcal{D}, \tag{8c}$$

$$0 \le h_{il} \le 1, \quad \forall i \in \mathcal{D}, \quad l = 1, \cdots, |p| \tag{8d}$$

where (8b) is the cache capacity constraint and (8c) is due to the variable exchanges under MCDP as discussed above. Problem (8) under MCDP has a unique global optimum.

3.4   Upper Bound (UB) on optimal Aggregate Utility

Constraint (8c) in (8) is enforced due to variable exchanges under MCDP as discussed above. Here we can define an upper bound on optimal aggregate utility by removing (8c) and solving the following optimization problem

$$\max \sum_{i \in \mathcal{D}} \sum_{l=1}^{|p|} \psi^{|p|-l} U_i(\lambda_i h_{il}), \quad \text{s.t., constraints (8b) and (8d).} \tag{9}$$

Note that the UB optimization problem is now independent of any timer driven caching policy and can be used as a performance benchmark for comparing various caching policies. Furthermore, it is easier to solve UB optimization problem (9) since it involves a smaller number of constraints compared to MCDP based optimization problem (8).

3.5   Distributed Algorithm

In Sect. 3.3, we formulated convex utility maximization problems with a fixed cache size. However, system parameters can change over time, so it is not feasible to solve the optimization offline and implement the optimal strategy. Thus, we need to design distributed algorithms to implement the optimal strategy and adapt to the changes in the presence of limited information.

Primal Algorithm: We aim to design an algorithm based on the optimization problem in (8), which is the primal formulation. Denote $\boldsymbol{h}_i = (h_{i1}, \cdots, h_{i|p|})$ and

$\boldsymbol{h} = (\boldsymbol{h}_1, \cdots, \boldsymbol{h}_n)$. We first define the following objective function.

$$Z(\boldsymbol{h}) = \sum_{i \in \mathcal{D}} \sum_{l=1}^{|p|} \psi^{|p|-l} U_i(\lambda_i h_{il}) - \sum_{l=1}^{|p|} C_l \left(\sum_{i \in \mathcal{D}} h_{il} - B_l \right)$$
$$- \sum_{i \in \mathcal{D}} \tilde{C}_i \left(\sum_{l=1}^{|p|} h_{il} - 1 \right) - \sum_{i \in \mathcal{D}} \sum_{l=1}^{|p|} \hat{C}_{il}(-h_{il}), \quad (10)$$

where $C_l(\cdot), \tilde{C}_i(\cdot)$ and $\hat{C}_{il}(\cdot)$ are convex and non-decreasing penalty functions denoting the cost for violating constraints (8b) and (8c).

Note that constraint (8c) ensures $h_{il} \leq 1 \; \forall i \in \mathcal{D}, \; l = 1, \cdots, |p|$, provided $h_{il} \geq 0$. One can assume that $h_{il} \geq 0$ holds in writing down (10). This would be true, for example, if the utility function is a β-fair utility function with $\beta > 0$ (Sect. 2.5 [26]). For other utility functions, it is challenging to incorporate constraint (8d) since it introduces $n|p|$ additional price functions. For all cases evaluated across various system parameters we found $h_{il} \geq 0$ to hold true. Hence we ignore constraint (8d) in the primal formulation and define the following objective function

$$Z(\boldsymbol{h}) = \sum_{i \in \mathcal{D}} \sum_{l=1}^{|p|} \psi^{|p|-l} U_i(\lambda_i h_{il}) - \sum_{l=1}^{|p|} C_l \left(\sum_{i \in \mathcal{D}} h_{il} - B_l \right) - \sum_{i \in \mathcal{D}} \tilde{C}_i \left(\sum_{l=1}^{|p|} h_{il} - 1 \right). \quad (11)$$

It is clear that $Z(\cdot)$ is strictly concave. Hence, a natural way to obtain the maximal value of (11) is to use the standard *gradient ascent algorithm* to move the variable h_{il} for $i \in \mathcal{D}$ and $l \in \{1, \cdots, |p|\}$ in the direction of gradient,

$$\frac{\partial Z(\boldsymbol{h})}{\partial h_{il}} = \lambda_i \psi^{|p|-l} U_i'(\lambda_i h_{il}) - C_l' \left(\sum_{j \in \mathcal{D}} h_{jl} - B_l \right) - \tilde{C}_i' \left(\sum_{m=1}^{|p|} h_{im} - 1 \right), \quad (12)$$

where $U_i'(\cdot), C_l'(\cdot), \tilde{C}_i'(\cdot)$ denote partial derivatives w.r.t. h_{il}.

Since h_{il} indicates the probability that content i is in cache l, $\sum_{j \in \mathcal{D}} h_{jl}$ is the expected number of contents currently in cache l, denoted by $B_{\mathrm{curr},l}$.

Therefore, the primal algorithm for MCDP is given by

$$T_{il}[k] \leftarrow \begin{cases} \frac{1}{\lambda_i} \log \left(1 + \frac{h_{il}[k]}{1 - \left(h_{i1}[k] + h_{i2}[k] + \cdots + h_{i|p|}[k] \right)} \right), & l = 1; \\ \frac{1}{\lambda_i} \log \left(1 + \frac{h_{il}[k]}{h_{i(l-1)}[k]} \right), & l = 2, \cdots, |p|, \end{cases} \quad (13a)$$

$$h_{il}[k+1] \leftarrow \max \left\{ 0, h_{il}[k] + \zeta_{il} \left[\lambda_i \psi^{|p|-l} U_i'(\lambda_i h_{il}[k]) \right. \right.$$
$$\left. \left. - C_l'(B_{\mathrm{curr},l} - B_l) - \tilde{C}_i' \left(\sum_{m=1}^{|p|} h_{im}[k] - 1 \right) \right] \right\}, \quad (13b)$$

where $\zeta_{il} > 0$ is the step-size parameter, and k is the iteration number incremented upon each request arrival.

Remark 2. Note that the primal formulation in (13) can be implemented distributively with respect to (w.r.t.) different contents and caches by some amount of book-keeping. For example in (13b), $\sum_{m=1}^{|p|} h_{im}[k]$ at cache l can be computed by first storing the value of $\sum_{m=1}^{|p|} h_{im}[k]$ at the edge cache in previous iteration and updating it during delivery of content i (from cache l) to the user.

4   General Edge Networks

In Sect. 3, we consider linear edge networks and characterize the optimal TTL policy for content when coupled with MCDP. Inspired by these results, we consider general edge networks in this section.

4.1   Contents, Servers and Requests

We consider the general edge network described in Sect. 2. Denote by $\mathcal{P}$ the set of all requests, and $\mathcal{P}^i$ the set of requests for content i. Suppose a cache in node v serves two requests (v_1, i_1, p_1) and (v_2, i_2, p_2), then there are two cases: (i) non-common requested content, i.e., $i_1 \neq i_2$; and (ii) common requested content, i.e., $i_1 = i_2$.

4.2   Non-common Requested Content

In this section, we consider the case that each network cache serves requests for different contents from each request (v, i, p) passing through it. Since there is no coupling between different requests (v, i, p), we can directly generalize the results for a particular path p in Sect. 3 to a tree network. Hence, given the utility maximization formulation in (8), we can directly formulate the optimization problem for MCDP as

$$\max \quad \sum_{i \in \mathcal{D}} \sum_{p \in \mathcal{P}^i} \sum_{l=1}^{|p|} \psi^{|p|-l} U_{ip}(\lambda_{ip} h_{il}^{(p)}) \tag{14a}$$

$$\text{s.t.} \quad \sum_{i \in \mathcal{D}} \sum_{p:l \in \{1,\cdots,|p|\}} h_{il}^{(p)} \leq B_l, \quad \forall l \in V, \tag{14b}$$

$$\sum_{l=1}^{|p|} h_{il}^{(p)} \leq 1, \quad \forall i \in \mathcal{D}, p \in \mathcal{P}^i, \tag{14c}$$

$$0 \leq h_{il}^{(p)} \leq 1, \quad \forall i \in \mathcal{D}, l \in \{1, \cdots, |p|\}, p \in \mathcal{P}^i. \tag{14d}$$

The optimization problem defined in (14) under MCDP has a unique global optimum.

4.3 Common Requested Contents

Now consider the case where different users share the same content, e.g., there are two requests (v_1, i, p_1) and (v_2, i, p_2). Suppose that cache l is on both paths p_1 and p_2, where v_1 and v_2 request the same content i. If we cache separate copies on each path, results from the previous section apply. However, maintaining redundant copies in the same cache decreases efficiency. A simple way to deal with that is to only cache one copy of content i at l to serve both requests from v_1 and v_2. Though this reduces redundancy, it complicates the optimization problem.

In the following, we formulate a utility maximization problem for MCDP with TTL caches, where all users share the same requested contents $\mathcal{D}$.

$$\max \quad \sum_{i \in \mathcal{D}} \sum_{p \in \mathcal{P}^i} \sum_{l=1}^{|p|} \psi^{|p|-l} U_{ip}(\lambda_{ip} h_{il}^{(p)}) \tag{15a}$$

$$\text{s.t.} \quad \sum_{i \in \mathcal{D}} \left(1 - \prod_{p:j \in \{1, \cdots, |p|\}} (1 - h_{ij}^{(p)}) \right) \le B_j, \quad \forall j \in V, \tag{15b}$$

$$\sum_{j \in \{1, \cdots, |p|\}} h_{ij}^{(p)} \le 1, \quad \forall i \in \mathcal{D}, p \in \mathcal{P}^i, \tag{15c}$$

$$0 \le h_{il}^{(p)} \le 1, \quad \forall i \in \mathcal{D}, j \in \{1, \cdots, |p|\}, p \in \mathcal{P}^i, \tag{15d}$$

where (15b) ensures that only one copy of content $i \in \mathcal{D}$ is cached at node j for all paths p that pass through node j. This is because the term $1 - \prod_{p:j \in \{1, \cdots, |p|\}} (1 - h_{ij}^{(p)})$ is the overall hit probability of content i at node j over all paths. (15c) is the cache capacity constraint and (15d) is the constraint from MCDP TTL cache policy as discussed in Sect. 3.2. (15) under MCDP is a non-convex optimization problem.

Example 1. Consider two requests (v_1, i, p_1) and (v_2, i, p_2) with paths p_1 and p_2 intersecting at j. Let the corresponding path perspective hit probability be $h_{ij}^{(p_1)}$ and $h_{ij}^{(p_2)}$. Then the term inside outer summation of (15b) is $1 - (1 - h_{ij}^{(p_1)})(1 - h_{ij}^{(p_2)})$, i.e., the hit probability of content i in j.

Remark 3. Note that we assume independence between different requests (v, i, p) in (15), e.g., in Example 1, if the insertion of content i in node j is caused by request (v_1, i, p_1), when request (v_2, i, p_2) comes, it is not counted as a cache hit from its perspective. Our framework still holds if we follow the logical TTL MCDP on a path. However, in that case, the utilities will be larger than the one we consider here.

In the following, we develop an optimization framework that handles the non-convexity issue in this optimization problem and provides a distributed solution.

To this end, we first introduce the Lagrangian function

$$L(\boldsymbol{h}, \boldsymbol{\nu}, \boldsymbol{\mu}) = \sum_{i \in \mathcal{D}} \sum_{p \in \mathcal{P}^i} \sum_{l=1}^{|p|} \psi^{|p|-l} U_{ip}(\lambda_{ip} h_{il}^{(p)}) - \sum_{i \in \mathcal{D}} \sum_{p \in \mathcal{P}^i} \mu_{ip} \left(\sum_{j \in \{1, \cdots, |p|\}} h_{ij}^{(p)} - 1 \right)$$

$$- \sum_{j \in V} \nu_j \left(\sum_{i \in \mathcal{D}} \left[1 - \prod_{p: j \in \{1, \cdots, |p|\}} (1 - h_{ij}^{(p)}) \right] - B_j \right), \tag{16}$$

where the Lagrangian multipliers (price vector and price matrix) are $\boldsymbol{\nu} = (\nu_j)_{j \in V}$, and $\boldsymbol{\mu} = (\mu_{ip})_{i \in \mathcal{D}, p \in \mathcal{P}}$. Constraint (15d) is ignored in the Lagrangian function due to the same reason stated for the primal formulation in Sect. 3.5.

The dual function can be defined as

$$d(\boldsymbol{\nu}, \boldsymbol{\mu}) = \sup_{\boldsymbol{h}} L(\boldsymbol{h}, \boldsymbol{\nu}, \boldsymbol{\mu}), \tag{17}$$

and the dual problem is given as

$$\min_{\boldsymbol{\nu}, \boldsymbol{\mu}} \quad d(\boldsymbol{\nu}, \boldsymbol{\mu}) = L(\boldsymbol{h}^*(\boldsymbol{\nu}, \boldsymbol{\mu}), \boldsymbol{\nu}, \boldsymbol{\mu}), \quad \text{s.t.} \quad \boldsymbol{\nu}, \boldsymbol{\mu} \geq 0, \tag{18}$$

where the constraint is defined pointwise for $\boldsymbol{\nu}, \boldsymbol{\mu}$, and $\boldsymbol{h}^*(\boldsymbol{\nu}, \boldsymbol{\mu})$ is a function that maximizes the Lagrangian function for given $(\boldsymbol{\nu}, \boldsymbol{\mu})$, i.e.,

$$\boldsymbol{h}^*(\boldsymbol{\nu}, \boldsymbol{\mu}) = \arg \max_{\boldsymbol{h}} L(\boldsymbol{h}, \boldsymbol{\nu}, \boldsymbol{\mu}). \tag{19}$$

The dual function $d(\boldsymbol{\nu}, \boldsymbol{\mu})$ is always convex in $(\boldsymbol{\nu}, \boldsymbol{\mu})$ regardless of the convexity of the optimization problem (15) [4]. Therefore, it is always possible to iteratively solve the dual problem using

$$\nu_l[k+1] = \nu_l[k] - \gamma_l \frac{\partial L(\boldsymbol{\nu}, \boldsymbol{\mu})}{\partial \nu_l}, \quad \mu_{ip}[k+1] = \mu_{ip}[k] - \eta_{ip} \frac{\partial L(\boldsymbol{\nu}, \boldsymbol{\mu})}{\partial \mu_{ip}}, \tag{20}$$

where γ_l and η_{ip} are the step sizes, and $\frac{\partial L(\boldsymbol{\nu}, \boldsymbol{\mu})}{\partial \nu_l}$ and $\frac{\partial L(\boldsymbol{\nu}, \boldsymbol{\mu})}{\partial \mu_{ip}}$ are the partial derivative of $L(\boldsymbol{\nu}, \boldsymbol{\mu})$ w.r.t. ν_l and μ_{ip}, respectively, satisfying

$$\frac{\partial L(\boldsymbol{\nu}, \boldsymbol{\mu})}{\partial \nu_l} = -\left(\sum_{i \in \mathcal{D}} \left[1 - \prod_{p: l \in \{1, \cdots, |p|\}} (1 - h_{il}^{(p)}) \right] - B_l \right),$$

$$\frac{\partial L(\boldsymbol{\nu}, \boldsymbol{\mu})}{\partial \mu_{ip}} = -\left(\sum_{j \in \{1, \cdots, |p|\}} h_{ij}^{(p)} - 1 \right). \tag{21}$$

Sufficient and necessary conditions for the uniqueness of $\boldsymbol{\nu}, \boldsymbol{\mu}$ are given in [19]. The convergence of primal-dual algorithm consisting of (19) and (20) is guaranteed if the original optimization problem is convex. However, our problem is not convex. Nevertheless, we next show that the duality gap is zero, hence (19) and (20) converge to the globally optimal solution. To begin with, we introduce the following results

Theorem 1 *[27] (Sufficient Condition). If the price based function $h^*(\nu, \mu)$ is continuous at one or more of the optimal lagrange multiplier vectors ν^* and μ^*, then the iterative algorithm consisting of (19) and (20) converges to the globally optimal solution.*

Theorem 2 *[27]. If at least one constraint of (15) is active at the optimal solution, the condition in Theorem 1 is also a necessary condition.*

Hence, if we can show the continuity of $h^*(\nu, \mu)$ and that constraints (15) are active, then given Theorems 1 and 2, the duality gap is zero, i.e., (19) and (20) converge to the globally optimal solution.

Take the derivative of $L(h, \nu, \mu)$ w.r.t. $h_{il}^{(p)}$ for $i \in \mathcal{D}$, $l \in \{1, \cdots, |p|\}$ and $p \in \mathcal{P}^i$, we have

$$\frac{\partial L(h, \nu, \mu)}{\partial h_{il}^{(p)}} = \psi^{|p|-l} \lambda_{ip} U'_{ip}(\lambda_{ip} h_{il}^{(p)}) - \mu_{ip} - \nu_l \left(\prod_{\substack{q:q \neq p, \\ j \in \{1, \cdots, |q|\}}} (1 - h_{ij}^{(q)}) \right). \quad (22)$$

Setting (22) equal to zero, we obtain

$$U'_{ip}(\lambda_{ip} h_{il}^{(p)}) = \frac{1}{\psi^{|p|-l} \lambda_{ip}} \left(\nu_l \left(\prod_{\substack{q:q \neq p, \\ j \in \{1, \cdots, |q|\}}} (1 - h_{ij}^{(q)}) \right) + \mu_{ip} \right). \quad (23)$$

Consider the utility function $U_{ip}(\lambda_{ip} h_{il}^{(p)}) = w_{ip} \log(1 + \lambda_{ip} h_{il}^{(p)})$, then $U'_{ip}(\lambda_{ip} h_{il}^{(p)}) = w_{ip}/(1 + \lambda_{ip} h_{il}^{(p)})$. Hence, from (23), we have

$$h_{il}^{(p)} = \frac{w_{ip} \psi^{|p|-l}}{\nu_l \left(\prod_{\substack{q:q \neq p, \\ j \in \{1, \cdots, |q|\}}} (1 - h_{ij}^{(q)}) \right) + \mu_{ip}} - \frac{1}{\lambda_{ip}}. \quad (24)$$

Lemma 1. *Constraints (15b) and (15c) cannot be both non-active, i.e., at least one of them is active.*

Proof. We prove this lemma by contradiction. Suppose both constraints (15b) and (15c) are non-active, i.e., $\nu = (0)$, and $\mu = (0)$. Then the optimization problem (14) achieves its maximum when $h_{il}^{(p)} = 1$ for all $i \in \mathcal{D}$, $l \in \{1, \cdots, |p|\}$ and $p \in \mathcal{P}^i$. If so, then the left hand size of (15b) equals $|\mathcal{D}|$ which is much greater than B_l for $l \in V$, which is a contradiction. Hence, constraints (15b) and (15c) cannot be both non-active.

From Lemma 1, we know that the feasible region for the Lagrangian multipliers satisfies $\mathcal{R} = \{\nu_l \geq 0, \mu_{ip} \geq 0, \nu_l + \mu_{ip} \neq 0, \forall i \in \mathcal{D}, l \in \{1, \cdots, |p|\}, p \in \mathcal{P}^i\}$.

Theorem 3. *The hit probability $h_{il}^{(p)}$ given in (24) is continuous in ν_l and μ_{ip} for all $i \in \mathcal{D}$, $l \in \{1, \cdots, |p|\}$ and $p \in \mathcal{P}^i$ in the feasible region $\mathcal{R}$.*

Proof. From Lemma 1, we know at least one of ν_l and μ_{ip} is non-zero, for all $i \in \mathcal{D}$, $l \in \{1, \cdots, |p|\}$ and $p \in \mathcal{P}^i$. Hence there are three cases, (i) $\nu_l \neq 0$ and $\mu_{ip} = 0$; (ii) $\nu_l = 0$ and $\mu_{ip} \neq 0$; and (iii) $\nu_l \neq 0$ and $\mu_{ip} \neq 0$.

For case (i), we have

$$
h_{il}^{(p)} = \frac{w_{ip}\psi^{|p|-l}}{\nu_l \left(\prod_{\substack{q:q \neq p, \\ j \in \{1, \cdots, |q|\}}} (1 - h_{ij}^{(q)}) \right)} - \frac{1}{\lambda_{ip}}, \tag{25}
$$

which is clearly continuous in ν_l, for all $i \in \mathcal{D}$, $l \in \{1, \cdots, |p|\}$ and $p \in \mathcal{P}^i$. Similarly for case (ii), we have

$$
h_{il}^{(p)} = \frac{w_{ip}\psi^{|p|-l}}{\mu_{ip}} - \frac{1}{\lambda_{ip}}, \tag{26}
$$

which is also clearly continuous in μ_{ip}, for all $i \in \mathcal{D}$, $l \in \{1, \cdots, |p|\}$ and $p \in \mathcal{P}^i$.

For case (iii), from (24), it is obvious that $h_{il}^{(p)}$ is continuous in ν_l and μ_{ip} for all $i \in \mathcal{D}$, $l \in \{1, \cdots, |p|\}$ and $p \in \mathcal{P}^i$. Therefore, we know that $h_{il}^{(p)}$ is is continuous in ν_l and μ_{ip} for all $i \in \mathcal{D}$, $l \in \{1, \cdots, |p|\}$ and $p \in \mathcal{P}^i$.

Remark 4. Note that similar arguments (by using Lemma 1) hold true for various other choice of utility functions such as: β- fair utility functions (Sect. 2 [26]). Therefore, the primal-dual algorithm consisting of (19) and (20) converges to the globally optimal solution for a wide range of utility functions.

Algorithm 1 summarizes the details of this algorithm.

Algorithm 1. Primal-Dual Algorithm

Input: $\forall \nu_0$, $\mu_0 \in \mathcal{R}$ and h_0
Output: The optimal hit probabilities h
Step 0: $t = 0$, $\nu[t] \leftarrow \nu_0$, $\mu[t] \leftarrow \mu_0$, $h[t] \leftarrow h_0$
Step $t \geq 1$
while Equation (21) $\neq 0$ **do**
    First, compute $h_{il}^{(p)}[t+1]$ for $i \in \mathcal{D}$, $l \in \{1, \cdots, |p|\}$ and $p \in \mathcal{P}^i$ through (24);
    Second, update $\nu_l[t+1]$ and $\mu_{ip}[t+1]$ through (20) given $h[t+1]$, $\nu[t]$ and $\mu[t]$ for $l \in V$, $i \in \mathcal{D}$ and $p \in \mathcal{P}^i$

---

5 Numerical Evaluation

In this Section we validate our analytical results with simulations for MCDP across different network topologies.

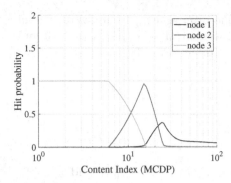

Fig. 2. Hit probability for MCDP in a three-node path.

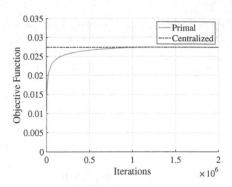

Fig. 3. Convergence of primal algorithm.

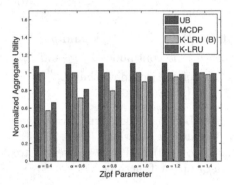

Fig. 4. Optimal aggregated utilities in a three-node path.

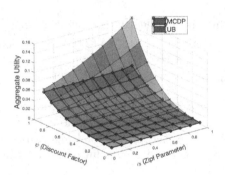

Fig. 5. Aggregate utility under MCDP and UB across different network parameters.

5.1   Linear Edge Network

First, we consider a three-node path with cache capacities $B_l = 10$, $l = 1, 2, 3$. The total number of unique contents considered in the system is $n = 100$. We consider the Zipf popularity distribution with parameter $\alpha = 0.8$. W.l.o.g., we consider log based utility function[3] $U_i(x) = \lambda_i \log(1 + x)$ [28], and discount factor $\psi = 0.1$. We assume that requests arrive according to a Poisson process with aggregate request rate $\Lambda = 1$.

We solve optimization problem (8) using a Matlab routine `fmincon`. From Fig. 2, we observe that popular contents are assigned higher hit probabilities at cache node 3, i.e. at the edge cache closest to the user as compared to other caches. The optimal hit probabilities assigned to popular contents at other caches

---

[3] One can also choose $U_i(x) = \lambda_i \log x$. However, $U_i(x)$ evaluated at $x = 0$ becomes negative infinity, which may produce undesired results while comparing the performance of MCDP with other caching policies.

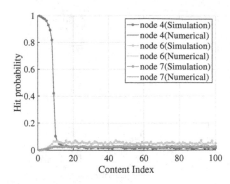
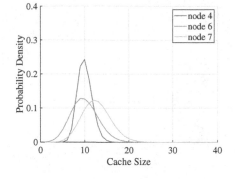

Fig. 6. Hit probability of MCDP under three-layer edge network with distinct contents.

Fig. 7. Cache size pdf of MCDP under three-layer edge network with distinct contents.

are almost negligible. However, the assignment is reversed for moderately popular contents. For non-popular contents, optimal hit probabilities at cache node 1 (closest to origin server) are the highest.

We then implement our primal algorithm given in (13), where we take the following penalty functions [26] $C_l(x) = \max\{0, x - B_l \log(B_l + x)\}$ and $\tilde{C}_i(x) = \max\{0, x - \log(1 + x)\}$. Figure 3 shows that the primal algorithm successfully converges to the optimal solution.

We also compare the performance of MCDP to other policies such as K-LRU (K=3), K-LRU with big cache abstraction: K-LRU(B) and the UB based bound. We plot the relative performance w.r.t. the optimal aggregated utilities of all above policies, normalized to that under MCDP shown in Fig. 4. We observe that MCDP significantly outperforms K-LRU and K-LRU(B) for low and moderate values of Zipf parameter. Furthermore, the performance gap between UB and MCDP increases in Zipf parameter.

Finally, we consider the effect of α and ψ on the performance gap (difference between optimal aggregate utility) between UB and MCDP. We present the simulation results in Fig. 5. Note that utility under both UB and MCDP increases in either α or ψ or both. When either α or ψ is small, irrespective of the value of the other, the performance gap is minor or negligible. However, the gap is considerably large when both α and ψ are high. We believe, due to high communication overhead between successive layers of caches, ψ has a small value, thus indicating a minor performance gap between MCDP and UB.

5.2   General Edge Network with Non-Common Requested Contents

We consider a three-layer edge network shown in Fig. 1 with node set $\{1, \cdots, 7\}$, which is consistent with the YouTube video delivery system [23]. Nodes 1-4 are edge caches, and node 7 is tertiary cache. There exist four paths $p_1 = \{1, 5, 7\}$, $p_2 = \{2, 5, 7\}$, $p_3 = \{3, 6, 7\}$ and $p_4 = \{4, 6, 7\}$. Each edge cache serves requests for 100 distinct contents, and cache size is $B_v = 10$ for $v \in \{1, \cdots, 7\}$. Assume

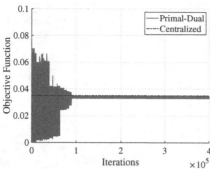
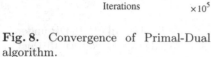

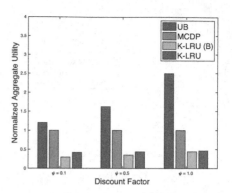

Fig. 8. Convergence of Primal-Dual algorithm.

Fig. 9. Optimal aggregated utilities under common requested contents.

that content follows a Zipf distribution with parameter $\alpha_1 = 0.2$, $\alpha_2 = 0.4$, $\alpha_3 = 0.6$ and $\alpha_4 = 0.8$, respectively. We consider utility function $U_{ip}(x) = \lambda_{ip} \log(1 + x)$, where λ_{ip} is the request arrival rate for content i on path p, and requests are described by a Poisson process with $\Lambda_p = 1$ for $p = 1, 2, 3, 4$. The discount factor $\psi = 0.1$.

Figure 6 shows results for path $p_4 = \{4, 6, 7\}$. From Fig. 6, we observe that our algorithm yields the exact optimal and empirical hit probabilities under MCDP. Figure 7 shows the probability density for the number of contents[4] in the edge network. As expected, the density is concentrated around their corresponding cache sizes.

5.3   General Edge Network with Common Requested Contents

We evaluate the performance of Algorithm 1 on a three-layer edge network shown in Fig. 1. We assume that there are totally 100 unique contents in the system requested from four paths. The cache size is given as $B_v = 10$ for $v = 1, \cdots, 7$. We consider the utility function $U_{ip}(x) = \lambda_{ip} \log(1 + x)$, and the popularity distribution over these contents is Zipf with parameter 0.8. W.l.o.g., the aggregate request arrival rate is one. The discount factor $\psi = 0.1$.

We solve the optimization problem in (15) using a Matlab routine `fmincon`. Then we implement our primal-dual algorithm given in Algorithm 1. The result for aggregate optimal utility is presented in Fig. 8. It is clear that the primal-dual algorithm successfully converges to the optimal solution.

Similar to Sect. 5.1, we compare MCDP to K-LRU, K-LRU(B) and UB. Figure 9 compares the performance of different eviction policies to our MCDP policy. We plot the relative performance w.r.t. the optimal aggregated utilities of all above policies, normalized to that under MCDP. We again observe a huge

---

[4] The constraint (14b) in problem (14) is on average cache occupancy. However it can be shown that if $n \to \infty$ and B_l grows in sub-linear manner, the probability of violating the target cache size B_l becomes negligible [8].

gain of MCDP w.r.t. K-LRU and K-LRU(B) across all values of discount factor. However, the performance gap between MCDP and UB increases with an increase in the value of discount factor.

5.4 Applications to General Graphs

Here we consider a general network topology with overlapping paths and common contents requested along different paths. We consider two network topologies: Grid and lollipop. *Grid* is a two dimensional square grid while a (a,b) *lollipop* network is a complete graph of size a, connected to a path graph of length b. Denote the network as $G = (V, E)$. For grid, we consider $|V| = 16$, while we consider a $(3, 4)$ lollipop topology with $|V| = 7$ and clique size 3. The library contain $|\mathcal{D}| = 100$ unique contents. We assign a weight to each edge in E, selected uniformly from the interval $[1, 20]$.

Next, we generate a set of requests in G as described in [16]. To ensure that paths overlap, we randomly select a subset $\tilde{V} \subset V$ nodes to generate requests. Each node in $\tilde{V}$ can generate requests for contents in $\mathcal{D}$ following a Zipf distribution with parameter $\alpha = 0.8$. Requests are then routed over the shortest path between the requesting node in $\tilde{V}$ and the node in V that caches the content. Again, we assume that the aggregate request rate at each node in $\tilde{V}$ is one and the discount factor to be $\psi = 0.1$.

Table 1. Optimal aggregate utilities under various network topologies.

Topology	α	MCDP	UB	% Gap
Grid	0.8	0.0923	0.1043	11.50
Grid	1.2	0.3611	0.4016	10.08
Lollipop	0.8	0.0908	0.1002	9.38
Lollipop	1.2	0.3625	0.4024	9.91

We evaluate the performance of MCDP over the graphs across various Zipf parameter in Table 1. It is clear that for both network topologies, aggregate utility obtained from our TTL-based framework with MCDP policy is higher for higher zipf parameter as compared to lower zipf parameter. With increase in Zipf parameter, the difference between request rates of popular and less popular contents increases. The aggregate request rate over all contents is the same in both cases. Thus popular contents get longer fraction of rates which in turn yields higher aggregate utility.

6 Related Work

There is a rich literature on the design, modeling and analysis of cache networks, including TTL caches [3,13,24] and optimal caching [16,20]. In particular, Rodriguez et al. [24] analyzed the advantage of pushing content upstream,

Berger et al. [3] characterized the exactness of TTL policy in a hierarchical topology. A unified approach to study and compare different caching policies is given in [14] and an optimal placement problem under a heavy-tailed demand has been explored in [12].

Dehghan et al. [9] as well as Abedini and Shakkottai [1] studied joint routing and content placement with a focus on a bipartite, single-hop setting. Both showed that minimizing single-hop routing cost can be reduced to solving a linear program. Ioannidis and Yeh [17] studied the same problem under a more general setting for arbitrary topologies.

An adaptive caching policy for a cache network was proposed in [16], where each node makes a decision on which item to cache and evict. An integer programming problem was formulated by characterizing the content transfer costs. Both centralized and complex distributed algorithms were designed with performance guarantees. This work complements our work, as we consider TTL cache and control the optimal cache parameters through timers to maximize the sum of utilities over all contents across the network. However, [16] proposed only approximate algorithms while our timer-based models enable us to design optimal solutions since content occupancy can be modeled as a real variable.

Closer to our work, a utility maximization problem for a single cache was considered under IRM [8, 21] and stationary requests [22], while [12] maximized the hit probabilities under heavy-tailed demands over a single cache. None of these approaches generalizes to edge networks, which leads to non-convex formulations (See Sect. 4.2 and Sect. 4.3); addressing this lack of convexity in its full generality, for arbitrary network topologies, overlapping paths and request arrival rates, is one of our technical contributions.

7   Conclusion

We constructed optimal timer-based TTL polices for content placement in edge networks through a unified optimization approach. We formulated a general utility maximization framework, which is non-convex in general. We identified the non-convexity issue and proposed efficient distributed algorithm to solve it. We proved that the distributed algorithms converge to the globally optimal solutions. We showed the efficiency of these algorithms through numerical studies. An analysis of the MCDP algorithm with more general request arrivals would remain our future work.

Acknowledgment. This research was sponsored by the U.S. ARL and the U.K. MoD under Agreement Number W911NF-16-3-0001 and by the NSF under Grant CNS-1617437 and CRII-CNS-2104880. This research was also supported by the U.S. Department of Energy's Office of Energy Efficiency and Renewable Energy (EERE) under the Solar Energy Technologies Office Award Number DE-EE0009341. The views and conclusions contained in this document are those of the authors and should not be interpreted as representing the official policies, either expressed or implied, of the National Science Foundation, U.S. ARL, the U.S. Government, the U.K. MoD or the U.K. Government.

A   Appendix

A.1   Stationary Behaviors of MCDP

[15] considered a caching policy LRU(m). Though the policy differ from MCDP, the stationary analysis is similar. Under IRM model, the request for content i arrives according a Poisson process with rate λ_i. As discussed earlier, for TTL caches, content i spends a deterministic time in a cache if it is not requested, which is independent of all other contents. We denote the timer as T_{il} for content i in cache l on the path p, where $l \in \{1, \cdots, |p|\}$.

Denote t_k^i as the k-th time that content i is either requested or moved from one cache to another. For simplicity, we assume that content is in cache 0 (i.e., server) if it is not in the cache network. Then we can define a discrete time Markov chain (DTMC) $\{X_k^i\}_{k \geq 0}$ with $|p| + 1$ states, where X_k^i is the cache index that content i is in at time t_k^i. Since the event that the time between two requests for content i exceeds T_{il} happens with probability $e^{-\lambda_i T_{il}}$, then the transition matrix of $\{X_k^i\}_{k \geq 0}$ is given as

$$\mathbf{P}_i^{\mathrm{MCDP}} = \begin{bmatrix} 0 & 1 & & & \\ e^{-\lambda_i T_{i1}} & 0 & 1 - e^{-\lambda_i T_{i1}} & & \\ & \ddots & \ddots & \ddots & \\ & & e^{-\lambda_i T_{i(|p|-1)}} & 0 & 1 - e^{-\lambda_i T_{i(|p|-1)}} \\ & & & e^{-\lambda_i T_{i|p|}} & 1 - e^{-\lambda_i T_{i|p|}} \end{bmatrix}. \quad (27)$$

Let $(\pi_{i0}, \cdots, \pi_{i|p|})$ be the stationary distribution for $\mathbf{P}_i^{\mathrm{MCDP}}$, we have

$$\pi_{i0} = \frac{1}{1 + \sum_{j=1}^{|p|} e^{\lambda_i T_{ij}} \prod_{s=1}^{j-1}(e^{\lambda_i T_{is}} - 1)}, \quad (28a)$$

$$\pi_{i1} = \pi_{i0} e^{\lambda_i T_{i1}}, \quad (28b)$$

$$\pi_{il} = \pi_{i0} e^{\lambda_i T_{il}} \prod_{s=1}^{l-1}(e^{\lambda_i T_{is}} - 1), \ l = 2, \cdots, |p|. \quad (28c)$$

The average time that content i spends in cache $l \in \{1, \cdots, |p|\}$ is given by

$$\mathbb{E}[t_{k+1}^i - t_k^i | X_k^i = l] = \int_0^{T_{il}} \left(1 - [1 - e^{-\lambda_i t}]\right) dt = \frac{1 - e^{-\lambda_i T_{il}}}{\lambda_i}, \quad (29)$$

and $\mathbb{E}[t_{k+1}^i - t_k^i | X_k^i = 0] = \frac{1}{\lambda_i}$. Given (28) and (29), the timer-average probability that content i is in cache $l \in \{1, \cdots, |p|\}$ is given by (2) with h_{il} being the hit probability for content i at cache l.

References

1. Abedini, N., Shakkottai, S.: Content caching and scheduling in wireless networks with elastic and inelastic traffic. IEEE/ACM Trans. Netw. **22**(3), 864–874 (2014)

2. Baccelli, F., Brémaud, P.: Elements of Queueing Theory: Palm Martingale Calculus and Stochastic Recurrences, vol. 26. Springer, Heidelberg (2013). https://doi.org/10.1007/978-3-662-11657-9
3. Berger, D.S., Gland, P., Singla, S., Ciucu, F.: Exact analysis of TTL cache networks. Perf. Eval. **79**, 2–23 (2014)
4. Boyd, S., Vandenberghe, L.: Convex Optimization. Cambridge University Press, Cambridge (2004)
5. Che, H., Tung, Y., Wang, Z.: Hierarchical web caching systems: modeling, design and experimental results. IEEE J. Sel. Areas Commun. **20**(7), 1305–1314 (2002)
6. Cisco, V.: Cisco Visual Networking Index: Forecast and Methodology 2014–2019 White Paper. Technical Report, Cisco (2015)
7. Coffman, E.G., Denning, P.J.: Operating Systems Theory, vol. 973. Prentice-Hall, Englewood Cliffs (1973)
8. Dehghan, M., Massoulie, L., Towsley, D., Menasche, D., Tay, Y.: A utility optimization approach to network cache design. In: Proceedings of IEEE INFOCOM (2016)
9. Dehghan, M., et al.: On the complexity of optimal routing and content caching in heterogeneous networks. In: Proceedings of IEEE INFOCOM, pp. 936–944 (2015)
10. Fagin, R.: Asymptotic miss ratios over independent references. J. Comput. Syst. Sci. **14**(2), 222–250 (1977)
11. Feldman, M., Chuang, J.: Service differentiation in web caching and content distribution. In: Proceedings of CCN (2002)
12. Ferragut, A., Rodríguez, I., Paganini, F.: Optimizing TTL caches under heavy-tailed demands. In: Proceedings of ACM SIGMETRICS (2016)
13. Fofack, N.C., Nain, P., Neglia, G., Towsley, D.: Performance evaluation of hierarchical TTL-based cache networks. Comput. Netw. **65**, 212–231 (2014)
14. Garetto, M., Leonardi, E., Martina, V.: A unified approach to the performance analysis of caching systems. ACM TOMPECS **1**(3), 12 (2016)
15. Gast, N., Houdt, B.V.: Asymptotically exact TTL-approximations of the cache replacement algorithms LRU(m) and h-LRU. In: ITC, vol. 28 (2016)
16. Ioannidis, S., Yeh, E.: Adaptive caching networks with optimality guarantees. In: Proceedings of ACM SIGMETRICS, pp. 113–124 (2016)
17. Ioannidis, S., Yeh, E.: Jointly optimal routing and caching for arbitrary network topologies. In: Proceedings of ACM ICN, pp. 77–87 (2017)
18. Jiang, W., Ioannidis, S., Massoulié, L., Picconi, F.: Orchestrating massively distributed CDNs. In: Proceedings of ACM CoNEXT, pp. 133–144 (2012)
19. Kyparisis, J.: On uniqueness of Kuhn-Tucker multipliers in nonlinear programming. Math. Program. **32**(2), 242–246 (1985)
20. Li, J., et al.: DR-cache: distributed resilient caching with latency guarantees. In: Proceedings of IEEE INFOCOM (2018)
21. Panigrahy, N.K., Li, J., Towsley, D.: Hit rate vs. hit probability based cache utility maximization. In: Proceedings of ACM MAMA (2017)
22. Panigrahy, N.K., Li, J., Towsley, D.: Network cache design under stationary requests: exact analysis and poisson approximation. In: Proceedings of IEEE MASCOTS (2018)
23. Ramadan, E., Narayanan, A., Zhang, Z.L., Li, R., Zhang, G.: Big cache abstraction for cache networks. In: Proceedings Of IEEE ICDCS (2017)
24. Rodríguez, I., Ferragut, A., Paganini, F.: Improving performance of multiple-level cache systems. In: SIGCOMM (2016)
25. Rosensweig, E.J., Kurose, J., Towsley, D.: Approximate models for general cache networks. In: Proceedings of IEEE INFOCOM (2010)

26. Srikant, R., Ying, L.: Communication Networks: an Optimization, Control, and Stochastic Networks Perspective. Cambridge University Press, Cambridge (2013)
27. Tychogiorgos, G., Gkelias, A., Leung, K.K.: A non-convex distributed optimization framework and its application to wireless ad-hoc networks. IEEE Trans. Wirel. Commun $12(9)$, 4286–4296 (2013)
28. Vecer, J.: Dynamic scoring: probabilistic model selection based on utility maximization. In: SSRN (2018)

A Two-Step Fitting Approach of Batch Markovian Arrival Processes for Teletraffic Data

Gang Chen[1], Li Xia[1(✉)], Zhaoyu Jiang[2], Xi Peng[3], Li Chen[3], and Bo Bai[3]

[1] Business School, Sun Yat-Sen University, Guangzhou 510275, China
{chengang5,xiali5}@sysu.edu.cn, xial@tsinghua.edu.cn
[2] Huawei Tech. Co., Ltd., No.156 Beiqing Rd., Hai-Dian District, Beijing, China
jiangzhaoyu2@huawei.com
[3] Huawei Tech. Investment Co., Ltd., Pak Shek Kok, Shatin, N.T., Hong Kong, China
{pancy.pengxi,chen.li7}@huawei.com

Abstract. Batch Markovian arrival process (BMAP) is a powerful stochastic process model for fitting teletraffic data since its MAP structure can capture the mode of packet interarrival times and its batch structure can capture packet size distributions. Compared with Poisson-related models and simple MAP models richly studied in the literature, the BMAP model and its fitting approach are much less investigated. Motivated by a practical project collaborated with Huawei company, we propose a new two-step parameter fitting approach of the BMAP model for teletraffic data generated from IP networks. The first step is the phase-type fitting for packet interarrival times, which is implemented by the framework of EM (expectation maximization) algorithms. The second step is the approximation of the lag correlation values of packet interarrival times and packet sizes, which is implemented by the framework of MM (moment matching) algorithms. The performance of our two-step EM-MM fitting approach is demonstrated by numerical experiments on both simulated and real teletraffic data sets, and compared with the MAP and MMPP (Markovian modulated Poisson process) models to illustrate the advantages of the BMAP model. Numerical examples also show that our proposed two-step fitting approach can obtain a good balance between the computation efficiency and accuracy.

Keywords: Batch Markovian arrival process · Traffic modeling · Fitting approach · Expectation maximization · Moment matching

Supported by the Guangdong Basic and Applied Basic Research Foundation (2021A1515011984, 2020A1515110824), the National Natural Science Foundation of China (62073346, 11671404, 61573206), and a collaborated project between the Sun Yat-Sen University and the Huawei Company.

1 Introduction

Along with the widely deployed infrastructure of the Internet, Internet protocol (IP) traffic has been becoming the main source of teletraffic data. It is important to study the statistical properties of IP traffic, such as the correlation, self-similarity, and burstyness. For IP traffic, the dilemma of the traffic modeling is to capture these statistical properties of the underlying measured trace data. A variety of traffic models have been developed in the past two decades. Most of the existing traffic models focus on two research directions. One direction is the consideration of non-analytically tractable models, e.g., fractional Gaussian noise model [15]. The other direction is the study of analytically tractable models, e.g., MMPP (Markovian modulated Poisson process) [9] and MAP (Markovian arrival process) [4].

As analytical models, the MAP and MMPP are widely used for traffic modeling, and they are easy to integrate in simulation models. A myriad of useful results has been achieved on these Markovian models for fitting the IP traffic. Some effective and stable fitting approaches of MAP and MMPP for traffic modeling are developed in the literature. These fitting approaches are mainly divided into moments matching (MM) [4,10,11] and expectation maximization (EM) [3,6,18]. Most of these available approaches fit the parameters of MAP or MMPP for the characterization of packet arrivals. Actually, packet arrivals and packet sizes are both observed in the measured traffic data. As we know, the characteristics of packet sizes also paly an important role in the performance analysis of the real traffic data. While, the MAP and MMPP are only used to model the packet arrivals of the traffic flow. As an extensive process of MAP, the stationary BMAP (batch Markovian arrival process) is capable of approximating any stationary batch point process. Moreover, one addition advantage of the BMAP model provides a more comprehensive tool for capturing the packet size distribution of IP traffic, while it still remains analytically tractable. The general BMAP is a highly parameterized model. So far, there have been only a limited number of studies on BMAP fitting. Breuer [1] studied a parameter estimation approach for a class of BMAP models. Then, he proposed an EM algorithm for the BMAP and its comparison to a simpler estimation procedure [2]. Klemm et al. [13] developed an efficient and numerically stable method for estimating the parameters of BMAP with an EM algorithm. Salvador et al. [21] introduced a fitting procedure for discrete-time BMAP which allows a simultaneous matching of joint characterization of packet arrival times and packet sizes. To our knowledge, only these above papers studied the parameter estimation of BMAP models. They all showed that BMAP is analytically tractable and it closely captures the statistics of the measured traffic data.

In this paper, we show how to utilize the EM and MM algorithms for the parameter estimation of BMAPs, by using a two-step fitting procedure. In the first step, the packet interarrival time is fitted by a phase-type distribution which corresponds to the embedded arrival process of BMAP. In the second step, the approximation of the lag correlation values of packet arrivals and packet sizes is computed through a non-linear optimization problem. As we know, the EM algorithm is accurate for Markovian models fitting of teletraffic data, while its computational cost is relatively high. A challenging issue for the teletraffic fitting, especially in practice, is to identify methods to reduce the computational cost. Our approach integrates the advantages of the EM and MM algorithms during the aforementioned two-step fitting procedure. Based on our proposed algorithm, we conduct some numerical experiments which illustrate the accuracy of our fitting method on real traffic data. Furthermore, in order to show the advantage of BMAP traffic model over other widely used analytically tractable models, we compare the BMAP with the MAP and MMPP models by means of visual inspection of sample paths from a real trace data. We demonstrate that BMAP model can capture the important statistical properties of packet arrivals and packet sizes. The numerical results also show that our two-step fitting approach can obtain a good balance between the computation efficiency and accuracy, which is an important feature for implementation in practice.

Our main contributions are summarized as follows. First, we introduce a traffic model and a fitting procedure that provide a detailed characterization of packet arrivals and packet sizes. Second, we show that the proposed traffic model BMAP and its parameters estimation algorithm are capable of matching closely the simulated and real traffic traces. Finally, we make a comparison study with the MAP, MMPP models fitted by the existing EM algorithm. The numerical results show that our proposed two-step EM-MM algorithm is both accurate and numerically efficient. The paper is organized as follows. The mathematical model and some properties of BMAP are introduced in Sect. 2. In Sect. 3, we analyze the implementation details of our proposed two-step fitting procedure. A number of numerical examples are conducted in Sect. 4, which demonstrate the efficiency of our proposed BMAP fitting approach. Finally, we draw concluding remarks with some future discussions in Sect. 5.

2   Basic Properties of BMAP

BMAP was first introduced by Neuts [17]. Its current and more tractable description can refer to the work by Lucantoni [16]. In this section, we firstly present a brief introduction of the BMAP model. The moments and autocorrelations characterization for the BMAP model are then provided.

2.1   Description of BMAP

We denote an m-state BMAP as symbol $BMAP_m(K)$, where K is the maximum batch size. With the model of $BMAP_m(K)$, a doubly stochastic process can be identified as $X(t) = \{J(t), N(t)\}$, where $J(t)$ represents an irreducible Markov process with finite state space $\mathcal{S} = \{1, 2, ..., m\}$, and $N(t)$ denotes the total number of arrivals up to time t. Moreover, $J(t)$ is normally called the *phase process* and $N(t)$ is the *counting process*.

For each state $i \in \mathcal{S}$, the process $J(t)$ spends an exponentially distributed amount of time in state i with rate λ_i. The transition that follows this sojourn can be one of two following types. For the first type, an arrival of batch size k $(k \geq 1)$ occurs, and the process transitions to state $j \in \mathcal{S}$ with probability $p_{ij}(k)$. For the second type, the batch size is 0 indicating no arrival and the process transitions to state $j \neq i$ with probability $p_{ij}(0)$. For each state $i \in \mathcal{S}$, the probabilities $p_{ij}(k)$ satisfy

$$\sum_{k=1}^{K}\sum_{j=1}^{m} p_{ij}(k) + \sum_{j \in \mathcal{S}\setminus\{i\}} p_{ij}(0) = 1.$$

In the context of a $BMAP_m(K)$, the matrices, $\{\mathbf{D}_0, \mathbf{D}_k, k = 1, ..., K\}$, are said to form a representation of the $BMAP_m(K)$, i.e., the $BMAP_m(K)$ is completely specified by these matrices. Here, $\mathbf{D}_0$ is an $m \times m$ matrix, with negative diagonal elements and non-negative off-diagonal elements, which represents the state transitions that correspond to no arrival occurring. $\mathbf{D}_k$ $(1 \leq k \leq K)$ is an $m \times m$ matrix, with non-negative elements, which represents the state transitions that correspond to a batch arrival of size k. For $k = 0, 1, ..., K$, these matrices $\mathbf{D}_k = [d_{i,j}(k)]_{i,j \in \mathcal{S}}$ are defined as follows,

$$d_{i,j}(0) = \begin{cases} -\lambda_i, & j = i, \\ \lambda_i p_{i,j}(0) & j \neq i, \end{cases}$$

$$d_{i,j}(k) = \lambda_i p_{i,j}(k), i, j \in \mathcal{S}.$$

The matrix $\mathbf{D}_0$ is assumed to be stable and nonsingular. The definition of the rate matrices implies that

$$\mathbf{Q} = \sum_{k=0}^{K} \mathbf{D}_k,$$

which is the infinitesimal generator of the underlying Markov process $J(t)$. The steady state probability vector $\boldsymbol{\pi}$ of the process $J(t)$ is the solution of the linear system $\boldsymbol{\pi}\mathbf{Q} = \mathbf{0}$, $\boldsymbol{\pi}\mathbf{e} = 1$ where $\mathbf{e}$ is a column vector of ones.

2.2   Performances of the Interarrival Times and Batch Sizes

Next, we provide some properties of the interarrival times and batch sizes of $BMAP_m(K)$. Its discrete time process embedded at arrival instants play an important role in the performance analysis of the $BMAP_m(K)$ model.

An important property of the $BMAP_m(K)$ concerns Markovian renewal theory. Let S_n denote the state of the process $J(t)$ at the time of the nth arrival event. Then $\{S_n\}_{n=0}^{\infty}$ is a Markov chain and its transition matrix is $\mathbf{P} = (-\mathbf{D}_0)^{-1}\mathbf{D}$ where $\mathbf{D} = \sum_{k=1}^{K} \mathbf{D}_k$. The steady state probability vector $\boldsymbol{\phi}$ of the embedded process is the solution of the linear equations $\boldsymbol{\phi}\mathbf{P} = \boldsymbol{\phi}$, $\boldsymbol{\phi}\mathbf{e} = 1$. From the results in [14], the steady state distributions of the original process $J(t)$ and the embedded process S_n are related as follows

$$\boldsymbol{\phi} = (\boldsymbol{\pi}\mathbf{D}\mathbf{e})^{-1}\boldsymbol{\pi}\mathbf{D}.$$

Interarrival Time. Let T_n denote the interarrival time between the $(n-1)$th and nth events in the $BMAP_m(K)$ model. The variables T_n's are phase-type distributed with representation $\{\phi, \mathbf{D}_0\}$ (see [22] for more details).
The distribution of the interarrival times T_n in the stationary case is

$$P(T < t) = 1 - \phi e^{\mathbf{D}_0 t} \mathbf{e}, \quad t > 0.$$

The moments of T_n in the stationary case are given by

$$\mu_r = E[T^r] = r! \phi(-\mathbf{D}_0)^{-r} \mathbf{e}, \quad r \geq 1.$$

The lag-r correlation function of the sequence of interarrival times is

$$\rho_T(r) = \rho(T_1, T_{r+1}) = \frac{\mu_1 \boldsymbol{\pi}[(-\mathbf{D}_0)^{-1}\mathbf{D}]^{r-1}(-\mathbf{D}_0)^{-1}\mathbf{e} - \mu_1^2}{\mu_2 - \mu_1^2}, \quad r \geq 1. \quad (1)$$

Batch Size. Let B_n denote the batch size of the nth arrival in the $BMAP_m(K)$ model. From the results in [20], we know that
The mass probability function of the stationary batch size, B, is

$$P(B = k) = \phi(-\mathbf{D}_0)^{-1}\mathbf{D}_k \mathbf{e}, \quad k = 1, 2, ..., K.$$

The moments of the stationary batch size B are obtained as

$$\beta_r = E[B^r] = \phi(-\mathbf{D}_0)^{-1}\mathbf{D}_r^* \mathbf{e}, \quad r \geq 1, \quad (2)$$

where $\mathbf{D}_r^* = \sum_{k=1}^{K} k^r \mathbf{D}_k$. Also, the lag-$r$ correlation function in the stationary version of the batch process is given by

$$\rho_B(r) = \rho(B_1, B_{r+1}) = \frac{\phi(-\mathbf{D}_0)^{-1}\mathbf{D}_1^*[(-\mathbf{D}_0)^{-1}\mathbf{D}]^{r-1}(-\mathbf{D}_0)^{-1}\mathbf{D}_1^*\mathbf{e} - \beta_1^2}{\beta_2 - \beta_1^2}. \quad (3)$$

By the joint Laplace Stieltjes transform (LST) of the interarrival times and batch sizes of a stationary $BMAP_m(K)$ given in the Lemma 1 of [20], we obtain $E[TB]$ as follows

$$\eta = E[TB] = \phi(-\mathbf{D}_0)^{-2}\mathbf{D}_1^*\mathbf{e}. \quad (4)$$

Moreover, the covariance between T and B is derived as

$$cov(T, B) = \phi(-\mathbf{D}_0)^{-2}\mathbf{D}_1^*\mathbf{e} - \phi(-\mathbf{D}_0)^{-1}\mathbf{e}\phi(-\mathbf{D}_0)^{-1}\mathbf{D}_1^*\mathbf{e}. \quad (5)$$

All the performances of the interarrival times and batch sizes for the BMAP model obtained above are very important. It will be used in our proposed algorithm for estimating BMAP parameters in the next section. Especially, the analytically tractable BMAP makes it possible for us to take the MM algorithm to consider the parameters estimation. That is, we can compute a BMAP that matches or approximates the performances of the observed processes including the interarrival times and packet sizes for a traffic data.

3   The Fitting Procedure

In terms of the observed information of the teletraffic trace, the sequences of packet interarrival times $\mathbf{t} = (t_1, t_2, ..., t_n)$ and packet sizes $\mathbf{b} = (b_1, b_2, ..., b_n)$ constitute the available observed samples. In this section, we summarize the theoretical issues of the fitting procedure for the $BMAP_m(K)$ model with the observing time series. Based on the properties of the $BMAP_m(K)$ (marginal distribution and lag correlation), we present a general framework of two-step $BMAP_m(K)$ fitting algorithm. Generally speaking, the main idea of the applied approach is that the matrix $\mathbf{D}_0$ and the matrices $\mathbf{D}_k$, $k = 1, 2, ..., K$ are constructed separately.

- In the first step, the interarrival time distribution is fitted by a phase-type distribution from the sequences of interarrival times, This procedure could determine the $\mathbf{D}_0$ matrix and the ϕ vector.
- Then, the $\mathbf{D}_k$, $k = 1, 2, ..., K$ matrices are constructed by solving a non-linear optimization problem, such that the interarrival time distribution of the resulting $BMAP_m(K)$ remains the same, and its lag correlation functions of the interarrival times and batch sizes approximate the traffic data.

3.1   Constructing the D_0 Matrix and the ϕ Vector

At the first step of the procedure, we try to fit a phase type distribution of arrival times. The interarrival time distribution of the original process can be given with its samples or by a given number of moments. Among a number of past research results on PH fitting, we find some fitting methods to solve the phase-type fitting problem [8, 23].

For the estimation of the $\mathbf{D}_0$ matrix and the ϕ vector, we can use the following software tools for analyzing the PH fitting, which are based on EM algorithm or MM algorithm [5]. By reviewing the relative literature, we find three fitting tools (KPC-Toolbox [7], BUTools [12], Mapfit [19]), which could solve the phase-type fitting problem. In our paper, we will use the fitting method from BUTools [12], which provides the fast EM algorithms for PH fitting with teletraffic data.

3.2   Constructing the D_k Matrices

In this subsection, we provide the detailed estimating process of the matrices $\mathbf{D}_k$, $k = 1, 2, ..., K$. The characterization of $BMAP_m(K)$ can be extended from the case $K = 2$ to the case with an arbitrary maximum batch size K. The generalization of such process is due to the fact that given a $BMAP_m(K)$ represented by $\mathbf{B}_K = \{\mathbf{D}_0, \mathbf{D}_1, ..., \mathbf{D}_K\}$, then K different $BMAP_m(2)$s can be obtained as

$$\mathbf{B}_2^i = \{\mathbf{D}_0, \mathbf{D}_i, \sum_{k \neq i} \mathbf{D}_k\}, \quad i = 1, 2, ..., K.$$

Then, we show the constructing process of $\mathbf{D}_1$ in the $BMAP_m(2)$ model, i.e., $\mathbf{B}_2^1 = \{\mathbf{D}_0, \mathbf{D}_1, \mathbf{D}_{-1}\}$ where $\mathbf{D}_{-1} = \mathbf{D}_2 + \mathbf{D}_3 + \cdots + \mathbf{D}_K$. It should be pointed out that, in order to compute the empirical moment β_r and lag-r correlation $\bar{\rho}_B(r)$, all batch sizes in $\mathbf{b}$ larger than 2 are considered as equal to 2.

Constraints of the $\mathbf{D}_1$ and $\mathbf{D}_{-1}$ Matrices. Once the matrix $\mathbf{D}_0$ and the vector ϕ are obtained by the first step in Sect. 3.1, we consider the constraints of the matrices $\mathbf{D}_1$ and $\mathbf{D}_{-1}$ based on the performances of the $BMAP_m(K)$ model. The matrices $\mathbf{D}_1$ and $\mathbf{D}_{-1}$ have to satisfy the following two constraints to maintain the interarrival time distribution determined in the first step:

- C1: $(\mathbf{D}_1 + \mathbf{D}_{-1})\mathbf{e} = -\mathbf{D}_0\mathbf{e}$,
- C2: $\phi(-\mathbf{D}_0)^{-1}(\mathbf{D}_1 + \mathbf{D}_{-1}) = \phi$.

Exact Performance Measures Fitting. From Sect. 2.2, we find that the moments of the batch size β_r and the joint expected of interarrival time and batch size $E[TB]$ can also be expressed as linear constraints. Specially, by Eq. (3), we know that the lag correlation function of the batch process depends on the first and second moments of the batch sizes. Here we provide the following linear constraints to exact β_1, β_2 (related to the batch size distribution) and $\eta = E[TB]$ (joint moments concerning interarrival times and batch sizes) fitting.

- C3: $\beta_1 = \phi(-\mathbf{D}_0)^{-1}(\mathbf{D}_1 + 2\mathbf{D}_{-1})\mathbf{e}$,
- C4: $\beta_2 = \phi(-\mathbf{D}_0)^{-1}(\mathbf{D}_1 + 4\mathbf{D}_{-1})\mathbf{e}$,
- C5: $\eta = \phi(-\mathbf{D}_0)^{-2}(\mathbf{D}_1 + 2\mathbf{D}_{-1})\mathbf{e}$.

We formulate these constrains $C1 \sim C5$ as a linear system of equations. To do so, we introduce a column vector $\mathbf{x}$ (of size $2m^2$), which is composed by the columns of the matrices $\mathbf{D}_1$ and $\mathbf{D}_{-1}$ as below.

$$\mathbf{x} = \begin{pmatrix} \{\mathbf{D}_1\}_1 \\ \vdots \\ \{\mathbf{D}_1\}_m \\ \{\mathbf{D}_{-1}\}_1 \\ \vdots \\ \{\mathbf{D}_{-1}\}_m \end{pmatrix}.$$

All possible $\mathbf{x}$ vectors (thus, $\mathbf{D}_1$ and $\mathbf{D}_{-1}$ matrices) satisfying constraints $C1 \sim C5$ are the solutions of the following linear equations with coefficient matrix $\mathcal{A}$

$$\begin{bmatrix} I & I & \cdots & I & I & I & \cdots & I \\ \omega & & & & \omega & & & \\ & \omega & & & & \omega & & \\ & & \ddots & & & & \ddots & \\ & & & \omega & & & & \omega \\ \omega & \omega & \cdots & \omega & 2\omega & 2\omega & \cdots & 2\omega \\ \omega & \omega & \cdots & \omega & 4\omega & 4\omega & \cdots & 4\omega \\ \varphi & \varphi & \cdots & \varphi & 2\varphi & 2\varphi & \cdots & 2\varphi \end{bmatrix}_{(2m+3)\times 2m^2} \cdot \begin{bmatrix} \\ \\ \mathbf{x} \\ \\ \\ \end{bmatrix}_{2m^2 \times 1} = \begin{bmatrix} d \\ \\ \phi \\ \\ \\ \beta_1 \\ \beta_2 \\ \eta \end{bmatrix}_{(2m+3)\times 1}, \quad (6)$$

where $\omega = \phi(-\mathbf{D}_0)^{-1}$, $\varphi = \phi(-\mathbf{D}_0)^{-2}$ and $d = -\mathbf{D}_0 e$. The first m lines of $\mathcal{A}$ correspond to constraint $C1$, and the second m lines of $\mathcal{A}$ are related to constraint $C2$. The last three lines of $\mathcal{A}$ correspond to constraint $C3 \sim C5$. Proper matrices $\mathbf{D}_1$ and $\mathbf{D}_{-1}$ (i.e., $\mathbf{x}$ vector) satisfy the following set of linear equations and inequalities:

$$\mathcal{A}\mathbf{x} = b, \quad \mathbf{x} \geq 0. \tag{7}$$

As we can see, the $\mathcal{A}\mathbf{x} = b$ equation is under-determined for $m \geq 2$, since we have $2m + 3$ equations and $2m^2$ unknowns. So that, the $\mathbf{D}_0$ matrix, the ϕ vector, and the performance measures $\{\beta_1, \beta_2, \eta\}$ cannot determine the $\mathbf{D}_1$ and $\mathbf{D}_{-1}$ matrices when $m \geq 2$. Moreover, the study in [20] explores the identifiability of the stationary two-state $BMAP_m(K)$. The results in [20] show that the $BMAP_m(K)$ is not identifiable if $m = 2$ and $K \geq 2$.

Fitting with Lag Correlations. In order to estimate the matrices $\mathbf{D}_1$ and $\mathbf{D}_{-1}$, we use the lag correlation functions of the interarrival times and batch sizes. To fit lag correlation values, we define an optimization problem with the linear constraints (7) such that a properly chosen goal function ensures the approximation of lag correlation values. The fitting of a given number of lag correlations is a linearly constrained nonlinear optimization problem.

We apply the objective function $c(\mathbf{x})$, which is the squared difference between the lag correlations of the original process for $(\overline{\rho}_T(r), \overline{\rho}_B(r))$ and the fitted $(\rho_T(r), \rho_B(r))$ of the $BMAP_m(K)$ model weighted with w_r and v_r, respectively. Then we can set up the following nonlinear optimization problem

$$c(\mathbf{x}) = \sum_{r=1}^{R_1} w_r \left[\frac{\rho_T(r) - \overline{\rho}_T(r)}{\overline{\rho}_T(r)}\right]^2 + \sum_{r=1}^{R_2} v_r \left[\frac{\rho_B(r) - \overline{\rho}_B(r)}{\overline{\rho}_B(r)}\right]^2. \tag{8}$$

In the objective function (8), $\rho_T(r)$ and $\rho_B(r)$ are the functions w.r.t. $\mathbf{x}$ (thus, $\mathbf{D}_1$ and $\mathbf{D}_{-1}$ matrices), presented in Eqs. (1) and (3), respectively. The largest lag correlation coefficient of interarrival times and batch sizes considered in this objective function are the lag-R_1 and lag-R_2 correlation coefficients, respectively. The weights w_r and v_r can be used to increase the importance of the accuracy of lower lag-r correlation with respect to higher ones or vice-versa.

Once the $\mathbf{D}_1$ matrix is obtained as the solutions of the nonlinear optimization problem with linear constraints (7), the approach will be repeated for estimating $\mathbf{D}_2$ (using the representation of $\mathbf{B}_2^2 = \{\mathbf{D}_0, \mathbf{D}_2, \mathbf{D}_{-2}\}$), $\mathbf{D}_3, ...,$ and finally $\mathbf{D}_K$. The computation process is summarized in Algorithm 1. It is important to comment that the optimization problems of (9) in Algorithm 1 for $k = 1, ..., K$ are straightforward problems in $2m^2$ variables each, solved by using standard optimization routines (fmincon in MATLAB), where a multistart with 100 randomly chosen starting points was executed.

4   Numerical Experiments

To illustrate the fitting properties of our proposed approach, we conduct two numerical experiments in this section. The first one approximates the sample data generated by a given $BMAP_m(K)$. We show how close the result of

---

Algorithm 1. A two-step EM-MM algorithm for fitting $BMAP_m(K)$.

---

1. **Input:**
Trace: the sequences of interarrival times $\mathbf{t} = (t_1, t_2, ..., t_n)$ and batch sizes $\mathbf{b} = (b_1, b_2, ..., b_n)$.

2. **A phase-type fitting for interarrival times**
• Obtain $\mathbf{D}_0$ and ϕ by the EM algorithm (from BUTools) on the trace of interarrival times $\mathbf{t} = (t_1, t_2, ..., t_n)$.

3. **The lag correlation fitting for interarrival time and batch sizes**
For $k = 1, ..., K - 1$ repeat:
• Compute the empirical moments $\{\beta_1, \beta_2, \eta\}$ and lag-r correlation $\{\rho_B(r), \rho_T(r)\}$ of the trace of interarrival times $\mathbf{t} = (t_1, t_2, ..., t_n)$ and batch sizes $\mathbf{b} = (b_1, b_2, ..., b_n)$ by Eqs. (1)–(4).
• From the obtained $\mathbf{D}_0, ..., \mathbf{D}_{k-1}$, consider the $BMAP_m(2)$: $\mathbf{B}_2^k = \{\mathbf{D}_0, \mathbf{D}_i, \sum_{i \neq k} \mathbf{D}_i\}$, and obtain $\mathbf{D}_k$ as the solutions of

$$\min \sum_{r=1}^{R_1} w_r \left[\frac{\rho_T(r) - \overline{\rho}_T(r)}{\overline{\rho}_T(r)} \right]^2 + \sum_{r=1}^{R_2} v_r \left[\frac{\rho_B(r) - \overline{\rho}_B(r)}{\overline{\rho}_B(r)} \right]^2 \tag{9}$$

$$s.t. \; \mathcal{A}\mathbf{x} = b, \quad \mathbf{x} \geq 0.$$

It can be solved by using standard optimization routines (fmincon in MATLAB), where a multistart with 100 randomly chosen starting points is executed.

4. **Output:**
The parameters of the fitted $BMAP_m(K)$ model: $M = \{\mathbf{D}_0, \mathbf{D}_1, ..., \mathbf{D}_K\}$.

---

our fitting algorithm is in terms of the performances on the interarrival times and packet sizes. In the second experiment, a real teletraffic data is fitted of a $BMAP_m(K)$ model by our proposed Algorithm 1. This teletraffic trace data is measured from a popular video game, which is provided from a practical project collaborated with Huawei company.

4.1 BMAP Traffic Modeling Framework

In our numerical experiments, the first data set was simulated from a $BMAP_3(2)$ model represented by the rate matrices $\{\mathbf{D}_0, \mathbf{D}_1, \mathbf{D}_2\}$ shown in the second column of the top part of Table 1. Based on this parameter set, a trace file with $n = 100,000$ arrivals and corresponding batch sizes is generated. This trace file is used as input for our proposed Algorithm 1 in order to derive the (known) parameter set of the BMAP model. This numerical experiment was conduct to demonstrate the accuracy of our proposed two-step EM-MM algorithm for fitting BMAP. The results obtained by our proposed approach are shown in Table 1. The second column in Table 1 presents the generator process and the characterizing theoretical performances according to Sect. 2.2. The third column in Table 1 presents the empirical moments from the simulated traces. The rest of columns present the estimated rate matrices and estimated characterizing performances by using the Algorithm 1. As can be seen in Table 1, very good results were obtained since the fitted BMAP matches closely the performances of the empirical data. Moreover, we should note that the estimated parameters of the fitted BMAP are different from that of the generator process. This is due to the fact that the $BMAP_m(K)$ is not identifiable when $K \geq 2$.

Table 1. Performances of the estimation method for a simulated trace from a $BMAP_3(2)$.

	Generator process	Empirical	Fitted process
$\mathbf{D}_0$	$\begin{pmatrix} -10 & 0 & 0 \\ 0 & -20 & 0 \\ 0 & 0 & -30 \end{pmatrix}$	–	$\begin{pmatrix} -8.6109 & 0 & 0 \\ 0 & -12.7691 & 0 \\ 0 & 0 & -26.1805 \end{pmatrix}$
$\mathbf{D}_1$	$\begin{pmatrix} 2 & 3 & 1 \\ 0 & 3 & 5 \\ 10 & 8 & 2 \end{pmatrix}$	–	$\begin{pmatrix} 0.5926 & 0.7304 & 2.5585 \\ 0.0000 & 3.0802 & 4.5875 \\ 0.5646 & 3.6611 & 8.9966 \end{pmatrix}$
$\mathbf{D}_2$	$\begin{pmatrix} 1 & 1 & 2 \\ 3 & 7 & 2 \\ 5 & 2 & 3 \end{pmatrix}$	–	$\begin{pmatrix} 1.1467 & 1.7818 & 1.8008 \\ 0.0000 & 3.2711 & 1.8304 \\ 3.1485 & 3.6417 & 6.1681 \end{pmatrix}$
μ_1	5.9935e−2	5.9796e−2	6.0122e−2
β_1	1.4207e−1	9.6640e−2	1.3730e−1
$\rho_B(1)$	1.4658	1.4680	1.4660
$\rho_T(1)$	5.5148e−5	7.8551e−3	6.3589e−4
$E[TB]$	8.7514e−2	8.7414e−2	8.7812e−2
$Cov(T,B)$	3.3882e−4	3.7050e−4	3.2825e−4

4.2 Comparative Study of IP Traffic Modeling

As a real example, we use the trace from a popular game which is provided from our project collaborated with Huawei company. The underlying trace traffic including 393206 packets, comprises packet interarrival times and the corresponding packet lengths. The analysis of the packet length distributions reveals that the packet lengths of the real traffic trace follow to a large extent of discrete values. Recalling the definition of $BMAP_m(K)$ in Sect. 2, the mapping process of packet lengths to $BMAP_m(K)$ rewards results in K kinds of different batch sizes. We map the packet lengths from a real traffic trace onto the discrete packet lengths s_k, for $1 \leq k \leq K$, where s_k is the average length of all packets of the considered trace comprising packet lengths between $\frac{L(k-1)}{K}$ bytes and $\frac{Lk}{K}$ bytes, where L denotes the maximum packet length of the real traffic trace. The aggregated traffic model utilizes these observations where different reward values of the $BMAP_m(K)$ represent different discrete packet lengths. In our numerical study, the parameter estimation procedure is applied for the $BMAP_m(K)$ model with $K = 2$ distinct batch sizes. The largest packet length is $L = 1506$ bytes in the original WZRY trace.

We consider a $BMAP_3(2)$ model to capture the packet arrival process of the considered trace by using Algorithm 1. We take the parameters $R_1 = R_2 = 5$ in our numerical experiments. We use weights $w_r = v_r = 10^{-(r-1)}$ for the fitting of lag-r correlations for interarrival times and batch sizes. The average packet lengths of our measurements are as follows: $s_1 = 77.4887$ bytes and $s_2 = 1113.3234$ bytes. To show the efficiency and accuracy of the BMAP traffic model and our proposed fitting method, we make a numerical comparison study with the other existing Markovian models, MAP and MMPP. Note that the parameter matrices $\mathbf{D}_0$ and $\mathbf{D}_1$ of the MAP, MMPP were estimated by means of the EM algorithm from BUTools [12]. For the BMAP, MAP, MMPP, the corresponding fitting time and estimated parameter sets are derived as follows:

(1) BMAP model: the fitting time is 292.32 s and the estimated parameters are as follows

$$\mathbf{D}_0 = \begin{pmatrix} -237.00 & 141.46 & 74.94 \\ 135.72 & -13792.60 & 6216.24 \\ 113.55 & 2032.53 & -14919.27 \end{pmatrix}, \mathbf{D}_1 = \begin{pmatrix} 8.34 & 9.18 & 3.10 \\ 0.89 & 6089.52 & 899.54 \\ 0.11 & 631.49 & 8.85 \end{pmatrix}.$$

$$\mathbf{D}_2 = \begin{pmatrix} 0 & 0 & 0 \\ 0.02 & 441.38 & 9.29 \\ 0 & 0 & 0 \end{pmatrix}.$$

(2) MAP model: the fitting time is 536.92 s and the estimated parameters are as follows

$$\mathbf{D}_0 = \begin{pmatrix} -17011.63 & 8.98 & 3573.99 \\ 1.46 & -335.49 & 1.06 \\ 43.97 & 0.73 & -2052.95 \end{pmatrix}, \mathbf{D}_1 = \begin{pmatrix} 9083.65 & 40.13 & 4304.88 \\ 1.89 & 329.97 & 1.10 \\ 738.95 & 0.68 & 1268.62 \end{pmatrix}.$$

(3) MMPP model: the fitting time is 433.13 s and the estimated parameters are as follows

$$\mathbf{D}_0 = \begin{pmatrix} -3042.07 & 2150.33 & 0.31 \\ 6511.04 & -16053.45 & 22.14 \\ 2.29 & 3.26 & -334.04 \end{pmatrix}, \mathbf{D}_1 = \begin{pmatrix} 891.43 & 0 & 0 \\ 0 & 9520.26 & 0 \\ 0 & 0 & 328.49 \end{pmatrix}.$$

Table 2. The moments of interarrival times of the fitted models and original process.

Moment ordinal	Raw trace	BMAP	MAP	MMPP
$E[T]$	$6.0254e^{-4}$	$6.0328e^{-4}$	$6.5187e^{-4}$	$6.1063e^{-4}$
$E[T^2]$	$2.8704e^{-6}$	$2.6430e^{-6}$	$2.4471e^{-6}$	$2.1487e^{-6}$
$E[T^3]$	$6.5608e^{-8}$	$2.9795e^{-8}$	$2.0150e^{-8}$	$1.6958e^{-8}$
$E[T^4]$	$6.6922e^{-9}$	$4.8607e^{-10}$	$2.4367e^{-10}$	$1.9352e^{-10}$

In this numerical experiment, the parameters of a BMAP model are estimated by our proposed two-step EM-MM algorithm, and the parameters of the MAP and MMPP models are estimated by the EM algorithm from [12]. The numerical results show that the fitting time of the BMAP model is almost half of the

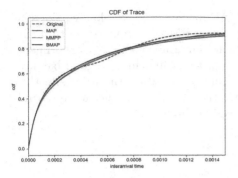

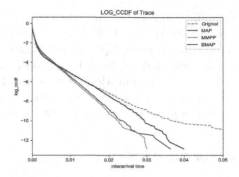

Fig. 1. The CDF of interarrival times of the fitted MMPP, MAP, BMAP and the original process.

Fig. 2. The Log(ccdf) of interarrival times of the fitted MMPP, MAP, BMAP and the original process.

fitting time of the MAP, MMPP models, which illustrates that our proposed two-step EM-MM algorithm for fitting the BMAP model is very efficient. Moreover, we present statistical properties for the measured traffic data, and the fitted BMAP, MAP, MMPP models in terms of the interarrival times. As we can see, Table 2 shows the moments of the interarrival times of the fitted MMPP, MAP, BMAP models and the original process. We observe that the mean and standard deviation of the interarrival times approximation are very accurate. Figure 1 presents the CDF (cumulative distribution function) of the interarrival times of the fitted BMAP, MAP, MMPP models and the original process. From Fig. 1, we see that the three statistic models perform very well w.r.t. the CDF of interarrival times. Moreover, Fig. 2 gives the Log of CCDF (complementary CDF) of the interarrival times of the fitted MMPP, MAP, BMAP models and the original process. It is obvious to find that the CCDF of the measured traffic and the BMAP, MAP, MMPP models perform different. The BMAP model is slightly better than the MAP and MMPP models.

5 Conclusions

In this paper, we present a two-step parameter fitting approach for BMAP to model real teletraffic traces. The approach is based on a two-step procedure combining EM and MM algorithms, where the first step is the phase-type fitting of the interarrival times and the second step is the lag correlation fitting of interarrival times and packet sizes. We evaluate our proposed BMAP fitting approach in two ways. We perform a fitting experiment for BMAP based on a trace data set generated by a given BMAP, such that we can evaluate the accuracy of our fitting method with theoretical values. We also conduct the second experiment fitting the models of BMAP, MAP, MMPP based on a real teletraffic trace, respectively. The numerical results illustrate the advantages of the BMAP modeling approach over other widely used analytically tractable models, MAP and MMPP. Such advantage mainly comes from the strong approximation capability of BMAP, since the class of BMAP includes the well-known Poisson process, MMPP, and MAP as special cases, and BMAP is able to capture the interarrival time and packet size distributions simultaneously. The experiment results also show that our two-step fitting approach can achieve a good balance between the computation efficiency and accuracy. Based on the results in this paper, one direction for future research is to study the performance analysis and optimization of the queueing system with BMAP inputs.

References

1. Breuer, L.: Parameter estimation for a class of BMAPs. In: Latouche, G., Taylor, P. (eds.) Advances in Algorithmic Methods for Stochastic Models: Proceedings of the 3rd International Conference on Matrix Analytic Methods, pp. 87–97. Notable Publications, Neshanic Station, New Jersey, USA, July 2000
2. Breuer, L.: An EM algorithm for batch Markovian arrival processes and its comparison to a simpler estimation procedure. Ann. Oper. Res. 112(1), 123–138 (2002)
3. Buchholz, P.: An EM-algorithm for MAP fitting from real traffic data. In: Kemper, P., Sanders, W.H. (eds.) TOOLS 2003. LNCS, vol. 2794, pp. 218–236. Springer, Heidelberg (2003). https://doi.org/10.1007/978-3-540-45232-4_14
4. Buchholz, P., Kriege, J.: A heuristic approach for fitting MAPs to moments and joint moments. In: 2009 Sixth International Conference on the Quantitative Evaluation of Systems, pp. 53–62. IEEE (2009)
5. Buchholz, P., Kriege, J., Felko, I.: Software tools. In: Input Modeling with Phase-Type Distributions and Markov Models. SM, pp. 111–114. Springer, Cham (2014). https://doi.org/10.1007/978-3-319-06674-5_7
6. Buchholz, P., Panchenko, A.: A two-step EM algorithm for MAP fitting. In: Aykanat, C., Dayar, T., Körpeoğlu, İ (eds.) ISCIS 2004. LNCS, vol. 3280, pp. 217–227. Springer, Heidelberg (2004). https://doi.org/10.1007/978-3-540-30182-0_23
7. Casale, G., Zhang, E.Z., Smirni, E.: KPC-toolbox: simple yet effective trace fitting using Markovian arrival processes. In: 2008 Fifth International Conference on Quantitative Evaluation of Systems, pp. 83–92. IEEE (2008)
8. Fiondella, L., Puliafito, A.: Principles of Performance and Reliability Modeling and Evaluation. Springer (2016). https://doi.org/10.1007/978-3-319-30599-8
9. Fischer, W., Meier-Hellstern, K.: The Markov-modulated Poisson process (MMPP) cookbook. Performance Eval. 18(2), 149–171 (1993)
10. Heindl, A., Horvath, G., Gross, K.: Explicit inverse characterizations of acyclic MAPs of second order. Formal methods, pp. 108–122 (2006)
11. Horvath, G., Buchholz, P., Telek, M.: A MAP fitting approach with independent approximation of the inter-arrival time distribution and the lag correlation, pp. 124–133 (2005)
12. Horváth, G., Telek, M.: BuTools 2: a rich toolbox for Markovian performance evaluation. In: VALUETOOLS (2016)
13. Klemm, A., Lindemann, C., Lohmann, M.: Modeling IP traffic using the batch Markovian arrival process. Perform. Eval. 54(2), 149–173 (2003)
14. Latouche, G., Ramaswami, V.: Introduction to matrix analytic methods in stochastic modeling. J. Am. Stat. Assoc. 95(452), 1379 (1999)

15. Ledesma, S., Liu, D.: Synthesis of fractional Gaussian noise using linear approximation for generating self-similar network traffic. ACM SIGCOMM Comput. Commun. Rev. **30**(2), 4–17 (2000)
16. Lucantoni, D.M.: New results on the single server queue with a batch Markovian arrival process. Commun. Stat. Stochas. Mod. **7**(1), 1–46 (1991)
17. Neuts, M.F.: A versatile Markovian point process. J. Appl. Probab. **16**(4), 764–779 (1979)
18. Okamura, H., Dohi, T.: Faster maximum likelihood estimation algorithms for Markovian arrival processes, pp. 73–82, September 2009
19. Okamura, H., Dohi, T.: Fitting phase-type distributions and Markovian arrival processes: algorithms and tools. In: Fiondella, L., Puliafito, A. (eds.) Principles of Performance and Reliability Modeling and Evaluation. SSRE, pp. 49–75. Springer, Cham (2016). https://doi.org/10.1007/978-3-319-30599-8_3
20. Rodriguez, J., Lillo, R.E., Ramirezcobo, P.: Nonidentifiability of the two-state BMAP. Methodol. Comput. Appl. Probab. **18**(1), 81–106 (2016)
21. Salvador, P., Pacheco, A., Valadas, R.: Modeling IP traffic: joint characterization of packet arrivals and packet sizes using BMAPs. Comput. Netw. **44**(3), 335–352 (2004)
22. Telek, M., Horvath, G.: A minimal representation of Markov arrival processes and a moments matching method. Perform. Eval. **64**(9), 1153–1168 (2007)
23. Thummler, A., Buchholz, P., Telek, M.: A novel approach for fitting probability distributions to real trace data with the EM algorithm. In: 2005 International Conference on Dependable Systems and Networks (DSN 2005), pp. 712–721. IEEE (2005)

Markov Chains and Hitting Times for Error Accumulation in Quantum Circuits

Long Ma[1(✉)] and Jaron Sanders[2]

[1] Faculty of Electrical Engineering, Mathematics, and Computer Science,
Delft University of Technology, Delft, Netherlands
`l.ma-2@tudelft.nl`
[2] Department of Mathematics and Computer Science,
Eindhoven University of Technology, Eindhoven, Netherlands
`jaron.sanders@tue.nl`

Abstract. We study a classical model for the accumulation of errors in multi-qubit quantum computations. By modeling the error process in a quantum computation using two coupled Markov chains, we are able to capture a weak form of time-dependency between errors in the past and future. By subsequently using techniques from the field of discrete probability theory, we calculate the probability that error quantities such as the fidelity and trace distance exceed a threshold analytically. The formulae cover fairly generic error distributions, cover multi-qubit scenarios, and are applicable to the randomized benchmarking protocol. To combat the numerical challenge that may occur when evaluating our expressions, we additionally provide an analytical bound on the error probabilities that is of lower numerical complexity. Besides this, we study a model describing continuous errors accumulating in a single qubit. Finally, taking inspiration from the field of operations research, we illustrate how our expressions can be used to decide how many gates one can apply before too many errors accumulate with high probability, and how one can lower the rate of error accumulation in existing circuits through simulated annealing.

Keywords: Markov chains · Error accumulation · Quantum circuits

1 Introduction

The development of a quantum computer is expected to revolutionize computing by being able to solve hard computational problems faster than any classical computer [37]. However, present-day state-of-the-art quantum computers are prone to errors in their calculations due to physical effects such as unwanted qubit–qubit interactions, qubit crosstalk, and state leakage [38]. Minor errors can be corrected, but error correction methods will still be overwhelmed once too

© ICST Institute for Computer Sciences, Social Informatics and Telecommunications Engineering 2021
Published by Springer Nature Switzerland AG 2021. All Rights Reserved
Q. Zhao and L. Xia (Eds.): VALUETOOLS 2021, LNICST 404, pp. 36–55, 2021.
https://doi.org/10.1007/978-3-030-92511-6_3

many errors occur [12,20,31]. Quantum circuits with different numbers of qubits and circuit depths have been designed to implement algorithms more reliably [16], and the susceptibility of a circuit to the accumulation of errors remains an important evaluation criterion. We therefore study now Markov chains that provide a model for the accumulation of errors in quantum circuits. Different types of errors [21] that can occur and are included in our model are e.g. Pauli channels [37], Clifford channels [23,34], depolarizing channels [37], and small rotational errors [7,26]. If the random occurrence of such errors only depends on the last state of the quantum mechanical system, then the probability that error quantities such as the fidelity and trace distance accumulate beyond a threshold can be related to different hitting time distributions of two coupled Markov chains [8]. These hitting time distributions are then calculated analytically using techniques from probability theory and operations research.

Error accumulation models that share similarities with the Markov chains under consideration here can primarily be found in the literature on randomized benchmarking [46]. From the modeling point of view, the dynamical description of error accumulation that we adopt is shared in [3,27,33,43]. These articles however do not explicitly tie the statistics of error accumulation to a hitting time analysis of a coupled Markov chain. Furthermore, while Markovianity assumptions on noise are common [14], the explicit mention of an underlying random walk is restricted to a few papers only [3,17]. From the analysis point of view, research on randomized benchmarking has predominantly focused on generalizing expressions for the expected fidelity over time. For example, the expected decay rates of the fidelity are analyzed for cases of randomized benchmarking with restricted gate sets [9], Gaussian noise with time-correlations [15], gate-dependent noise [43], and leakage errors [45]; and the expected loss rate of a protocol related to randomized benchmarking is calculated in [10,33,39,43,44]. In this article, we focus instead on the probability distributions of both the error and maximum error in the Markov chain model – which capture the statistics in more detail than an expectation – for arbitrary distance measures, and in random as well as nonrandom quantum circuits. Finally, [3,33,43,45] resort to perturbation or approximate analyses (via e.g. Taylor expansions, and independence or decorrelation assumptions) to characterize the fidelity, whereas here we provide the exact, closed-form expressions for the distributions using the theory of Markov chains.

To be precise: this article first studies a model for discrete Markovian error accumulation in a multi-qubit quantum circuit. We suppose for simplicity that both the quantum gates and errors belong to a finite unitary group $\mathcal{G}_n \subseteq \mathcal{U}(2^n)$, where $\mathcal{U}(2^n)$ is the unitary group for n qubits. The group $\mathcal{G}_n$ can e.g. be the generalized Pauli group (i.e., the discrete Heisenberg–Weyl group), or the Clifford group. By modeling the quantum computation with and without errors as two coupled Markov chains living on the state space consisting of pairs of elements from these groups, we are able to capture a weak form of time-dependency within the process of error accumulation. To see this, critically note that the assumption of a Markov property does not imply that the past and the future in the quantum computation are independent given any information concerning the present [8]. We must also note that while the individual elements of our two-dimensional

Markov chain belong to a group, the two-dimensional Markov chain itself, here, is generally not a random walk on a group. Lastly, our Markov chain model works for an arbitrary number of qubits. These model features are all relevant to the topic of error modeling in quantum computing, and since the Markov property is satisfied in randomized benchmarking, the model has immediate application. The method is generic in the sense that any measure of distance between two pure quantum states may be used to quantify the error, and that it allows for a wide range of error distributions. The method can handle nonuniform, gate-, and time-dependent errors. Concretely, for arbitrary measure of distance and a wide range of error distributions, we will calculate (i) the expected error at time t, (ii) the probability that an error is larger than a threshold δ at time t, and (iii) the probability that the error has *ever* been larger than a threshold δ before time t, and we do so both for random and nonrandom circuits.

In addition to studying a model for discrete Markovian error accumulation in quantum circuits, we also briefly study a random walk model on the three-dimensional sphere [40]. This model is commonly used to describe the average dephasing of a single qubit (or spin) [24]. Using this model, we characterize the distribution and expectation of the trace distance measuring the error that is accumulated over time. These derivations are, essentially, refinements that provide information about the higher-order statistics of the error accumulation in a single qubit.

The approach taken in this article is a hybrid between classical probability theory and quantum information theory. This hybridization allows us to do quite detailed calculations, but not every quantum channel will satisfy the necessary assumptions such as Markovianity of the error distribution. On the other hand, in cases where one introduces their own source of randomness (such as in randomized benchmarking), the assumptions are met naturally. It should furthermore be noted that the numerical complexity of the exact expressions we provide is high for large quantum circuits. The precise difficulty of evaluating our expressions depends on the particulars of the quantum circuit one looks at. For practical purposes, we therefore also provide an explicit bound on the maximum error probability that is of lower numerical complexity. Reference [27] is relevant to mention here, because similar to our observations, these authors also note the generally high computational complexity of error analysis in quantum circuits. The issue is approached in [27] differently and in fact combinatorially by converting circuits into directed graphs, tracing so-called fault-paths through these graphs, and therewith estimating the success rates of circuits.

Finally, we use the expressions that describe how likely it is that errors accumulate to answer two operational questions that will help advance the domain of practical quantum computing [29]. First, we calculate and bound analytically how many quantum gates $t^\star_{\delta,\gamma}$ one can apply before an error measure of your choice exceeds a threshold δ with a probability above γ. This information is useful for deciding how often a quantum computer should perform repairs on qubits, and is particularly opportune at this moment since quantum gates fail $O(0.1–1\%)$ of the time [29]. Related but different ideas can be found in [21, §2.3], where the accumulation of bit-flips and rotations on a repetition code is studied and a time to failure is derived, and in [25, §V], where an upper bound on the number of necessary measurements for a randomized benchmarking protocol is derived. Second, using techniques from optimization, we design a simulated annealing method that improves existing circuits by swapping out gate pairs to achieve lower rates of error accumulation. There is related literature where the aim is to reduce the circuit depth [2,28,35], but an explicit expression for error accumulation has not yet been leveraged in the same way. Moreover, we also discuss conditions under which this tailor-made method is guaranteed to find the best possible circuit. Both of these excursions illustrate how the availability of an analytical expression for the accumulation of errors allows us to proceed with second-tier optimization methods to facilitate quantum computers in the long-term. We further offer an additional proof-of-concept that simulated annealing algorithms can reduce error accumulation rates in existing quantum circuits when taking error distributions into account: we illustrate that the misclassification probability in a circuit that implements the Deutsch–Jozsa Algorithm for one classical bit [11,13] can be lowered by over 40%. In this proof of concept we have chosen an example error distribution that is gate-dependent and moreover one that is such that *not* applying a gate gives the lowest error rate in this model; applying a single-qubit gate results in a medium error rate; and applying a two-qubit gate gives the largest probability that an error may occur.

This paper is structured as follows. In Sect. 2, we give the model aspects pertaining to the quantum computation (gates, error dynamics, and error measures) and we introduce the coupled Markov chain to describe error accumulation. In Sect. 3, we provide the relation between the probability of error and the hitting time distributions, and we derive the error distributions as well as its bound. We also calculate the higher-order statistics of an error accumulation model for a single qubit that undergoes (continuous) random phase kicks and depolarization. In Sect. 4, we illustrate our theoretical results by comparing to numerical results of a quantum simulator we wrote for this article. In Sect. 5, we discuss the simulated annealing scheme. Finally, we conclude in Sect. 6.

2   Model and Coupled Markov Chain

2.1   Gates and Errors in Quantum Computing

It is generally difficult to describe large quantum systems on a classical computer for the reason that the state space required increases exponentially in size with the number of qubits [36]. However, the stabilizer formalism is an efficient tool to analyze such complex systems [18]. Moreover, the stabilizer formalism covers many paradoxes in quantum mechanics [1], including the Greenberger–Horne–Zeilinger (GHZ) experiment [22], dense quantum coding [5], and quantum teleportation [4]. Specifically, the stabilizer circuits are the smallest class of quantum circuits that consist of the following four gates: $\omega = e^{i\pi/4}$, $H = (1/\sqrt{2})((1,1);(1,-1))$, $S = ((1,0);(0,i))$, and $Z_c = ((1,0,0,0);(0,1,0,0);(0,0,1,0);(0,0,0,-1))$. These four gates are closed under the operations of tensor product and composition [42]. As a consequence of the Gottesman–Knill theorem, stabilizer circuits can be efficiently simulated on a classical computer [19].

Unitary stabilizer circuits are also known as the Clifford circuits; the Clifford group $\mathcal{C}_n$ can be defined as follows. First: let $P \triangleq \{I, X, Y, Z\}$ denote the Pauli matrices, so $I = ((1,0); (0,1))$, $X = ((0,1); (1,0))$, $Y = ((0,-i); (i,0))$, and $Z = ((1,0); (0,-1))$, and let $P_n \triangleq \{\sigma_1 \otimes \cdots \otimes \sigma_n \mid \sigma_i \in P\}$ denote the Pauli matrices on n qubits. The Pauli matrices are commonly used to model errors that can occur due to the interactions of the qubit with its environment [41]. In the case of a single qubit, the matrix I represents that there is no error, the matrix X that there is a bit-flip error, the matrix Z that there is a phase-flip error, and the matrix Y that there are both a bit-flip and a phase-flip error. The multi-qubit case interpretations follow analogously. Second: let $P_n^* = P_n / I^{\otimes n}$. We now define the Clifford group on n qubits by $\mathcal{C}_n \triangleq \{U \in \mathcal{U}(2^n) \mid \sigma \in \pm P_n^* \Rightarrow U\sigma U^\dagger \in \pm P_n^*\}/\mathcal{U}(1)$.

The fact that $\mathcal{C}_n$ is a group can be verified by checking the two necessary properties (see our extended version [32]). The Clifford group on n qubits is finite [30], and we will ignore the global phase throughout this paper for convenience; its size is then $|\mathcal{C}_n| = 2^{n^2+2n} \prod_{i=1}^{n} (4^i - 1)$. Moreover, for a single qubit, a representation for the Clifford group $\mathcal{C}_1 = \{C_1, C_2, \cdots, C_{24}\}$ can then be enumerated and its elements are for example shown in [46] and [3].

2.2 Dynamics of Error Accumulation

Suppose that we had a faultless, perfect quantum computer. Then a faultless quantum mechanical state ρ_t at time t could be calculated under a gate sequence $\mathcal{U}_\tau = \{U_1, \ldots, U_\tau\}$ from the initial state $\rho_0 \triangleq |\psi_0\rangle \langle\psi_0|$. Here $\tau < \infty$ denotes the sequence length, and $t \in \{0, 1, \cdots, \tau\}$ enumerates the intermediate steps. On the other hand, with an imperfect quantum computer, a possibly faulty quantum mechanical state σ_t at time t would be calculated under both $\mathcal{U}_t$ and some (unknown) noise sequence $\mathcal{E}_t = \{\Lambda_1, \ldots, \Lambda_t\}$ starting from an initial state $\sigma_0 \triangleq |\Psi_0\rangle \langle\Psi_0|$ possibly different from ρ_0. We define the set of all pure states for n qubits as $\mathcal{S}^n$ and consider the situation that $|\psi_0\rangle, |\Psi_0\rangle \in \mathcal{S}^n$.

To be precise, define for the faultless quantum computation

$$\rho_t \triangleq |\psi_t\rangle \langle\psi_t| = U_t |\psi_{t-1}\rangle \langle\psi_{t-1}| U_t^\dagger$$

for times $t = 1, 2, \ldots, \tau$. Let $X_t \triangleq U_t U_{t-1} \cdots U_1$ be shorthand notation such that $\rho_t = X_t \rho_0 X_t^\dagger$. For the possibly faulty quantum computation, define

$$\sigma_t \triangleq |\Psi_t\rangle \langle\Psi_t| = \Lambda_t U_t |\Psi_{t-1}\rangle \langle\Psi_{t-1}| U_t^\dagger \Lambda_t^\dagger$$

for times $t = 1, 2, \ldots, \tau$, respectively. Introduce also the shorthand notation $Y_t \triangleq \Lambda_t U_t \Lambda_{t-1} U_{t-1} \cdots \Lambda_1 U_1$ such that $\sigma_t = Y_t \sigma_0 Y_t^\dagger$. The analysis in this paper can immediately be extended to the case where errors (also) precede the gate. The error accumulation process is also illustrated in Fig. 1.

2.3 Distance Measures for Quantum Errors

The error can be quantified by any measure of distance between the faultless quantum-mechanical state ρ_t and the possibly faulty quantum-mechanical

a) Faultless computation:

$$\rho_0 \xrightarrow{\ U_1\ } \rho_1 \xrightarrow{\ U_2\ } \cdots \xrightarrow{\ U_{\tau-1}\ } \rho_{\tau-1} \xrightarrow{\ U_\tau\ } \rho_\tau$$

b) Potentially faulty computation:

$$\sigma_0 \xrightarrow{\ \Lambda_1 U_1\ } \sigma_1 \xrightarrow{\ \Lambda_2 U_2\ } \cdots \xrightarrow{\ \Lambda_{\tau-1} U_{\tau-1}\ } \sigma_{\tau-1} \xrightarrow{\ \Lambda_\tau U_\tau\ } \sigma_\tau$$

Fig. 1. Schematic depiction of the coupled quantum mechanical states ρ_t and σ_t for times $t = 0, 1, \cdots, \tau$. a) Faultless computation. The state ρ_t is calculated based on a gate sequence $\mathcal{U}_t = \{U_1, \ldots, U_t\}$ from the initial state ρ_0. b) Potentially faulty computation. The state σ_t is calculated using *the same* gate sequence $\mathcal{U}_t = \{U_1, \ldots, U_t\}$ and an additional error sequence $\mathcal{E}_t = \{\Lambda_1, \ldots, \Lambda_t\}$. The final state σ_τ can depart from the faultless state ρ_τ because of errors.

state σ_t for steps $t = 0, 1, \ldots, \tau$. For example, we can use the fidelity $F_t \triangleq \mathrm{Tr}\sqrt{\rho_t^{1/2} \sigma_t \rho_t^{1/2}}$ [37], or the Schatten d-norm [6] defined by

$$D_t \triangleq \|\sigma_t - \rho_t\|_d = \tfrac{1}{2} \mathrm{Tr}\left[\left\{(\sigma_t - \rho_t)^\dagger (\sigma_t - \rho_t)\right\}^{\frac{d}{2}}\right]^{\frac{1}{d}}$$

for any $d \in [1, \infty)$. The Schatten d–norm reduces to the trace distance for $d = 1$, the Frobenius norm for $d = 2$, and the spectral norm for $d = \infty$. In the case of one qubit, the trace distance between quantum-mechanical states ρ_t and σ_t equals half of the Euclidean distance between ρ_t and σ_t when representing them on the Bloch sphere [37]. It is well known that the trace distance is invariant under unitary transformations [37]; a fact that we leverage in Sect. 3.

In this paper, we are going to analyze the statistical properties of some arbitrary distance measure (one may choose) between the quantum mechanical states ρ_t and σ_t for times $t = 0, 1, \ldots, \tau$. For illustration, we will state the results in terms of the Schatten d–norm, and so are after its expectation $\mathbb{E}[D_t]$, as well as the probabilities $\mathbb{P}[D_t \le \delta]$, $\mathbb{P}[\max_{0 \le s \le t} D_s \le \delta]$. Throughout this paper, the operator $\mathbb{P}$ and thus also $\mathbb{E}$ are with respect to a sufficiently rich probability space $(\Omega, \mathbb{P}, \mathcal{F})$ that each time can describe the Markov chain being considered.

3   Error Accumulation

3.1   Discrete, Random Error Accumulation (Multi-Qubit Case)

Following the model described in Sect. 2 and illustrated in Fig. 1 and Fig. 2a, we define the gate pairs $Z_t \triangleq (X_t, Y_t)$ for $t = 0, 1, 2, \ldots, \tau$, and suppose that $Z_0 = z_0$ with probability one where $z_0 = (x_0, y_0)$ is deterministic and given a priori. Note in particular that if the initial state is prepared without error, then $\rho_0 = \sigma_0$ and consequently $z_0 = (I^{\otimes n}, I^{\otimes n})$. If on the other hand the initial state is prepared incorrectly as $y_0 |\psi_0\rangle$ instead of $|\psi_0\rangle$, then $z_0 = (I^{\otimes n}, y_0)$.

The Case of Random Circuits. We consider first the scenario that each next gate is selected randomly and independently from everything but the last system state. This assumption is satisfied in the randomized benchmarking protocol [3, 9, 14, 15, 17, 27, 33, 43–46]. The probabilities $\mathbb{P}_{z_0}[D_t > \delta]$ and $\mathbb{P}_{z_0}[\max_{0 \le s \le t} D_s \le \delta]$ can then be calculated once the initial states $|\psi_0\rangle$, $|\Psi_0\rangle$ and the *transition matrix* are known. Here, the subscript z_0 reminds us of the initial state the Markov chain is started from.

Let the transition matrix of the Markov chain $\{Z_t\}_{t \ge 0}$ be denoted element-wise by $P_{z,w} \triangleq \mathbb{P}[Z_{t+1} = w | Z_t = z]$ for $z = (x, y), w = (u, v) \in \mathcal{G}_n^2$. The transition matrix satisfies $P \in [0, 1]^{|\mathcal{G}_n|^2 \times |\mathcal{G}_n|^2}$ and the elements of each of its rows sum to one. Let $P_{z_0, w}^{(t)} \triangleq \mathbb{P}[Z_t = w | Z_0 = z_0] = (P^t)_{z_0, w}$ stand in for the probability that the process is at state w at time t starting from $Z_0 = z_0$. Note that the second equality follows from the Markov property [8].

We are now after the probability that the distance D_t is larger than a threshold δ. We define thereto the set of δ-*bad gate pairs* by

$$\mathcal{B}_{|\psi_0\rangle, \delta}^{|\Psi_0\rangle} \triangleq \left\{ (x, y) \in \mathcal{G}_n^2 \, \big| \, \|x\rho_0 x^\dagger - y\sigma_0 y^\dagger\|_d > \delta \right\} \tag{1}$$

for $|\psi_0\rangle, |\Psi_0\rangle \in \mathcal{S}^n, \delta \ge 0$, as well as the *hitting time* of any set $\mathcal{A} \subseteq \mathcal{G}_n^2$ by

$$T_\mathcal{A} \triangleq \inf\{t \ge 0 | Z_t \in \mathcal{A}\} \tag{2}$$

with the convention that $\inf \phi = \infty$. Note that $T_\mathcal{A} \in \mathbb{N}_0 \cup \{\infty\}$ and that it is random. With Definitions (1), (2), we have the convenient representation

$$\mathbb{P}_{z_0}[\max_{0 \le s \le t} D_s \le \delta] = 1 - \mathbb{P}_{z_0}[\max_{0 \le s \le t} D_s > \delta] = 1 - \mathbb{P}_{z_0}[T_{\mathcal{B}_{|\psi_0\rangle, \delta}^{|\Psi_0\rangle}} \le t] \tag{3}$$

for this homogeneous Markov chain. As a consequence of (3), the analysis comes down to an analysis of the hitting time distribution for this coupled Markov chain (Fig. 2b).

Results. Define the matrix $B_{|\psi_0\rangle, \delta}^{|\Psi_0\rangle} \in [0, 1]^{|\mathcal{G}_n|^2 \times |\mathcal{G}_n|^2}$ element-wise by

$$\left(B_{|\psi_0\rangle, \delta}^{|\Psi_0\rangle}\right)_{z, w} \triangleq \begin{cases} P_{z,w} & \text{if} \quad w \notin \mathcal{B}_{|\psi_0\rangle, \delta}^{|\Psi_0\rangle}, \\ 0 & \text{otherwise.} \end{cases}$$

Let the initial state vector be denoted by e_{z_0}, a $|\mathcal{G}_n|^2 \times 1$ vector with just the z_0-th element 1 and the others 0. Also let $1_\mathcal{A}$ denote the $|\mathcal{G}_n|^2 \times 1$ vector with ones in every coordinate corresponding to an element in the set $\mathcal{A}$. Let the transpose of an arbitrary matrix A be denoted by A^T and defined element-wise $(A^\mathrm{T})_{i,j} = A_{j,i}$. Finally, we define a $|\mathcal{G}_n|^2 \times 1$ vector $d_{|\psi_0\rangle}^{|\Psi_0\rangle} = \left(\|x\rho_0 x^\dagger - y\sigma_0 y^\dagger\|_d \right)_{(x,y) \in \mathcal{G}_n^2}$ enumerating all possible Schatten d-norm distances. We now state our first result, and defer to [32] for its proof:

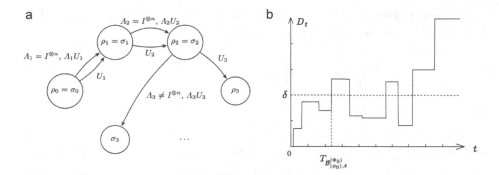

Fig. 2. a) Coupled chain describing the quantum circuit with errors. In this depiction, we start from *the same* initial state for simplicity. Here an error $\Lambda_3 \neq I^{\otimes n}$ occurs as the third gate is applied. Note that the coupled chain ρ_t, σ_t separates. b) Schematic diagram of the hitting time $T_{\mathcal{B}^{|\Psi_0\rangle}_{|\psi_0\rangle,\delta}}$.

Proposition 1 (Error accumulation in random circuits). *For any $z_0 \in \mathcal{G}_n^2$, $\delta \geq 0$, $t = 0,1,\ldots,\tau < \infty$: the expected error is given by $\mathbb{E}_{z_0}[D_t] = e_{z_0}^{\mathrm{T}} P^t d_{|\psi_0\rangle}^{|\Psi_0\rangle}$. Similarly, the probability of error is given by*

$$\mathbb{P}_{z_0}[D_t > \delta] = e_{z_0}^{\mathrm{T}} P^t 1_{\mathcal{B}^{|\Psi_0\rangle}_{|\psi_0\rangle,\delta}} , \qquad (4)$$

and is nonincreasing in δ. Furthermore; if $z_0 \notin \mathcal{B}^{|\Psi_0\rangle}_{|\psi_0\rangle,\delta}$, the probability of maximum error is given by

$$\mathbb{P}_{z_0}[\max_{0\leq s\leq t} D_s > \delta] = \sum_{s=1}^{t} e_{z_0}^{\mathrm{T}} \left(B^{|\Psi_0\rangle}_{|\psi_0\rangle,\delta}\right)^{s-1} \left(P - B^{|\Psi_0\rangle}_{|\psi_0\rangle,\delta}\right) 1_{\mathcal{B}^{|\Psi_0\rangle}_{|\psi_0\rangle,\delta}} , \qquad (5)$$

and otherwise it equals one. Lastly, (5) is nonincreasing in δ, and nondecreasing in t.

The probability in (5) is a more stringent error measure than (4) is. The event $\{\max_{0\leq s\leq t} D_s < \delta\}$ implies after all that the error D_t has always been below the threshold δ up to and including at time t. The expected error $\mathbb{E}_{z_0}[D_t]$ and probability $\mathbb{P}_{z_0}[D_t > \delta]$ only concern the error *at* time t. Additionally, (5) allows us to calculate the maximum number of gates that can be performed. That is, $\mathbb{P}_{z_0}[\max_{0\leq s\leq t} D_s > \delta] \leq \gamma$ as long as $t \leq t_{\delta,\gamma}^\star \triangleq \max\{t \in \mathbb{N}_0 | \mathbb{P}_{z_0}[\max_{0\leq s\leq t} D_s > \delta] \leq \gamma\}$. In words: at most $t_{\delta,\gamma}^\star$ gates can be applied before an accumulated error of size at least δ occurred with probability at least γ.

For general $\mathcal{B}^{|\Psi_0\rangle}_{|\psi_0\rangle,\delta}$, the explicit calculation of (5) can be numerically intensive. It is however possible to provide a lower bound of lower numerical complexity via the expected hitting time of the set $\mathcal{B}^{|\Psi_0\rangle}_{|\psi_0\rangle,\delta}$.

Lemma 1 (Lower bound for random circuits). *For any set $\mathcal{A} \subseteq \mathcal{G}_n^2$, the expected hitting times of a homogeneous Markov chain are the solutions to the* *linear system of equations* $\mathbb{E}_z[T_{\mathcal{A}}] = 0$ *for* $z \in \mathcal{A}$, $\mathbb{E}_z[T_{\mathcal{A}}] = 1 + \sum_{w \notin \mathcal{A}} P_{z,w} \mathbb{E}_w[T_{\mathcal{A}}]$ *for* $z \notin \mathcal{A}$. *Furthermore; for any* $z_0 \in \mathcal{G}_n^2$, $\delta \geq 0$, $t = 0, 1, \ldots, \tau < \infty$:

$$\mathbb{P}_{z_0}[\max_{0 \leq s \leq t} D_s > \delta] \geq 0 \vee \left(1 - \frac{\mathbb{E}_{z_0}[T_{\mathcal{B}_{|\psi_0\rangle,\delta}^{|\Psi_0\rangle}}]}{t+1}\right). \tag{6}$$

Here $a \vee b \triangleq \max\{a, b\}$.

A proof of (6) can be found in our extended version [32]. As a consequence of Lemma 1, $\mathbb{P}_{z_0}[\max_{0 \leq s \leq t} D_s > \delta] \geq \gamma$ when $t \geq \mathbb{E}_{z_0}[T_{\mathcal{B}_{|\psi_0\rangle,\delta}^{|\Psi_0\rangle}}]/(1-\gamma) - 1$, and in particular $\mathbb{P}_{z_0}[\max_{0 \leq s \leq t} D_s > 0] > 0$ when $t \geq \mathbb{E}_{z_0}[T_{\mathcal{B}_{|\psi_0\rangle,0}^{|\Psi_0\rangle}}]$. The values in the right-hand sides are thus upper bounds to the number of gates $t_{\delta,\gamma}^{\star}$ one can apply before δ error has occurred with probability γ:

$$t_{\delta,\gamma}^{\star} \leq \mathbb{E}_{z_0}[T_{\mathcal{B}_{|\psi_0\rangle,0}^{|\Psi_0\rangle}}] \wedge \left(\frac{\mathbb{E}_{z_0}[T_{\mathcal{B}_{|\psi_0\rangle,\delta}^{|\Psi_0\rangle}}]}{1-\gamma} - 1\right)$$

for $\delta \geq 0, \gamma \in [0, 1]$. Here, $a \wedge b \triangleq \min\{a, b\}$.

Limitations of the Method: Types of Quantum Noise Channels. The approach taken in this article is a hybrid between classical probability theory and quantum information theory. The results of this article are therefore not applicable to all quantum channels, and it is important that we signal you the limitations.

As an illustrative example, consider the elementary circuit of depth $\tau = 1$ with $n = 1$ qubit, in which the one gate is restricted to the Clifford group $\{C_1, \ldots, C_{24}\}$, say. For such an elementary circuit, this article describes a classical stochastic process that chooses one of twenty-four quantum noise channel $\mathcal{F}^{(1)}, \ldots \mathcal{F}^{(24)}$ say according to some arbitrary classical probability distribution $\{p_i(\rho)\}$, i.e.,

$$\rho_0 \to \rho_1 = \mathcal{F}(\rho_0) = \begin{cases} \mathcal{F}^{(1)}(\rho_0) = C_1 \rho_0 C_1^{\dagger} & \text{w.p. } p_1(\rho_0), \\ \mathcal{F}^{(2)}(\rho_0) = C_2 \rho_0 C_2^{\dagger} & \text{w.p. } p_2(\rho_0), \\ \cdots \\ \mathcal{F}^{(24)}(\rho_0) = C_{24} \rho_0 C_{24}^{\dagger} & \text{w.p. } p_{24}(\rho_0). \end{cases} \tag{7}$$

Here, the classical probability distribution $\{p_i(\rho)\}$ may be chosen arbitrarily, and depend on the initial quantum state ρ_0 as indicated. For this elementary quantum circuit of depth $\tau = 1$ with $n = 1$ qubit, (7) characterizes the set of stochastic processes covered by our results in its entirety.

For example, Proposition 1 cannot be applied to the deterministic process

$$\rho_0 \to \rho_1 = \left\{\mathcal{E}^{(1)}(\rho_0) = (1-p)\rho_0 + pY\rho_0 Y^{\dagger} \text{ w.p. } 1,\right.$$

nor to the deterministic process

$$\rho_0 \to \rho_1 = \left\{\mathcal{E}^{(2)}(\rho_0) = (1-p)\rho_0 + \frac{p}{2}U\rho_0 U^{\dagger} + \frac{p}{2}U^{\dagger}\rho_0 U \text{ w.p. } 1.\right.$$

Here, $p \in (0, 1)$ can be chosen arbitrarily and $U = e^{-i\pi Y/4}$ is a Clifford gate. The reason is that $(\mathcal{F}^{(1)} \neq \mathcal{F}^{(2)} \neq \cdots \neq \mathcal{F}^{(24)}) \neq (\mathcal{E}^{(1)} = \mathcal{E}^{(2)})$ by the unitary freedom in the operator-sum representation [37, Thm. 8.2]. A meticulous reader will now note that the example quantum channels $\mathcal{E}^{(1)}, \mathcal{E}^{(2)}$ are however *averages* of two particular stochastic processes $\mathcal{F}$. That is: if $p_I = 1 - p, p_Y = p$, then $\mathcal{E}^{(1)}(\rho) = \mathbb{E}[\mathcal{F}(p)]$; or if $p_I = 1 - p, p_U = p_{U^\dagger} = \frac{p}{2}$, then $\mathcal{E}^{(2)}(\rho) = \mathbb{E}[\mathcal{F}(\rho)]$.

The Case of Nonrandom Circuits. Suppose that the gate sequence $\mathcal{U}_\tau = \{U_1, ..., U_\tau\}$ is fixed *a priori* and that it is not generated randomly. Because the gate sequence is nonrandom, we have now that the faultless state $\rho_t = X_t \rho_0 X_t^\dagger$ is deterministic for times $t = 0, 1, \ldots, \tau$. On the other hand the potentially faulty state $\sigma_t = Y_t \rho_0 Y_t^\dagger$ is still (possibly) random.

We can now use a lower dimensional Markov chain to represent the system. To be precise: we will now describe the process $\{Y_t\}_{t \geq 0}$ (and consequently $\{\sigma_t\}_{t \geq 0}$) as an *inhomogeneous Markov chain*. Its transition matrices will now be time-dependent and given element-wise by $Q_{y,v}(t) = \mathbb{P}[Y_{t+1} = v | Y_t = y]$ for $y, v \in \mathcal{G}_n, t \in \{0, 1, \ldots, \tau - 1\}$. Letting $Q_{y,v}^{(t)} \triangleq \mathbb{P}[Y_t = v | Y_0 = y]$ stand in for the probability that the process $\{Y_t\}_{t \geq 0}$ is at state v at time t starting from y, we have by the Markov property [8] that $Q_{y,v}^{(t)} = \left(\prod_{s=1}^{t} Q(s)\right)_{y,v}$ for $y, v \in \mathcal{G}_n$. Note that the Markov chain modeled here is inhomogeneous, which is different from Sect. 3.1. In particular, the time-dependent transition matrix $Q(t)$ here cannot be expressed in terms of a power P^t of a transition matrix P on the same state space as in Sect. 3.1.

Results. Now define the sets of (δ, t)-bad gate pairs by $\mathcal{B}_{|\psi_0\rangle,\delta}^{|\Psi_0\rangle,t} \triangleq \{x \in \mathcal{U}_n | \|\rho_t - x\sigma_0 x^\dagger\|_d > \delta\}$ for $|\psi_0\rangle, |\Psi_0\rangle \in \mathcal{S}^n$, $t \in \{0, 1, \ldots, \tau\}$, $\delta \geq 0$. Also define the matrices $B_{|\psi_0\rangle,\delta}^{|\Psi_0\rangle,t} \in [0, 1]^{|\mathcal{G}_n| \times |\mathcal{G}_n|}$ element-wise by

$$\left(B_{|\psi_0\rangle,\delta}^{|\Psi_0\rangle,t}\right)_{y,v} \triangleq \begin{cases} Q_{y,v}(t) & \text{if } v \notin \mathcal{B}_{|\psi_0\rangle,\delta}^{|\Psi_0\rangle,t}, \\ 0 & \text{otherwise}, \end{cases}$$

for $t = 0, 1, \ldots, \tau$. Recall the notation introduced above Proposition 1. Similarly enumerate in the vector d_{ρ_t} the Schatten d-norms between any of the possibles states of σ_t and the faultless state ρ_t. We state our second result; see [32] for a proof:

Proposition 2 (Error accumulation in nonrandom circuits). *For any* $y_0 \in \mathcal{G}_n$, $\delta \geq 0$, $t = 0, 1, \ldots, \tau < \infty$: *the expected error is given by* $\mathbb{E}_{y_0}[D_t] = e_{y_0}^{\mathrm{T}} \left(\prod_{k=1}^{t} Q(k)\right) d_{\rho_t}$. *Similarly, the probability of error is given by*

$$\mathbb{P}_{y_0}[D_t > \delta] = e_{y_0}^{\mathrm{T}} \left(\prod_{k=1}^{t} Q(k)\right) 1_{\mathcal{B}_{|\psi_0\rangle,\delta}^{|\Psi_0\rangle,t}}. \tag{8}$$

Furthermore; if $y_0 \notin \mathcal{B}^{|\Psi_0\rangle,0}_{|\psi_0\rangle,\delta}$, the probability of maximum error is given by

$$\mathbb{P}_{y_0}[\max_{0\leq s\leq t} D_s > \delta] = \sum_{s=0}^{t-1}\left(e_{y_0}^{\mathrm{T}}(\prod_{r=0}^{s} B^{|\Psi_0\rangle,r}_{|\psi_0\rangle,\delta}) \times (Q(s+1)-B^{|\Psi_0\rangle,s+1}_{|\psi_0\rangle,\delta})1_{\mathcal{B}^{|\Psi_0\rangle,s+1}_{|\psi_0\rangle,\delta}}\right), \quad (9)$$

and otherwise it equals one.

For illustration, we have written a script that will generate a valid P and Q matrices after a user inputs a vector describing (gate-dependent) error probabilities. The code is available on TU/e's GitLab server at https://gitlab.tue.nl/20061069/markov-chains-for-error-accumulation-in-quantum-circuits.

3.2   Continuous, Random Error Accumulation (One-Qubit Case)

In this section, we analyze the case where a single qubit:

1. receives a random perturbation on the Bloch sphere after each s-th unitary gate according to a continuous distribution $p_s(\alpha)$, and
2. depolarizes to the completely depolarized state $I/2$ with probability $q \in [0,1]$ after each unitary gate,

by considering it an absorbing random walk on the Bloch sphere. The key point leveraged here is that the trace distance is invariant under rotations. Hence a rotationally symmetric perturbation distribution will still allow us to calculate the error probabilities.

Model. Let R_0 be an initial point on the Bloch sphere. Every time a unitary quantum gate is applied, the qubit is rotated and receives a small perturbation. This results in a random walk $\{R_t\}_{t\geq0}$ on the Bloch sphere for as long as the qubit has not depolarized. Because the trace distance is invariant under rotations and since the rotations are applied both to ρ_t and σ_t, we can ignore the rotations. We let ν denote the random time at which the qubit depolarizes. With the usual independence assumptions, $\nu \sim \mathrm{Geometric}(q)$.

Define $\mu_t(r)$ for $t < \nu$ as the probability that the random walk is in a solid angle Ω about r (in spherical coordinates) conditional on the qubit not having depolarized yet. That is,

$$\mathbb{P}[R_t \in \mathcal{S}|\nu > t] \triangleq \int_{\mathcal{S}} \mu_t(r)\mathrm{d}\Omega(r).$$

We assume without loss of generality that $R_0 = \hat{z}$. From [40], the initial distribution is then given by

$$\mu_0 = \sum_{n=0}^{\infty} \frac{2n+1}{4\pi}P_n(\cos\theta).$$

Here, the $P_n(\cdot)$ denote the Legendre polynomials. Also introduce the shorthand notation

$$\Lambda_{n,t} \triangleq \prod_{s=1}^{t} \int_{0}^{\pi} P_n(\cos\alpha)dp_s(\alpha)$$

for convenience. As we will see in Proposition 3 in a moment, these constants will turn out to be the coefficients of an expansion for the expected trace distance (see (10)). Recall that here, $p_s(\alpha)$ denotes the probability measure of the angular distance for the random walk on the Bloch sphere at time t (see (i) above). In particular: if $p_t(\alpha) = \delta(\alpha)$ for all $t \geq 0$ meaning that each step is taken into a random direction but exactly of angular length α, then $\Lambda_{n,t} = (P_n(\cos\alpha))^t$. From [40], it follows that after t unitary quantum gates have been applied without depolarization having occurred,

$$\mu_t = \sum_{n=0}^{\infty} \frac{2n+1}{4\pi} \Lambda_{n,t} P_n(\cos\theta).$$

Results. In this section we specify D_t as the trace distance. We are now in position to state our findings; proofs can be found in [32]:

Proposition 3 (Single qubit). *For $0 \leq \delta \leq 1$, $t \in \mathbb{N}_+$: the expected trace distance satisfies*

$$\mathbb{E}[D_t] = \tfrac{1}{2} - (1-q)^t \left(\tfrac{1}{2} + 2 \sum_{n=0}^{\infty} \frac{\Lambda_{n,t}}{(2n-1)(2n+3)} \right). \tag{10}$$

The probability of the trace distance deviating is given by

$$\mathbb{P}[D_t \leq \delta] = \mathbb{1}[\tfrac{1}{2} \in [0,\delta]]\left(1 - (1-q)^t\right)$$
$$+ (1-q)^t \sum_{n=0}^{\infty} (2n+1)\Lambda_{n,t} \sum_{r=1}^{n+1} (-1)^{r+1} \delta^{2r} C_{r-1} \binom{n+r-1}{2(r-1)}.$$

Here, the C_r denote the Catalan numbers. Finally; the probability of the maximum trace distance deviating is lower bounded by

$$\mathbb{P}[\max_{0 \leq s \leq t} D_s \leq \delta | \nu > t] \geq 0 \vee \left(1 - t + \delta^2 \sum_{s=1}^{t} \sum_{n=0}^{\infty} (2n+1)\Lambda_{n,s} \frac{n!}{(2)_n} P_n^{(1,-1)}(1 - 2\delta^2) \right).$$

4   Simulations

We will now briefly illustrate and validate our results numerically. For a more indepth numerical investigation, see [32].

Consider two nonrandom circuits: the first is a periodical single-qubit circuit that repeats a Hadamard, Pauli-X, Pauli-Y and Pauli-Z gate $k = 25$ times, and the second a two-qubit circuit that is repeated $k = 5$ times; see also Fig. 3. Here the controlled-NOT gate CNOT $= \big((1,0,0,0); (0,1,0,0); (0,0,0,1); (0,0,1,0)\big)$. Consider also the following two error models in which the errors depend on the gates:

(i) For the single-qubit circuit, presume $\mathbb{P}[\Lambda = I] = 0.990, \mathbb{P}[\Lambda = Z] = 0.010$.

(ii) For the two-qubit circuit, when labeling the qubits by A and B, suppose

$$\mathbb{P}[\Lambda_A = I] = 0.990,\ \mathbb{P}[\Lambda_A = X] = 0.006, \mathbb{P}[\Lambda_A = Y] = 0.003,\ \mathbb{P}[\Lambda_A = Z] = 0.001;$$
$$\mathbb{P}[\Lambda_B = I] = 0.980,\ \mathbb{P}[\Lambda_B = X] = 0.002, \mathbb{P}[\Lambda_B = Y] = 0.014,\ \mathbb{P}[\Lambda_B = Z] = 0.004.$$

In order to evaluate Proposition 2, we set the error threshold $\delta = 1/10$.

The theoretical and simulation results on the two circuits are shown in Fig. 3. Note that the simulation curves almost coincide with the theoretical curves; the deviation is only due to numerical limits. Furthermore, because different gates influence error accumulation to different degrees, the periodical ladder shape

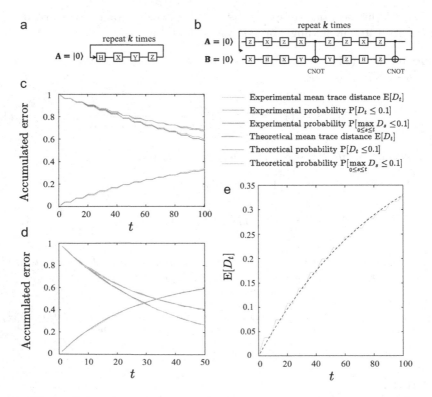

Fig. 3. Theoretical and simulation results for error accumulation on a single-qubit circuit (figures a, c, and e) and a two-qubit circuit (figures b and d). The numerical results are calculated from 2000 independent runs, and almost indistinguishable from the formulae. The dashed, black curve in figure e is a fit of $D_t \approx \frac{1}{2}(1 - (1 - \mu)^t)$ to the data—this formula describes the so-called *depolarization channel*, a basic model of depolarization of one qubit. The fit parameter is $\mu^{\text{fit}} \approx 0.011$.

occurs in Fig. 3. Observe furthermore that this periodical ladder shape is not captured by a fit method that only takes into account the decay of t applications of a single depolarizing channel.

5 Minimizing Errors in Quantum Circuit Through Optimization

The rate at which errors accumulate may be different for different quantum circuits that can implement the same algorithm. Using techniques from optimization and (9), we can therefore search for the quantum circuit that has the lowest error rate accumulation while maintaining the same final state. To see this, suppose we are given a circuit $\mathcal{U}_\tau = \{U_1, U_2, \ldots, U_\tau\}$. For given ρ_0 this brings the quantum state to some quantum state ρ_τ. Other circuits may go to the same final state and have a lower probability of error at time τ. We will therefore aim to

$$\underset{G_1,\ldots,G_\tau \in \mathcal{G}_n}{\text{minimize}} \quad u(\{G_1,\ldots,G_\tau\})$$
$$\text{subject to} \quad G_\tau \cdots G_1 = U_\tau \cdots U_1. \tag{11}$$

Here, one can for example choose for the objective function $u(\cdot)$ the probability of error (8), or probability of maximum error (9). To solve (11), we design a simulated annealing algorithm in Sect. 5.1 to improve the quantum circuit.

The minimization problem in (11) is well-defined and has a few attractive features. For starters, the minimization problem automatically detects shorter circuits if the probability of error when applying the identity operator $I^{\otimes n}$ is relatively small. The optimum may then for example occur at a circuit of the form

$$G_\tau G_{\tau-1} G_{\tau-2} \cdots G_2 G_1 = I^{\otimes n} G_{\tau-1} I^{\otimes n} \cdots I^{\otimes n} G_1,$$

which effectively means that only the two gates $G_{\tau-1} G_1$ are applied consecutively. The identity operators in this solution essentially describe the passing of time. Now, critically, note that while the minimization problem does consider all shorter circuits of depth at most τ, this does not necessarily mean that the physical application of one specific group element $G \in \mathcal{G}_n$ is always the best. Concretely, in spite of the fact that any quantum circuit of the form $G_\tau \cdots G_1 = G \in \mathcal{G}_n$ performs the single group element $G \in \mathcal{G}_n$, it is not necessarily true that

$$u(\{G, I^{\otimes n}, \ldots, I^{\otimes n}\}) < u(\{G_1, \ldots, G_\tau\}).$$

The reason for this is that the error distribution on the direct group element G may be worse than using a circuit utilizing multiple other group elements. In other words, the optimal circuit need not always be the 'direct' circuit, but of course it can be. (In Sect. 5.2 we also consider the situation in which an experimentalist can only apply a subset $\mathcal{A} \subseteq \mathcal{G}_n$ that need not necessarily be a group, and in such a case the direct group element G may not even be a viable solution to the experimentalist if $G \notin \mathcal{A}$.) Typically, the minimization problem will prefer shorter circuits if the probability of error when applying the identity operator $I^{\otimes n}$ is relatively small and the error distributions of all gate distributions are relatively homogeneous.

5.1   Simulated Annealing

We will generate candidate circuits as follows. Let $\{G_1^{[\eta]}, \ldots, G_\tau^{[\eta]}\}$ denote the circuit at iteration η. Choose an index $I \in [\tau - 1]$ uniformly at random, choose $G \in \mathcal{G}$ uniformly at random. Then set

$$G_i^{[\eta+1]} = \begin{cases} G & \text{if } i = I, \\ G_{I+1}^{[\eta]} G_I^{[\eta]} G^{\leftarrow} & \text{if } i = I + 1, \\ G_i^{[\eta]} & \text{otherwise.} \end{cases}$$

Here, $G^{\leftarrow}$ denotes the (left) inverse group element, i.e., $G^{\leftarrow} G = I^{\otimes n}$. The construction thus ensures that

$$G_{I+1}^{[\eta+1]} G_I^{[\eta+1]} = (G_{I+1}^{[\eta]} G_I^{[\eta]} G^{\leftarrow}) G = G_{I+1}^{[\eta]} G_I^{[\eta]}$$

so that the circuit's intent does not change: $G_\tau^{[\eta+1]} \cdots G_1^{[\eta+1]} = G_\tau^{[\eta]} \cdots G_1^{[\eta]}$.

We will use the Metropolis algorithm. Let

$$E = \left\{ \{G_1, \ldots, G_\tau\} \mid G_\tau \cdots G_1 = U_\tau \cdots U_1 \right\}$$

denote the set of all viable circuits. For two arbitrary circuits $i, j \in E$, let

$$\Delta(i,j) \triangleq \sum_{s=1}^{\tau-1} \mathbb{1}[i_s \neq j_s, i_{s+1} \neq j_{s+1}]$$

denote the number of consecutive gates that differ between both circuits. Under this construction, the *candidate-generator matrix* of the Metropolis algorithm is given by

$$q_{ij} = \begin{cases} \frac{1}{(\tau-1)|\mathcal{G}|} & \text{if } \Delta(i,j) \leq 1 \\ 0 & \text{otherwise.} \end{cases}$$

Since the candidate-generator matrix is symmetric, this algorithm means that we set $\alpha_{i,j}(T) = \exp\left(-\frac{1}{T} \max\{0, u(j) - u(i)\}\right)$ as the *acceptance probability* of circuit j over i. Here $T \in (0, \infty)$ is a positive constant. Finally, we need a cooling schedule. Let $M \triangleq \sup_{\{i,j \in E \mid \Delta(i,j) \leq 1\}} \{u(j) - u(i)\}$. Based on [8], if we choose a cooling schedule $\{T_\eta\}_{\eta \geq 0}$ that satisfies $T_\eta \geq \frac{\tau M}{\ln \eta}$, then the Metropolis algorithm will converge to the set of global minima of the minimization problem in (11).

Lemma 2. *Algorithm 1 converges to the global minimizer of* (11) *whenever* $T_\eta \geq \tau M / \ln \eta$ *for* $\eta = 1, 2, \cdots$.

Input: A group $\mathcal{G}$, a circuit $\{U_1, \ldots, U_\tau\}$, and number of iterations w
Output: A revised circuit $\{G_1^{[w]}, \ldots, G_\tau^{[w]}\}$
begin
    Initialize $\{G_1^{[0]}, \ldots, G_\tau^{[0]}\} = \{U_1, \ldots, U_\tau\}$;
    for $\eta \leftarrow 1$ **to** w **do**
        Choose $I \in [\tau - 1]$ uniformly at random;
        Choose $G \in \mathcal{G}$ uniformly at random;
        Set $J_I = G, J_{I+1} = G_{I+1}^{[\eta]} G_I^{[\eta]} G^{\leftarrow}, J_i = G_i^{[\eta]} \forall_{i \neq I, I+1}$;
        Choose $X \in [0, 1]$ uniformly at random;
        if $X \leq \alpha_{G^{[\eta]}, J}(T_\eta)$ **then**
            Set $G^{[\eta+1]} = J$;
        else
            Set $G^{[\eta+1]} = G^{[\eta]}$;
        end
    end
end

Algorithm 1: Pseudo-code for the simulated annealing algorithm described in Section 5.1.

5.2 Examples

Gate-Dependent Error Model. We are going to improve the one-qubit circuit in Fig. 3 using Algorithm 1. The gates are limited to the Clifford group $\mathcal{C}_1$ and the errors will be limited to the Pauli channel. The error probabilities considered here are gate-dependent and written out explicitly in [32, Appendix H]. The cooling schedule used here will be set as $T_\eta = C / \ln(\eta + 1)$, and the algorithm's result when using $C = 0.004$ is shown in Fig. 4a. Figure 4a illustrates that the improved circuit can indeed lower the error accumulation rate. The circuit with the lowest error accumulation rate that was found is shown in [32, Appendix H].

Gates in a Subset of One Group. The gates that are available in practice may be restricted to some subset $\mathcal{A} \subseteq \mathcal{G}$ not necessarily a group. Under such constraint, we could generate candidate circuits as follows: Let $\{G_1^{[\eta]}, \ldots, G_\tau^{[\eta]}\}$ denote the circuit at iteration η. In each iteration, two neighboring gates will be considered to be replaced by two other neighboring gates. There are $m \leq (\tau - 1)$ neighboring gate pairs $(G_1^{[\eta]}, G_2^{[\eta]}), \ldots, (G_{m-1}^{[\eta]}, G_m^{[\eta]})$ that can be replaced by two different neighboring gates. Choose an index $I \in [m - 1]$ uniformly at random, and replace $(G_I^{[\eta]}, G_{I+1}^{[\eta]})$ by any gate pair from $\{(\tilde{G}_1, \tilde{G}_2) \in \mathcal{A}^2 \mid G_I^{[\eta]} G_{I+1}^{[\eta]} = \tilde{G}_1 \tilde{G}_2\}$ uniformly at random. Pseudo-code for this modified algorithm can be found in [32, Algorithm 2]. It must be noted that this algorithm is not guaranteed to converge to the global minimizer of (11) (due to limiting the gates available); however, it may still find use in practical scenarios where one only has access to a restricted set of gates.

We now aim to decrease the probability of maximum error (9) by changing the two-qubit circuit shown in Fig. 3. The error model is the same as that in Sect. 4–B. The set of gates available for improving the circuit is here limited to $\{I, X, Y, Z, H, CNOT\}$. The result here for the two-qubit circuit is obtained by again using the cooling schedule $T_\eta = C/\ln(\eta + 1)$ but now letting the parameter $C = 0.002$. Figure 4b shows that a more error-tolerant circuit can indeed be found using this simulated annealing algorithm. The improved circuit is shown in [32, Appendix H].

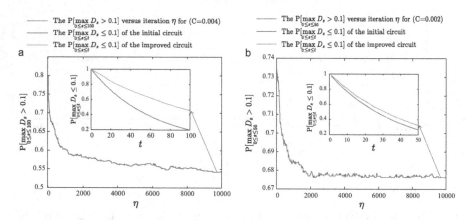

Fig. 4. a) Circuit optimization when using Algorithm 1. The error probabilities are gate-dependent. Note that the probability of maximum error (9) decreases as the number of iterations η increases when using Algorithm 1 ($C = 0.004$). b) Circuit optimization when the available gates are limited. The set of gates available is chosen limited to $\{I, X, Y, Z, H, CNOT\}$. Here we started from the two-qubit circuit shown in Fig. 3.

Deutsch–Jozsa Algorithm. Let us give further proof of concept through the Deutsch–Jozsa Algorithm for one classical bit [11,13]. This quantum algorithm determines if a function $f : \{0,1\} \to \{0,1\}$ is constant or balanced, i.e., if $f(0) = f(1)$ or $f(0) \neq f(1)$. It is typically implemented using the quantum circuit in Fig. 5. If no errors occur in this quantum circuit, then the first qubit would measure $|0\rangle$ or $|1\rangle$ w.p. one if f constant or balanced, respectively. If errors occur in this quantum circuit, then there is a strictly positive probability that the first qubit measures $|1\rangle$ or $|0\rangle$ in spite of f being constant or balanced, respectively, and thus for the algorithm to incorrectly output that f is constant or balanced. This *misclassification probability* ν of the algorithm depends on the underlying error distributions, and can be calculated by adapting (8)'s derivation.

We suppose now that errors occur according to a distribution in which two-qubit Clifford gates are more error prone than single-qubit gates. We can then revise the quantum circuit in Fig. 5 using a simulated annealing algorithm that aims at minimizing (11) by randomly swapping out poor gate pairs for better gate pairs. This simulated annealing algorithm, like any other, is sensitive to the choice of *cooling schedule* [8], here set as $T_\eta = C(\gamma/\eta + (1 - \gamma)/\ln(\eta + 1))$

with $C > 0$, $\gamma \in [0, 1]$; the integer η indexes the iterations. Figure 5 shows the ratio $\Theta \triangleq \nu_{\text{original circuit}}/\nu_{\text{revised circuit}}$ as a function of C, γ for $f_a(x) = x, f_b(x) = 1 - x, f_c(x) = 0, f_d(x) = 1$ where $x \in \{0, 1\}$. Note that $\Theta \geq 1$ always, ≥ 1.60 commonly, and sometimes even ≥ 2.20.

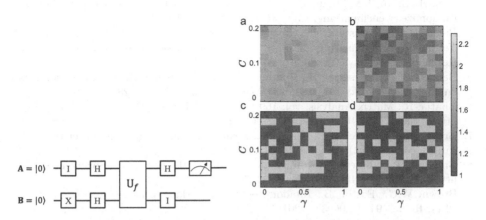

Fig. 5. (left) The Deutsch–Jozsa Algorithm for one classical bit in quantum circuit form. (right) Relative improvement when using Algorithm 1. For every pair (C, γ) here, Θ was calculated using a Monte Carlo simulation with 10^5 independent repetitions for the best circuit found throughout $w = 10^3$ iterations of the annealing algorithm. $u(\cdot)$ was set to the misclassification probability for a, c; and to (9) for b, d.

6 Conclusion

In conclusion; we have proposed and studied a model for discrete Markovian error accumulation in a multi-qubit quantum computation, as well as a model describing continuous errors accumulating in a single qubit. By modeling the quantum computation with and without errors as two coupled Markov chains, we were able to capture a weak form of time-dependency, allow for fairly generic error distributions, and describe multi-qubit systems. Furthermore, by using techniques from discrete probability theory, we could calculate the probability that error measures such as the fidelity and trace distance exceed a threshold analytically. To combat the numerical challenge that may occur when evaluating our expressions, we additionally provided an analytical bound on the error probabilities that is of lower numerical complexity. Finally, we showed how our expressions can be used to decide how many gates one can apply before too many errors accumulate with high probability, and how one can lower the rate of error accumulation in existing circuits by using techniques from optimization.

Acknowledgments. We are grateful to Bart van Schooten, who contributed the code on TU/e's GitLab server. Finally, this research received financial support from the Chinese Scholarship Council (CSC) in the form of a CSC Scholarship.

References

1. Aaronson, S., Gottesman, D.: Improved simulation of stabilizer circuits. Phys. Rev. A **70**(5), 052328 (2004)
2. Amy, M.: Formal methods in quantum circuit design (2019)
3. Ball, H., Stace, T.M., Flammia, S.T., Biercuk, M.J.: Effect of noise correlations on randomized benchmarking. Phys. Rev. A **93**(2), 022303 (2016)
4. Bennett, C.H., Brassard, G., Crépeau, C., Jozsa, R., Peres, A., Wootters, W.K.: Teleporting an unknown quantum state via dual classical and Einstein-Podolsky-Rosen channels. Phys. Rev. Lett. **70**(13), 1895 (1993)
5. Bennett, C.H., Wiesner, S.J.: Communication via one- and two-particle operators on Einstein-Podolsky-Rosen states. Phys. Rev. Lett. **69**(20), 2881 (1992)
6. Bhatia, R.: Matrix Analysis, vol. 169. Springer Science & Business Media (2013). https://doi.org/10.1007/978-1-4612-0653-8
7. Bravyi, S., Englbrecht, M., König, R., Peard, N.: Correcting coherent errors with surface codes. npj Quantum Inf. **4**(1), 55 (2018)
8. Brémaud, P.: Discrete Probability Models and Methods, vol. 78. Springer (2017). https://doi.org/10.1007/978-3-319-43476-6
9. Brown, W.G., Eastin, B.: Randomized benchmarking with restricted gate sets. Phys. Rev. A **97**(6), 062323 (2018)
10. Carignan-Dugas, A., Boone, K., Wallman, J.J., Emerson, J.: From randomized benchmarking experiments to gate-set circuit fidelity: how to interpret randomized benchmarking decay parameters. New J. Phys. **20**(9), 092001 (2018)
11. Cleve, R., Ekert, A., Macchiavello, C., Mosca, M.: Quantum algorithms revisited. Proc. Royal Soc. London. Ser. A: Math. Phys. Eng. Sci. **454**(1969), 339–354 (1998)
12. Cramer, J., et al.: Repeated quantum error correction on a continuously encoded qubit by real-time feedback. Nat. Commun. **7**, 11526 (2016)
13. Deutsch, D., Jozsa, R.: Rapid solution of problems by quantum computation. Proc. Royal Soc. London. Ser. A: Math. Phys. Sci. **439**(1907), 553–558 (1992)
14. Epstein, J.M., Cross, A.W., Magesan, E., Gambetta, J.M.: Investigating the limits of randomized benchmarking protocols. Phys. Rev. A **89**(6), 062321 (2014)
15. Fong, B.H., Merkel, S.T.: Randomized benchmarking, correlated noise, and ising models. arXiv preprint arXiv:1703.09747 (2017)
16. Fowler, A.G., Hollenberg, L.C.: Scalability of Shor's algorithm with a limited set of rotation gates. Phys. Rev. A **70**(3), 032329 (2004)
17. França, D.S., Hashagen, A.: Approximate randomized benchmarking for finite groups. J. Phys. A: Math. Theoret. **51**(39), 395302 (2018)
18. Fujii, K.: Stabilizer formalism and its applications. In: Quantum Computation with Topological Codes, pp. 24–55. Springer (2015). https://doi.org/10.1007/978-981-287-996-7
19. Gottesman, D.: The Heisenberg representation of quantum computers. arXiv preprint quant-ph/9807006 (1998)
20. Gottesman, D.: Efficient fault tolerance. Nature **540**, 44 (2016)
21. Greenbaum, D., Dutton, Z.: Modeling coherent errors in quantum error correction. Quantum Sci. Technol. **3**(1), 015007 (2017)
22. Greenberger, D.M., Horne, M.A., Zeilinger, A.: Going beyond Bell's theorem. In: Bell's Theorem, Quantum Theory and Conceptions of the Universe, pp. 69–72. Springer (1989). https://doi.org/10.1007/978-94-017-0849-4_10
23. Gutiérrez, M., Svec, L., Vargo, A., Brown, K.R.: Approximation of realistic errors by Clifford channels and Pauli measurements. Phys. Rev. A **87**(3), 030302 (2013)

24. Gutmann, H.: Description and control of decoherence in quantum bit systems. Ph.D. thesis, lmu (2005)
25. Harper, R., Hincks, I., Ferrie, C., Flammia, S.T., Wallman, J.J.: Statistical analysis of randomized benchmarking. Phys. Rev. A **99**(5), 052350 (2019)
26. Huang, E., Doherty, A.C., Flammia, S.: Performance of quantum error correction with coherent errors. Phys. Rev. A **99**(2), 022313 (2019)
27. Janardan, S., Tomita, Yu., Gutiérrez, M., Brown, K.R.: Analytical error analysis of Clifford gates by the fault-path tracer method. Quantum Inf. Process. **15**(8), 3065–3079 (2016). https://doi.org/10.1007/s11128-016-1330-z
28. Kliuchnikov, V., Maslov, D.: Optimization of Clifford circuits. Phys. Rev. A **88**(5), 052307 (2013)
29. Knill, E.: Quantum computing with realistically noisy devices. Nature **434**(7029), 39 (2005)
30. Koenig, R., Smolin, J.A.: How to efficiently select an arbitrary Clifford group element. J. Math. Phys. **55**(12), 122202 (2014)
31. Linke, N.M., et al.: Fault-tolerant quantum error detection. Sci. Adv. **3**(10), e1701074 (2017)
32. Ma, L., Sanders, J.: Markov chains and hitting times for error accumulation in quantum circuits. arXiv preprint arXiv:1909.04432 (2021)
33. Magesan, E., Gambetta, J.M., Emerson, J.: Scalable and robust randomized benchmarking of quantum processes. Phys. Rev. Lett. **106**(18), 180504 (2011)
34. Magesan, E., Puzzuoli, D., Granade, C.E., Cory, D.G.: Modeling quantum noise for efficient testing of fault-tolerant circuits. Phys. Rev. A **87**(1), 012324 (2013)
35. Maslov, D., Dueck, G.W., Miller, D.M., Negrevergne, C.: Quantum circuit simplification and level compaction. IEEE Trans. Comput.-Aided Des. Integrated Circ. Syst. **27**(3), 436–444 (2008)
36. Moll, N., et al.: Quantum optimization using variational algorithms on near-term quantum devices. Quantum Sci. Technol. **3**(3), 030503 (2018)
37. Nielsen, M.A., Chuang, I.L.: Quantum Computation and Quantum Information: 10th Anniversary Edition, 10th edn. Cambridge University Press, New York (2011)
38. Preskill, J.: Quantum computing: pro and con. Proc. Royal Soc. London. Ser. A: Math. Phys. Eng. Sci. **454**(1969), 469–486 (1998)
39. Proctor, T., Rudinger, K., Young, K., Sarovar, M., Blume-Kohout, R.: What randomized benchmarking actually measures. Phys. Rev. Lett. **119**(13), 130502 (2017)
40. Roberts, P.H., Ursell, H.D.: Random walk on a sphere and on a Riemannian manifold. Philos. Trans. Royal Soc. London. Ser. A, Math. Phys. Sci. **252**(1012), 317–356 (1960)
41. Ruskai, M.B.: Pauli exchange errors in quantum computation. Phys. Rev. Lett. **85**(1), 194 (2000)
42. Selinger, P.: Generators and relations for n-qubit Clifford operators. Logical Meth. Comput. Sci. **11** (2013)
43. Wallman, J.J.: Randomized benchmarking with gate-dependent noise. Quantum **2**, 47 (2018). https://doi.org/10.22331/q-2018-01-29-47
44. Wallman, J.J., Barnhill, M., Emerson, J.: Robust characterization of loss rates. Phys. Rev. Lett. **115**(6), 060501 (2015)
45. Wood, C.J., Gambetta, J.M.: Quantification and characterization of leakage errors. Phys. Rev. A **97**(3), 032306 (2018)
46. Xia, T., et al.: Randomized benchmarking of single-qubit gates in a 2D array of neutral-atom qubits. Phys. Rev. Lett. **114**(10), 100503 (2015)

Performance Evaluation of ZooKeeper Atomic Broadcast Protocol

Said Naser Said Kamil[1]([✉]), Nigel Thomas[2], and Ibrahim Elsanosi[3]

[1] Faculty of Science-Khoms, Elmergib University, Khoms, Libya
said.kamil@elmergib.edu.ly
[2] School of Computing, Newcastle University, Newcastle upon Tyne, UK
nigel.thomas@ncl.ac.uk
[3] Faculty of Science, Sebha University, Sebha, Libya
i.elsanosi@sebhau.edu.ly

Abstract. In this paper, we present a performance model of the Zab protocol formally specified using the Markovian process algebra PEPA. The model is parameterised from measurements taken from a real deployment of Zookeeper and is evaluated to derive estimates for average latency and throughput at various loads. These estimates are then compared against further measurements from the real system. Although the model is highly abstract and ignores much implementation detail, it is shown to give qualitative predictions for system behaviour, most notably for estimating the saturation point.

Keywords: Cloud computing · Performance evaluation · ZooKeeper · Zab protocol · Scalability · PEPA

1 Introduction

Large scale systems have witnessed dramatic increases for high performance, availability and coordinations. More recently, ZooKeeper has proposed a system for high performance and availability coordination service. ZooKeeper is a distributed coordination service for cloud computing applications and it is providing fundamental coordination primitives, for instance, synchronisation and group services for distributed applications [1,15,16]. Additionally, ZooKeeper is an open source Apache project providing a fault-tolerant coordination service and it is maintained by Apache Software Foundation and Yahoo. A key feature of ZooKeeper is offering high-performance services and coordination of the large scale systems. Cloud computing applications also, can leverage fundamental services that provided by ZooKeeper through the encapsulation of the distributed coordination algorithms and preserving a simple database [20,26]. According to Hunt et al. [15], "ZooKeeper can handle tens to hundreds of thousands of transactions per second". Indeed, ZooKeeper is used as a standard coordination kernel for several distributed systems, for instance, Facebook, Yahoo, Netflix

© ICST Institute for Computer Sciences, Social Informatics and Telecommunications Engineering 2021
Published by Springer Nature Switzerland AG 2021. All Rights Reserved
Q. Zhao and L. Xia (Eds.): VALUETOOLS 2021, LNICST 404, pp. 56–71, 2021.
https://doi.org/10.1007/978-3-030-92511-6_4

and Twitter [4,9,25]. Furthermore, a high availability is provided by ZooKeeper through the replication of data on each server in an ensemble of servers.

A collection of services is provided by the ZooKeeper; group services, configuration and synchronisation are provided in a form of centralised coordination service [1]. An additional characteristic offered by ZooKeeper interface is wait-free data objects [15] that provide a simple and powerful coordination service. ZooKeeper not only guarantees the ordering FIFO for the execution of clients' requests, but also guarantees linearizability for all write requests that update the state of the ZooKeeper. ZooKeeper guarantees liveness and durability of services. The design of ZooKeeper uses some ideas from previous coordination services, for example, Chubby [5] and Boxwood [19]. ZooKeeper does not use locks in its API, but allows a client to use them. While Chubby only allowed the client to connect to the leader, ZooKeeper allows a client to connect to any server in an ensemble (leader and/or followers). Therefore, ZooKeeper provides more flexibility and higher performance requirements.

The performance of ZooKeeper atomic broadcast protocol Zab will be examined using the PEPA Eclipse Plug-in tool [24]. Specifically, the broadcast phase will be modelled and an approximation of its behaviour will be shown taking into account several performance metrics (Latency and Throughput). Furthermore, the model will be presented considering the cyclic behaviour of the system, by means of the scalable analysis provided by the Eclipse Plug-in tool using Ordinary Differential Equations (ODEs) to derive measures of throughput and latency.

The outline of this paper is as follows. An overview for the ZooKeeper is given in the next section. Then the atomic broadcast protocol (Zab) is presented. This is followed by an illustration of the Zab PEPA model. Next, the experiments and results are shown and discussed. Finally, the paper ends with some conclusions and future work.

2   ZooKeeper

ZooKeeper service consists of an ensemble of servers, which use replication in order to attain high performance and availability. Several functions are provided through the ZooKeeper API, (create, delete, exists, getData, setData, getChildren and sync); these methods have both synchronous (an application use it in the case of executing a single operation) and asynchronous (can be used for both executing tasks and for multiple outstanding operations). That is to say, using an asynchronous operation technique allows the client to have multiple outstanding operations at one time. ZooKeeper uses the atomic broadcast protocol (Zab) to guarantee linearizability and the watches mechanism to allow clients to watch any updates and receive notification about their data object without requiring polling. Furthermore, ZooKeeper provides data nodes (znodes) in a hierarchical namespace that allow clients to manipulate these data objectives while taking into account a set of primitives, for instance, group membership and configuration management (for details, see [15]). A new approach in the coordination is

presented through implementing the powerful primitives at the client side using
the ZooKeeper API, which will provide a high performance. Figure 1 displays
the high level components of the ZooKeeper service.

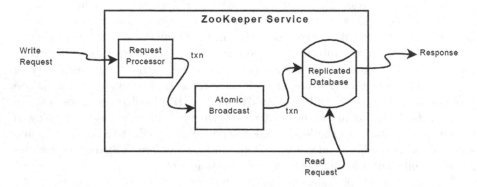

Fig. 1. Components of the ZooKeeper service [15]

As ZooKeeper is comprised of an ensemble of servers, it uses majority quo-
rums, which means that it will proceed only when the majority of servers work
and are consistent. Servers can crash and recover many times without affecting
the service as long as Zab has a majority. In the case of leader failure, Zab con-
trols the election of a new leader and guarantees the continuity of service. The
leader replicates requests consistently to followers, thus eventually they finish up
with the same state. Additionally, timeouts have been used in order to detect
failures. For recovering from faults, such as, power outages, it is essential to
record messages on the disk media of a quorum of followers before the delivery
process takes place [21]. Moreover, ZooKeeper uses a database in-memory and
stores periodic snapshots and transactions log on disk, as writing every single
transaction on the disk would result in the disk I/O bottleneck [21].

Clients can connect to one ZooKeeper server to submit its requests. Read
requests are serviced from a local replica of the database of the contacted server,
this provides high read throughput. However, write requests that change the
state of the service, are transformed to idempotent transactions and forwarded to
the leader using Zab protocol, thus write throughput is limited by the throughput
of Zab protocol [21]. Furthermore, the server that received the client request, will
respond to that client with the corresponding state update.

3   Atomic Broadcast Protocol Zab

ZooKeeper uses the atomic broadcast protocol (Zab) for the coordination of write
requests that update the state of ZooKeeper service. In short, Zab is a high-
performance protocol for primary-backup systems [16]. Zab provides efficient
mechanisms for process recovery, where it enables multiple outstanding state

changes [16]. Additionally, Zab keeps the replicas of the ZooKeeper state at each server consistent [21]. Furthermore, Zab protocol has two modes: broadcast and recovery [21], where it transitions to a recovery mode once the service starts or after a failure of a leader. There are two requirements to end the recovery mode: a leader emerges and then synchronising the state of a quorum of servers with the leader. The leader can start broadcasting once the synchronisation phase successfully finished, the broadcast mode ends if the leader fails or it lost the support of a quorum of followers.

The Zab differs from the other atomic broadcast protocols, such as Paxos [18], whereas Zab enables multiple outstanding operations and guarantees the execution order of the clients' requests. ZooKeeper maintaining ordered broadcast FIFO using Zab, where it guarantees to deliver the transactions in the same order to all servers, and guarantees that transactions cannot be skipped by servers [17]. ZooKeeper developers have proposed a Primary Order property that maintains the right order of the state change over time, which is extremely important for the primary-backup systems. As stated by Hunt *et al.* [15], the implementation of Zab showing that it can meet the high performance and low latency demands of broadcasts. Additionally, the TCP channel connections are used for the message exchange between servers (Zab processes), and it has been relied on the nature of the TCP to maintain the right order.

According to Junqueira *et al.* [16], Zab consists of three phases: discovery, synchronisation, and broadcast. There are two types of servers (Leader and Follower), which can be performed by Zab process. Phase 1, is used to determine which server is the leader and phase 2 to establish the leadership and phase 3 is the broadcast phase. In this work, we will only investigate the Zab broadcast phase, however, for more details about the phases 1 and 2 we refer the interested reader to [16].

As stated by Junqueira and Reed [17], there are additional checks that need to be performed by the follower before the acknowledgement of a proposal take place. Firstly, checking that the proposal is from the current leader that the follower f is following. Secondly, checking the order of proposals and transactions to be in the same order, as the leader broadcasts them.

4 Zab PEPA Model

A PEPA model has been created for the Zab protocol, specifically, the broadcast phase using the Performance Evaluation Process Algebra (PEPA) modelling paradigm. The primary objective is to inspect and evaluate the behaviour of the Zab protocol (broadcast phase). Validating the outcomes of the model against a real implementation of Zab is another objective. PEPA is a high-level quantitative modelling language developed by Hillston [13], and one of its essential usage is for modelling the distributed systems. Models built on a high level of abstraction using stochastic process algebra (SPA), for instance (PEPA [13], EMPA [3] and SPADES [10]), and stochastic Petri net (SPN) [7]. PEPA offers several significant features in performance modelling, such as compositionality, formality and abstraction [14]. The PEPA Eclipse Plug-in tool [24] is a supporting tool has been used for developing and analysing the performance of systems, offering a variety of analysis techniques, for example, continuous time Markov chain (CTMC), Stochastic Simulation and Ordinary Differential Equations (ODEs). Such approaches allowing the observation of a system as it evolves from an initial state over a period of time [24]. Furthermore, each action within the PEPA model has a rate which is the reciprocal of the average duration, or delay, to undertake by the action. Each of the model components has been modelled as a set of sequential actions to simulate the protocol behaviour. Furthermore, several performance metrics, for instance, throughput and latency have been used to evaluate the performance of the Zab PEPA model.

The model is analysed using Steady State Analysis CTMC, and also, its behaviour is approximated using the Ordinary Differential Equations (ODEs) analysis for both throughput and population, which supports the numerical calculation of a large scale model with a large number of *Client* and *Request* components. ODE solvers are continuous and deterministic [24] and have been used by [12] and [11] as a solution to the CTMC state space explosion problem. The number of clients and requests has been varied (from 1 to 10), every single client generates (20) threads and sends requests to the *Request* component. Accordingly, the number of requests that sent to a server equals to the number of clients times (20) threads (e.g. 10 clients * 20 threads = 200 requests sent in parallel). In order to model the Zab protocol in PEPA, several simplifications have been applied with the objective to get a scalable model, that can process a large number of requests, and also, to allow to explore further implications of some actions (c.g. cpu action). The Zab broadcast PEPA model is represented in an abstraction level with the following modifications: Firstly, it has been created *Request* component which has a sequence of all actions that specified by the Zab protocol. So, the *Request* component has been used in the model to simulate the overall performance, i.e. allowing to evaluate the overload of the *Leader* in another way. Additionally, the other components have been used in the model are (*Client* and *Leader*). Whereas, the actions of these components are manipulated in parallel. In other words, these components are simultaneously cooperating their actions with the primary component (*Request*). As well, the *Client* component is used to represent a client who sends a write request or update existing data.

Furthermore, the sequence of the actions is preserved through the *Request* component. These changes not only simplify the model, but also, allow the model to scale up without any restrictions. Hence, the overall performance will not be affected, i.e. quantitative analysis. Moreover, the model follows a regularly repeated sequence of events. For example, a client generates a request, wait for processing this request and once the processed request is received (i.e. *getRequest* action) by the client then the model returns to the initial state (i.e. generate request) to send a new request and so on, i.e. which is a cyclic model.

Some local actions of the *Leader*, such as, (*excuteRequest, processingCommit* and *leaderDeliver*) have been renamed to *cpu* action in the model and the rate of the *cpu* action equals to the average duration rate of these three actions. Whereas, we consider two different options for the interaction between the leader and the clients in ZooKeeper. Namely, we have made an assumption that the model will be examined into two different versions, where the action *leaderDeliver* will be used as an independent action, and then it will be used as a centralised action. Firstly, we have used the model with (2 cpu) actions (*excuteRequest and processingCommit*), and will be referred to as (*Assumption 1*). Secondly, it has been analysed by using (3 cpu) actions, where added the action (*leaderDeliver*), which is the action of the component $Request_{10}$. The main reason for using a different number of *cpu* actions is to examine the model with various performance scenarios and to explore which of these actions will give rise to overload the leader. Additionally, the *cpu* action is used because PEPA does not allow to directly limit the action across multiple actions, i.e. limits the rate across multiple actions. That is to say, it will be a single action (*cpu*) but we are limiting the total rate of this action, it is a combination of all those actions. So, the *cpu* cannot run faster than those all actions (bounded capacity).

The following is the Zab PEPA model:

$$Client \stackrel{def}{=} (generateRequest, r_1).(getRequest, r_1).Client$$

$$Request \stackrel{def}{=} (generateRequest, r_1).Request_1$$

$$Request_1 \stackrel{def}{=} (requestToL, r_1).Request_2$$

$$Request_2 \stackrel{def}{=} (cpu, c).Request_3$$

$$Request_3 \stackrel{def}{=} (sndProposalA, r_4).Request_4$$

$$Request_4 \stackrel{def}{=} (sndProposalB, r_4).Request_5$$

$$Request_5 \stackrel{def}{=} (processingAcknowledgeA, r_9).Request_6$$
$$+ (processingAcknowledgeB, r_9).Request_6$$

$$Request_6 \stackrel{def}{=} (acknowledgeA, r_5).Request_7$$
$$+ (acknowledgeB, r_5).Request_7$$

$$Request_7 \stackrel{def}{=} (cpu, c).Request_8$$

$$Request_8 \stackrel{def}{=} (commitA, r_7).Request_9$$

$$Request_9 \stackrel{def}{=} (commitB, r_7).Request_{10}$$

$$Request_{10} \stackrel{def}{=} (leaderDeliver, r_8).Request_{11}$$

$$Request_{11} \stackrel{def}{=} (getRequest, r_1).Request$$

$$Leader \stackrel{def}{=} (cpu, c).Leader$$

$$System \stackrel{def}{=} Client[N] \underset{L_1}{\bowtie} Request[N] \underset{L_2}{\bowtie} Leader$$

The components of the model are cooperated concurrently as exhibited in the system equation, where *Client* cooperate with *Request* over the set L_1. Indeed, the *getRequest* action has been used to restrict a client who sent a request from sending a new request until get a response, consequently, a client will be able to send a new request. Also, the *Request* and *Leader* cooperate over the set L_2. Whereas, N has been varied from 20 to 200, and the cooperation set $L_1 = \{generateRequest, getRequest\}$ and the set $L_2 = \{cpu\}$.

Table 1. The Rates of Zab PEPA model

Rate	Value	Rate	Value
r_1	7.42	r_3	61.96657616
r_4	7.42	r_5	7.42
r_6	437.1975906	r_7	7.42
r_8	26.33486286	r_9	1330.250132
c	3/(1/r3+1/r6+1/r8)		

It is worth noting that, the model parameters representing a real-time implementation, which is in milliseconds. However, the rates of $r1$, $r4$, $r5$ and $r7$ are not part of the real measurement of the system. That is because of the well-known problem in the real-time distributed systems (clock synchronisation). In fact, there are many solutions for the clock synchronisation problem, but they have not been augmented in the system that we have taken the measurements from. Therefore, we have calculated those rates as follows:

$$rate_i = \frac{L + \sum r_i}{N} \tag{1}$$

Where, L is the average latency of the real measurements, r_i representing the real time in ms for the actions associated with these rates (r_3, r_6, r_8 and r_9). Also, N is representing the repetition of times that the rates (r_1, r_4, r_5 and r_7) have been used in the model. The rate of each action equals $(1/Time(ms))$ as shown in the Table 1.

5   Performance Metrics

Two performance benchmarks will be considered (latency and throughput) with the intention of inspecting the model behaviour and to compare the PEPA model with the real implementation of Zab protocol.

Latency is a performance benchmark, whereby it means the total time for a data transmitted to a destination and then returned back to its source (measuring the round trip). In other terms, it used to refer to a delay. Based on the derived ODEs analysis results the latency has been calculated as follows:

– **Assumption 1:**

$$(1 + Pop(Req2 + Req7))\frac{2}{c} + \sum_{\forall a \in \sigma} \frac{1}{rate_a} \tag{2}$$

– **Assumption 2:**

$$(1 + Pop(Req2 + Req7 + Req10))\frac{3}{c} + \sum_{\forall a \in \sigma} \frac{1}{rate_a} \tag{3}$$

Where $\sigma = \{generateRequest,\ requestToL,\ sndProposalA,\ sndProposalB,\ processing\ AcknowledgeA,\ processingAcknowledgeB,\ acknowledgeA,\ acknowledgeB,\ commitA,\ commitB\ and\ getRequest\}$. Also, a is the type action for all actions in the set σ. It is clear that the model here is a closed queuing network. The reason of calculating the latency for the non cpu actions $1/rate$ because they are not subject to queuing, just the time it takes to do that action. But for the cpu actions, they have to queue (competitive actions) and they take serving time and the average waiting time. Hence, the population average queue length equals 1 plus the queue length for 1 less entity in it, so the population is for the system with $N - 1$ requests.

Throughput is the second performance metric used. It has been calculated using the PEPA Eclipse Plug-in tool scalable analysis (Throughput), which gives how many times a specific action happen in a millisecond.

6   Experiments and Results

In this section, a number of assumptions have been made to analyse the protocol, for instance, all experiments are processed in the absence of failure and the latency is only measured from the server side. The model which is illustrated in the Sect. 4 has been analysed and evaluated using PEPA Eclipse Plug-in tool, i.e. steady state analysis (CTMC) and the fluid flow analysis (ODEs). Basically, the time which is used in the model and illustrated in the Table 1, is taken from a real implementation of the Zab protocol. Specifically, it is representing the average of three implementations of Zab for each number of clients. However, in the experiments which follow the rates are calculated by taking the average of those times, which means that the rates that have been used in the Zab PEPA model are representing the average of (30) experiments for the number of clients (from 1 to 10). As each client generates (20) threads, in our model number of clients have been varied (from 20 to 200) in order to simulate the number of threads for each number of clients.

In the first set of experiments, we have analysed the model by the means of CTMC (i.e. compare *Assumption 1* and *Assumption 2*), and then the Zab performance (*Assumption 1* and *Assumption 2*) have been evaluated using the ODEs. Also, we have investigated the performance issue that arises in the *Assumption 2*. Finally, it has been compared the latency of Zab *Assumption 1* against the real implementation of Zab.

6.1   CTMC Analysis

The results of the CTMC analysis will be displayed in this section. The main reason for using the CTMC is to figure out the scale of the model, specifically, the number of clients and requests that can be analysed with this type of analyses. The following figures illustrate the outcomes of the model *Assumption 1* and *Assumption 2*, in terms of performance metrics (throughput and the populations), where the maximum number of clients can be derived using the CTMC is (4). As seen in Fig. 2, the throughput of *cpu* action of the *Assumption 2* is higher than the *Assumption 1* that is because we have used 3 *cpu* action in the *Assumption 2* and the number of clients and requests is small. So, the leader has enough resources to manipulate this loads. However, the throughput of the *getRequests* action of both *Assumption 1* and *Assumption 2* are identical, due to the use of small load (i.e. limited by the CTMC state space problem). In Fig. 3, the population of *Assumption 1* and *Assumption 2* are showing the same behaviour, and also, there is a very small waiting time. It is clearly that the utilisation is very low, as the CTMC only allowed us to analyse the model using (4 clients) before the state space problem arises.

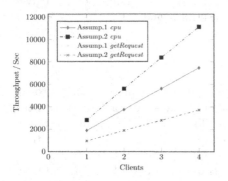

Fig. 2. Throughput of Zab (Assump.1 & Assump.2) using CTMC (*cpu* and *getRequest* actions).

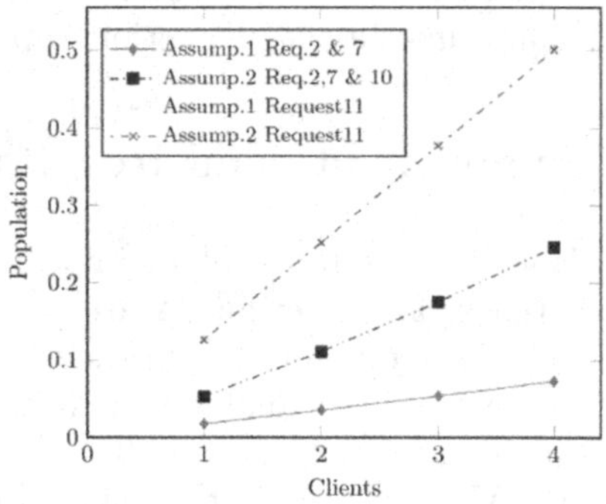

Fig. 3. Population of Zab (Assump.1 and Assump.2) using CTMC (Request2,7,10 and Request11).

6.2   Zab PEPA Model ODEs Analysis

The latency and the throughput of the Zab protocol PEPA models *Assumption 1* and *Assumption 2* are shown respectively in Fig. 4 and Fig. 5. Obviously, *Assumption 1* of Zab performs much better than the Zab *Assumption 2* in both latency and throughput. In Fig. 4 the latency of the Zab *Assumption 1* is displayed as a flat line at the beginning up to the point where $N = 60$. Then it has increased gradually to reach its maximum latency of 6.2538 ms, when the number of clients equals 200. This means that the Leader has the capacity to handle the coming requests up to $N = 60$, then the queuing time will start to increase with the increased load on the system. On the other hand, the latency of Zab *Assumption 2* is linearly increased from the very beginning and it is continued up to the end with the increase of the load, where the latency is extremely high.

It is obvious that introducing the extra *cpu* action indeed affects the scalability; overloading the Leader and saturation is reached with a very low number of clients. With 2 *cpu* actions that saturation does not occur until 60 clients, and shows a more scalable approach. These because the load on the Leader is not so great the latency does not rise as significantly.

The throughput shown in Fig. 5 illustrates that the performance of Zab *Assumption 2* is poor and it is just shown as a flat line. However, Zab *Assumption 1* presents a higher throughput, where it reaches its peak when the number of clients is 60 (i.e. 54273 requests per second). The poor performance exemplified by the Zab *Assumption 2* (Fig. 5) is because the system is saturated where the use of 3 *cpu* results in more contention on the resources. Hence, the rate which is used for the *cpu* action (*c*) is much lower than the same rate that has been used for the *Assumption 1* for the same action. The average duration rate (*c*) for the *Assumption 1* is calculated as:

$$c_1 = \frac{2}{\frac{1}{r_3} + \frac{1}{r_6}} \tag{4}$$

For the *Assumption 2*, it is calculated as:

$$c_2 = \frac{3}{\frac{1}{r_3} + \frac{1}{r_6} + \frac{1}{r_8}} \tag{5}$$

As $1/r_3 + 1/r_6 < 2/r_8$ using the parameters in Table 1, the value of c used in *Assumption 2* is less than in *Assumption 1*. In fact c is less than half as fast in *Assumption 2* than *Assumption 1*.

Thus, the performance of the model is significantly affected. Therefore, in the next section we will investigate further the behaviour of the Zab *Assumption 2* in order to tackle the bottleneck caused by the *cpu* action.

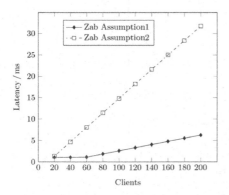

Fig. 4. Latency of Zab (*Assump.1 & Assump.2*)

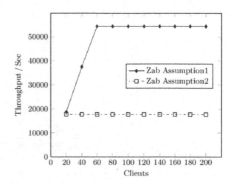

Fig. 5. Throughput of Zab (*Assump.1 & Assump.2*)

6.3   Investigating Zab Assumption 2 Bottleneck

The system is overloaded quite rapidly by the use of *Assumption 2*, as depicted in the graphs of the Sect. 6.2. As a result, an additional research has been conducted in order to solve the scalability issue of the Zab *Assumption 2*. In the model Zab *Assumption 2*, we have made an assumption that the action (*leaderDeliver*) will be used as a *cpu* action. Indeed, *leaderDeliver* is an action that occurs after the *commit*, where the *Leader* is proceeded to write physically on its memory. The *leaderDeliver* action is considered as one of the actions that extensively use the resources of the *Leader*. Consequently, it is renamed to *cpu* and its rate r_8 is used to calculate the average duration rate (c) of the *cpu* action, as illustrated in the Eq. 5. Due to the rate r_8 is the slowest rate among the other rates that used to calculate the rate of the *cpu*, we have varied the rate r_8 and observe how this will influence the behaviour of the Zab *Assumption 2* model.

Figure 6 and Fig. 7 are respectively displayed the latency and the throughput of the Zab *Assumption 2* model by means of varying the rate r_8. In Fig. 6 it is noticeable that the latency of the model is extremely decreased, with the increase of the rate r_8 value. Namely, it is declined from (31.74 ms) to (10.51 ms) when the active number of clients is equal to (200). The throughput also, has risen significantly in response to the change of the value of the r_8 rate as shown in Fig. 7. Whereas, the maximum reached throughput is (47588) request per second. Clearly, the model outcomes have illustrated that the overall performance is improved significantly by only increasing the rate r_8, which was limiting the *Leader* behaviour and causing the overload.

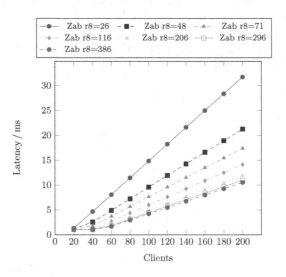

Fig. 6. Latency of Zab *Assump.2* varying (r_8)

6.4   Comparing Zab Model and Real Implementation of Zab

In this section, a comparison will be given between the performance of the Zab *Assumption 1* (2 *cpu*) model and a real implementation of Zab. This comparison aims to validate the model outcomes, specifically, the latency of the Zab

Assumption 1. Figure 8 showing the latency of both the Zab PEPA model and the implementation of Zab. As the load is relatively small at the beginning, it is noticeable that the latency of the Zab PEPA model is shown as a flat line, where the system has received (20, 40 and 60) requests. This is because there are enough resources and the coming requests are served without any waiting time.

However, the latency starts to increase linearly as the number of clients and requests are increased beyond (60), where the coming requests have to wait to be served. Although, the latency of the Zab model a bit higher (i.e. 6.25 ms to handle 200 requests) than the latency of the Zab implementation (i.e. 5 ms for the same number of requests); however, both are shown some consistency, in particular, the saturation point is the same for both, which is (60) clients. Then the latency of Zab model has risen linearly.

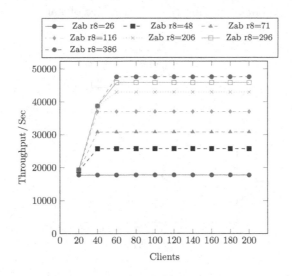

Fig. 7. Throughput of Zab *Assump.2* varying (r_8)

7   Related Work

High performance and scalability are desired requirements for cloud computing services providers and customers. A significant number of researches and studies have put forward this goal and not only presented practical solutions, but also, proposed designs of scalable systems. One example of using ZooKeeper coordination services is the study conducted by Skeirik *et al.* in [23], where they formally analysed a model of a ZooKeeper-based group key management service for fault-tolerance and performance. This formal analysis is based on rewriting logic model Maude [6] and the statistical model checking tool PVeStA [2]. This study also explored the reliability of ZooKeeper-based group key management service for handling faults and providing a scalable performance with low latency. Although, the authors have considered security aspects of infrastructure in the cloud using formal modelling and they are shown that ZooKeeper is capable to tolerate with fault, but also authors stated that they have only considered one aspect of failure, i.e. crash-failures.

More recently, El-Sanosi and Ezhilchelvan [8] have presented an approach for improving the performance of ZooKeeper atomic broadcast protocol by coin tossing. This study presents a practical solution for the performance problem of a leader-based protocol; which raises with the increase of the load on the leader. A condition has been made that followers can broadcast only if the outcome of a coin toss is head, which reduces the communication traffic at the leader. However, there is a case where switching back to Zab would be the better choice, if one of followers fails or cannot compute the probability of coin toss. Nonetheless, significant performance improvement can be gained with the adoption of ZabCT proposed by [8] without impacting the performance of Zab.

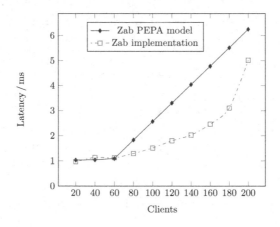

Fig. 8. Latency of Zab PEPA model (*Assump. 1*) and Zab implementation

8   Conclusion

This research has presented an approach to the evaluation of the performance of the ZooKeeper atomic broadcast protocol, specifically the Zab broadcast phase. A Zab model has been created using PEPA Eclipse Plug-in tool, and an extensive performance analysis is conducted for the developed Zab model. This research is motivated by inspecting a high-performance coordination service system by the use of Performance Evaluation Process Algebra PEPA. In addition, validating the Zab PEPA model results against a real implementation of Zab.

At first, we have explored the scaling limit of the CTMC analysis, whereas the maximum number of clients that can be derived is (4), because of the well-known state space issue. On the other hand, the fluid flow analysis (ODEs) has been used for a large scale system to derive two performance metrics (latency and throughput). Two versions of the Zab model with a slight difference (i.e. Zab *Assumption 1* and Zab *Assumption 2*) have been evaluated. The *Assumption 1* model has presented a high performance (low latency and high throughput). However, the system is overloaded with the use of *Assumption 2*. This is because of the *leaderDeliver* action, which is used as a *cpu* action in the Zab *Assumption 2* that has a relatively slow rate, hence, limited the behaviour of the Leader. Therefore, we have investigated further the scalability issue that shown in Fig. 4 and Fig. 5 of the model Zab *Assumption 2* by varying the rate r_8. Consequently, the performance of the model has improved significantly by increasing the rate r_8 as illustrated in Fig. 6 and Fig. 7, which is allowed to free up the Leader, thus performing much better for handling the increased loads.

Moreover, we have compared the latency of the model Zab *Assumption 1* with the latency of a real implementation of Zab, where the outcomes showing that they are noticeably consistent as depicted in Fig. 8; although the model gives a more idealised (smoother) performance due to the simplifications made.

In spite of the fact that we have based our analysis on the parameters that measured in a real system, but different real system can have slightly different parameters rates and that might affect the comparison between Zab *Assumption 1* and Zab *Assumption 2*, such that we may see better scalability on one than the other in our case. Furthermore, we have only looked at certain sets of parameters governed by a measurement, and also, we have only looked at one leader, if we have more leaders potentially can generate more capacity; but if there are coordinations between those leaders then it will limit that extra capacity. The model also accurately predicts the saturation point of the system (the point when throughput ceases to increase). Overall, the approach described in this study is highlighting a methodology to inspect a high-performance coordination services system, precisely, ZooKeeper (Zab protocol). Herein, we have only considered one phase of Zab (broadcast), therefore our future work is to model all phases of the Zab; taking into account further investigation in different performance aspects.

More recently, a new scalable distributed coordination service called Giraffe is released, which supposed to be an alternative to the coordination services, such as Zookeeper and Chubby. Shi *et al.* [22] claimed that Giraffe is 300% faster than Zookeeper. However, there is no independence analysis to support this study. So, analysing the performance of this system and compare it with the ZooKeeper would be an interesting point of research.

References

1. ZooKeeper, A.: https://cwiki.apache.org/confluence/display/ZOOKEEPER/Index (2015). Accessed 16 Dec 2015
2. AlTurki, M., Meseguer, J.: PVESTA: a parallel statistical model checking and quantitative analysis tool. In: Corradini, A., Klin, B., Cîrstea, C. (eds.) CALCO 2011. LNCS, vol. 6859, pp. 386–392. Springer, Heidelberg (2011). https://doi.org/10.1007/978-3-642-22944-2_28

3. Bernardo, M., Gorrieri, R.: A tutorial on EMPA: a theory of concurrent processes with nondeterminism, priorities, probabilities and time. Theoret. Comput. Sci. **202**, 1–54 (1998)
4. Borthakur, D., et al.: Apache Hadoop goes realtime at Facebook. In: Proceedings of the 2011 ACM SIGMOD International Conference on Management of Data. SIGMOD 2011, pp. 1071–1080. ACM, New York, NY, USA (2011)
5. Burrows, M.: The chubby lock service for loosely-coupled distributed systems. In: Proceedings of the 7th Symposium on Operating Systems Design and Implementation, pp. 335–350. OSDI 2006, USENIX Association (2006)
6. Clavel, M., et al.: All About Maude - A High-Performance Logical Framework. LNCS, vol. 4350. Springer, Heidelberg (2007). https://doi.org/10.1007/978-3-540-71999-1
7. Donatelli, S.: Superposed generalized stochastic petri nets: definition and efficient solution. In: Valette, R. (ed.) ICATPN 1994. LNCS, vol. 815, pp. 258–277. Springer, Heidelberg (1994). https://doi.org/10.1007/3-540-58152-9_15
8. EL-Sanosi, I., Ezhilchelvan, P.: Improving ZooKeeper atomic broadcast performance by coin tossing. In: Reinecke, P., Di Marco, A. (eds.) EPEW 2017. LNCS, vol. 10497, pp. 249–265. Springer, Cham (2017). https://doi.org/10.1007/978-3-319-66583-2_16
9. Fan, W., Bifet, A.: Mining big data: current status, and forecast to the future. SIGKDD Explor. Newsl. **14**(2), 1–5 (2013)
10. Harrison, P.G., Strulo, B.: Spades - a process algebra for discrete event simulation. J. Logic Comput. **10**(1), 3–42 (2000)
11. Hayden, R.A., Bradley, J.T.: Fluid-flow solutions in PEPA to the state space explosion problem. In: 6th Workshop on Process Algebra and Stochastically Timed Activities (PASTA), p. 25 (2007)
12. Hillston, J.: Fluid flow approximation of PEPA models. In: Second International Conference on the Quantitative Evaluation of Systems (QEST 2005), pp. 33–42 (2005)
13. Hillston, J.: A compositional Approach to Performance Modelling. Cambridge University Press, Cambridge (2008)
14. Hillston, J., Gilmore, S.: Performance evaluation process algebra. http://www.dcs.ed.ac.uk/pepa/about/ (2011). Accessed 05 April 2016
15. Hunt, P., Konar, M., Junqueira, F.P., Reed, B.: Zookeeper: wait-free coordination for internet-scale systems. In: USENIX Annual Technical Conference, vol. 8, p. 9 (2010)
16. Junqueira, F.P., Reed, B.C., Serafini, M.: Zab: high-performance broadcast for primary-backup systems. In: 2011 IEEE/IFIP 41st International Conference on Dependable Systems Networks (DSN), pp. 245–256 (2011)
17. Junqueira, F., Reed, B.: ZooKeeper: Distributed Process Coordination. O'Reilly Media, Inc. (2013)
18. Lamport, L.: The part-time parliament. ACM Trans. Comput. Syst. **16**(2), 133–169 (1998)
19. MacCormick, J., Murphy, N., Najork, M., Thekkath, C.A., Zhou, L.: Boxwood: abstractions as the foundation for storage infrastructure. In: Proceedings of the 6th Conference on Symposium on Operating Systems Design and Implementation - Volume 6. OSDI 2004, USENIX Association (2004)
20. Medeiros, A.: Zookeeper's atomic broadcast protocol: Theory and practice. Technical report (2012). Accessed 07 Oct 2015

21. Reed, B., Junqueira, F.P.: A simple totally ordered broadcast protocol. In: proceedings of the 2nd Workshop on Large-Scale Distributed Systems and Middleware. ACM (2008)
22. Shi, X., et al.: GIRAFFE: a scalable distributed coordination service for large-scale systems. In: 2014 IEEE International Conference on Cluster Computing (CLUSTER), pp. 38–47, September 2014
23. Skeirik, S., Bobba, R.B., Meseguer, J.: Formal analysis of fault-tolerant group key management using ZooKeeper. In: Proceedings of the 13th IEEE/ACM International Symposium on Cluster, Cloud, and Grid Computing 2013, pp. 636–641, May 2013
24. Tribastone, M., Duguid, A., Gilmore, S.: The PEPA Eclipse Plugin. SIGMETRICS Perform. Eval. Rev. **36**(4), 28–33 (2009)
25. Zimmerman, J.: Apache zookeeper in netflix. http://techblog.netflix.com/2011/11/introducing-curator-netflix-zookeeper.html (2011). Accessed 19 Jan 2016
26. Zookeeper, A.: Zookeeper: a distributed coordination service for distributed applications. https://zookeeper.apache.org/doc/trunk/zookeeperOver.html (2014). Accessed 16 Feb 2017

ComBench: A Benchmarking Framework for Publish/Subscribe Communication Protocols Under Network Limitations

Stefan Herrnleben[1]([⊠])[ID], Maximilian Leidinger[1], Veronika Lesch[1][ID],
Thomas Prantl[1], Johannes Grohmann[1][ID], Christian Krupitzer[2][ID],
and Samuel Kounev[1]

[1] University of Wuerzburg, 97074 Wuerzburg, Germany
{stefan.herrnleben,maximilian.leidinger,veronika.lesch,thomas.prantl,
johannes.grohmann,samuel.kounev}@uni-wuerzburg.de
[2] University of Hohenheim, 70599 Stuttgart, Germany
christian.krupitzer@uni-hohenheim.de

Abstract. Efficient and dependable communication is a highly relevant aspect for Internet of Things (IoT) systems in which tiny sensors, actuators, wearables, or other smart devices exchange messages. Various publish/subscribe protocols address the challenges of communication in IoT systems. The selection process of a suitable protocol should consider the communication behavior of the application, the protocol's performance, the resource requirements on the end device, and the network connection quality, as IoT environments often rely on wireless networks. Benchmarking is a common approach to evaluate and compare systems, considering the performance and aspects like dependability or security. In this paper, we present our IoT communication benchmarking framework *ComBench* for publish/subscribe protocols focusing on constrained networks with varying quality conditions. The benchmarking framework supports system designers, software engineers, and application developers to select and investigate the behavior of communication protocols. Our benchmarking framework contributes to (i) show the impact of fluctuating network quality on communication, (ii) compare multiple protocols, protocol features, and protocol implementations, and (iii) analyze scalability, robustness, and dependability of clients, networks, and brokers in different scenarios. Our case study demonstrates the applicability of our framework to support the decision for the best-suited protocol in various scenarios.

Keywords: IoT · Publish/subscribe · Benchmarking · Load testing

1 Introduction

The road to success of the Internet of Things (IoT) leads to an enormous increase of devices exchanging data over the internet [27]. Many popular IoT communication protocols follow the publish/subscribe communication pattern [1], which

Q. Zhao and L. Xia (Eds.): VALUETOOLS 2021, LNICST 404, pp. 72–92, 2021.
https://doi.org/10.1007/978-3-030-92511-6_5

decouples space, time, and synchronization and is well-suited for upscaled distributed systems due to its loose coupling [11]. MQTT, AMQP, and CoAP are well-known examples of publish/subscribe protocols which exchange their messages via a central message broker [3,15,17,22,28]. All participating clients can be both publishers and subscribers. Whoever wants to receive a message subscribes to a specific topic at the broker. For sending a message, a client publishes the message with a specific topic to the broker, which delivers the message to all subscribers of the topic.

The argument to apply a publish/subscribe protocol in an IoT system is straightforward due to the characteristics of those systems [10,20,30]. Performance, scalability, and overhead of a protocol often play a significant role when looking for a suitable protocol for a particular use case [3,17,22,28]. Selecting the most suitable protocol and a performant implementation can be challenging due to the wide variety of existing protocols, each with different broker and client implementations [10,20].

A special challenge in IoT is the robustness of a communication protocol against the variation of network quality [5,6]. While data centers typically use wired connections, IoT devices often communicate via wireless networks, which are often faced with bandwidth limitations, high latencies, and packet loss [4,16]. In addition to WiFi and cellular networks, low-power wireless network protocols such as LoRaWAN can also be used. In addition to already challenging aspects of standardization, availability of implementations, and licensing, developers should also be aware of protocol performance even in stressed environments and under unstable connections [6].

Existing benchmarking tools can test load and scalability [7,9,13], but do not limit the network communication by, e.g., adding artificial packet loss or latency.

To support designers, developers, and operators of IoT systems, we present in this paper our benchmarking framework *ComBench* coupled with a methodology for evaluating and comparing communication-related characteristics like performance, dependability, and security [19]. The key attributes of ComBench can be summarized as follows:

- **Multi-protocol client** supporting the protocols MQTT, AMQP, and CoAP out of the box.
- **Customized load profiles** with configurable message size and frequency for each client or group of clients.
- **Virtually unlimited clients** for testing large-scaled environments.
- **Fine-grained communication configuration** per client or group of clients. Messages can be published after a fixed delay, following a statistical distribution, or as a (delayed) response to an incoming message.
- **Configurable network quality** in terms of bandwidth limit, transmission delay, and packet loss. Each can be configured per client or group of clients, statically or using a time series, e.g., for emulating a moving device.
- A **Benchmarking controller** serving as a single configuration point for managing the experiment, collecting all measurements, and making them available for visualization and export.

We demonstrate the wide range of ComBench's applicability in five exemplary use cases by varying protocols, configurations, workloads, network conditions, and setups. The case study shows how our benchmarking framework and measurement methodology analyzes and compares communication protocols in different IoT environments. ComBench allows developers and system designers to investigate the performance of protocols and the resource consumption on clients and networks to select an appropriate protocol or optimize their application. Artificial variation of network quality allows statements about the protocol's performance even under limited network connectivity.

The remainder of this paper is structured as follows. In Sect. 2, we describe the characteristics of our benchmark, its components, the supported metrics and give an insight into its usage. Section 3 presents the case study demonstrating its applicability to various evaluation objectives. In Sect. 4, we discuss weaknesses of our benchmarking framework that an operator should be aware of. Section 5 shows related work that deals with benchmarking and performance evaluation of publish/subscribe systems. Section 6 summarizes the paper and provides ideas for future work.

2   Benchmark

This section introduces our benchmarking framework, including the specified requirements, the measurement methodology, the captured metrics, and some design and implementation details. Section 2.1 presents the requirements from which the capabilities of the framework are derived. In Sect. 2.2, the metrics based on the measured values are formally defined. Section 2.3 describes the design and architecture of our ComBench, while Sect. 2.4 provides implementation and configuration details.

2.1   Requirements

For the development of a benchmarking framework for IoT communication protocols, a number of requirements arise. The general requirements for a benchmark [18] are supplemented by additional requirements due to the large heterogeneity of protocols, clients, and communication behavior, as well as the usually wireless connection [5,26]. This section presents the essential requirements to be met by our benchmarking framework for IoT communication protocols.

Support of Multiple Protocols. Various publish/subscribe communication protocols are omnipresent in IoT systems. While some protocols are similar, other protocols are more suitable for specific communication scenarios and requirements. A benchmarking framework should address this heterogeneity and support various protocols to investigate scenarios under different protocols.

Variation of Network Conditions. IoT devices often communicate over wireless links exposed to bandwidth fluctuations, transmission delays, and packet loss. A benchmarking framework for IoT systems should take such varying network conditions into account by artificially influencing the network quality. This allows investigating and comparing protocol behavior and features under different workloads.

Heterogeneous Clients. IoT systems often consist of clients with different message types (e.g., messages with different payload sizes), different communication behavior (e.g., rarely arising vs. high frequent), and different network quality (e.g., GSM vs. LTE). A benchmarking framework should reproduce this heterogeneity by configuring bandwidth limit, transmission delay, and packet loss per device or device group.

Protocol Features. Protocols like MQTT support different QoS levels or security features like encryption via *Transport Layer Security (TLS)*. Such features often influence the performance and resource consumptions of the device as well as the network load. A benchmarking framework should support easy activation of these features to investigate the impact of such protocol features on CPU, RAM, and network.

Scalability. IoT systems often scale over a massive amount of devices, and the number of IoT devices is continuously increasing [27]. The number of emulated IoT devices should not be limited to the hardware capabilities of a single node; instead, the experiment should be scaleable to multiple servers or a cloud. Scaling the number of clients, messages, and network load enable investigating brokers in stressed environments.

Flexibility at Broker Selection. Various message broker implementations exist, which differ in the supported protocols, programming language, licensing, performance, stability, vendor, and how they are developed (e.g., community-driven, proprietary). Operators of IoT systems need to be aware of the broker's performance and reliability. Therefore, a benchmarking framework should not be fixed to a specific broker; instead, the broker, configuration, and deployment should be freely choosable.

Metrics. Metrics are used in benchmarks to provide insights into the performance, reliability, or security of a tested system [19]. The supported metrics should not be limited to a specific evaluation objective. Instead, the metrics and measurements like client resource consumption, i.e., CPU and RAM utilization, network throughput, latency, packet loss, and protocol efficiency contribute to investigate and compare different objectives. Metrics are formally defined in the next section.

2.2   Metrics

Before defining the metrics, we introduce the meaning of the used sets and symbols. The set C containing all clients, the set T refers to all topics (for topic-based protocols like MQTT), and the set M contains all published messages by any client to any arbitrary topic. $P_c = \{m \mid m \in M$ and message m was published by client $c\}$ represents the set of messages which have been published by client c. The subscribed messages of each client c are described by $S_c = \{m \mid m \in M$ and topic t of message m was subscribed by client $c\}$. Since messages can be lost, $S'_c = \{m \mid m \in S_c$ and message m was received by client $c\}$ with $S'_c \subseteq S_c$ only considers messages for client c that were actually delivered. From these definitions and the introduced measurements, we derive the following metrics:

Message Loss Ratio. This metric defines the ratio of lost messages to expected received messages at the application layer. This metric provides information about the reliability of a protocol and the success rate of packet loss compensation mechanisms, e.g., retransmissions. The message loss ratio is defined for each client c in Eq. (1) as well as for all clients in Eq. (2).

$$MsgLoss_c = 1 - \frac{|S'_c|}{|S_c|} \qquad (1) \qquad MsgLoss = 1 - \frac{\sum_{c \in C} |S'_c|}{\sum_{c \in C} |S_c|} \qquad (2)$$

In both cases, the numerator represents the number of messages subscribed and successfully received by a client. The denominator depicts the messages which have been subscribed by the client, regardless if they were received or not. It is essential to sum the messages per client for the total message loss. Considering only the number of total messages would include messages without a subscription and count messages with multiple subscriptions only once.

Latency of Received Messages. This metric refers to the mean latency of received messages. Equation (3) defines the mean latency for each client, while Eq. (4) defines it for all involved clients. $\sigma_{m,c} \in \mathbb{R}$ states the spent time for message m from sender to recipient c, including the broker processing time.

$$\overline{Lat_c} = \frac{\sum_{m \in S'_c} \sigma_{m,c}}{|S'_c|} \qquad (3) \qquad \overline{Lat} = \frac{\sum_{c \in C} \sum_{m \in S'_c} \sigma_{m,c}}{\sum_{c \in C} |S'_c|} \qquad (4)$$

In both definitions, the numerator sums up the latency of each received message. The denominator represents the total number of received messages per client, respectively, for all clients.

Protocol Efficiency. This metric indicates the ratio between the transferred payload and the number of bytes sent through the network interface. While the payload only considers the bytes of the message content, the bytes at the network interface include the complete protocol stack, i.e., payload, headers, checksums, and trailers. The protocol efficiency metric reflects the proportion of the payload to the transferred bytes. This metric further reflects additional bytes for retransmissions in case of packet loss. Moreover, the overhead for encryption or optional header fields are also included. Equation (5) shows the protocol efficiency for an individual client c while Eq. (6) considers all clients. $\psi_m \in \mathbb{N}$ refers to the payload of message m, and Ψ_c denotes the total transferred bytes for client c.

$$PE_c = \frac{\sum_{m \in P_c} \psi_m}{\Psi_c} \qquad (5) \qquad PE = \frac{\sum_{c \in C} \sum_{m \in P_c} \psi_m}{\sum_{c \in C} \Psi_c} \qquad (6)$$

In both definitions, the numerator sums up the payload of each published message. The denominator depicts the amount of sent bytes, measured at the network interface, for one client, respectively, for all clients.

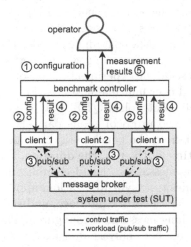

Fig. 1. Benchmark architecture with involved participants.

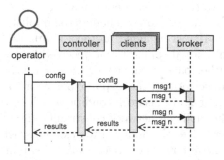

Fig. 2. Sequence diagram of a benchmark run.

2.3 Harness Design

Section 2.1 defined the requirements for a benchmarking framework for IoT communication protocols. For the software design, additional requirements arise addressing correctness, ease of use, modularity, and extensibility. This section presents the design, the entities involved, and the process of a benchmark run. Figure 1 depicts the involved participants, i.e., the benchmark operator, the benchmark controller as central management instance, and the *system under test (SUT)* which consist of the multiple clients and the broker. The benchmark harness contains the controller and the clients, which emulate a freely configurable application. Any message broker, including cloud brokers that supports the desired protocol, can be used as a message broker.

The *benchmark operator* configures the benchmark by providing a configuration to the central controller. This configuration includes, amongst others, the start time and duration of the benchmark, some global settings like the selected publish/subscribe protocol, and the definition of the communication behavior as well as network constraints of the clients. After the experiment, the operator retrieves the measurement results from the central benchmark controller. Instead of a human operator, scripts or external software artifacts can use our API to configure and run the benchmark.

The *benchmark controller* is responsible for the overall control of the benchmark. It is instantiated as a separate software artifact in a container and must be deployed once. The benchmark controller receives the global configuration from the operator via a REST API, connects to the clients via an additional API, and configures them independently. The controller knows the benchmark timing and waits for the duration of the experiment before collecting the raw measurement data from the clients, aggregating and evaluating them, and generating reports.

A *client* represents a communicating entity and acts as a workload generator using a publish/subscribe protocol. The client runs as a container, likewise the central controller. Clients can coexist on the same machine as long as their performance does not negatively affect each other. To scale the experiment or to isolate performance, the deployment can also be spread over multiple hosts. Our client implementation can handle the publish/subscribe protocols MQTT, AMQP, and CoAP via included protocol adapters. The selected protocol is one part of the individual configuration received by each client from the benchmark controller. Other parts of the configuration are the start time of the run, the duration, the subscriptions, the response behavior to incoming messages, scheduled publishing events, and network conditions. The client verifies time synchronization, applies the artificial network constraints on the container's network interface, and subscribes to the assigned topics, and starts publishing messages if configured. The central controller collects all measurements for further analysis after each run.

A central *message broker* is required by several publish/subscribe protocols like MQTT, AMQP, and CoAP. The message brokers manage the client subscriptions and deliver messages based on the message topics and subscriptions.

For an experiment, the benchmark operator provides a global configuration to the controller as shown in Fig. 2. This configuration contains the global specifications and client configurations, including communication behavior, load profiles, and network constraints. The central controller sends an individual configuration file to each client, as depicted in Fig. 1. After a client receives a configuration, it can operate autonomously, i.e., no further control traffic during the experiment is necessary. This time-decoupling allows using the same network connection for configuration and the workload. The clients configure themselves independently, start the experiment at a predefined time and collect measurement data. After a benchmark run, the controller collects and aggregates the measurement results from all clients, processes them, and generates summaries and graphs. This report is accessible from the controller, depicted in the last step in the sequence diagram of Fig. 2.

2.4   Implementation

The implementation of ComBench consists of the artifacts controller and client, as introduced in Sect. 2.3. The controller manages the benchmark, configures the clients, and evaluates the measurements. The clients exchange messages with the chosen publish/subscribe protocol and conduct the measurements. This section describes the implementation of these artifacts, which are written in Python.

The *controller* provides a REST-API for configuration. The benchmark operator configures the benchmark via the API and retrieves the measurement results. The controller sends the preprocessed configuration again as an HTTP request to each client and collects the raw measurement results.

The benchmark *client* application currently includes three protocol adapters for MQTT (*asyncio-mqtt*[1] library), AMQP (*pika* library[2]), and CoAP (*aiocoap*[3] library). The REST API on the client receives the configuration commands and returns the measurements. To influence the network conditions and retrieve specific system parameters, the client has some dependencies on Linux tools and is, therefore, only executable on Linux.

Both artifacts, controller and client, are published as open-source under the Apache 2.0 license. The source code, a manual, and some evaluation examples are available at ComBench's GitHub repository [29]. The Docker images can be pulled from Docker Hub[4,5].

The setup of an experiment is specified in a single configuration file to facilitate repeatability and sharing of different scenarios since all configuration options are stored in one file. The configuration contains some global settings like the chosen communication protocol, the start time, the runtime of the benchmark, and the IP address/hostname of the broker given as a JSON file. Furthermore, the configuration defines the client roles, like group settings, subscriptions, the associated publishing events, and network conditions. The network conditions like packet loss rates, bandwidth limits, and transmission delays can be specified either statically or as a time series so that the values change during the experiment and thus, e.g., emulate moving devices. The last part of the configuration is the individual clients. Each client is assigned to a group, simplifying large-scale experiments.

To calculate the metrics defined in Sect. 2.2, measurements must be performed. On the client, our benchmark collects (i) message logs, (ii) system performance parameters, and (iii) network performance parameters. The message log records every published message with its unique message ID, the topic, and the transmission timestamp. Each received message also creates a log entry at the subscriber with the message ID and arrival timestamp. The topic of received messages is derived using the unique message ID.

---

[1] https://pypi.org/project/asyncio-mqtt/.
[2] https://pika.readthedocs.io/.
[3] https://github.com/chrysn/aiocoap.
[4] https://hub.docker.com/r/descartesresearch/iot-pubsub-benchmark-controller.
[5] https://hub.docker.com/r/descartesresearch/iot-pubsub-benchmark-client.

For the system performance parameters, the client periodically queries the utilization of CPU (total and per core) and RAM as well as the numbers of sent packets, received packets, sent bytes, and received bytes from the operating system's network interface. This measurement is repeated every 1000 ms, and the measured values and timestamps are logged into a file. The measured values are stored on each client to avoid additional network traffic during each run. The benchmark controller collects all clients' measurements after the runs, as described in Sect. 2.3.

3   Case Study

ComBench addresses multiple evaluation objectives related to performance, scalability, reliability, and security, which can be investigated with different network conditions. The captured measurement data and derived metrics introduced in Sect. 2.1 enable a wide range of investigations and comparisons. This section presents the universal applicability of ComBench in a case study consisting of five exemplary IoT scenarios. We show how ComBench contributes to analyzing specific concerns for each of those scenarios, such as broker resilience, protocol efficiency, and the impact of network limitations on communication. Each scenario includes a short motivation, the associated study objectives (SO), a description of the benchmarking experiment, and a brief interpretation of results. We like to emphasize that the focus of this case study is to demonstrate the capabilities of ComBench and how it can be applied and not the discussion of the observations itself. Hence, we do not present a detailed description and interpretation of the concrete measurement results.

The experiments are executed on a node with a four-core Intel Core i7-4710HQ processor and 8 GB RAM using Ubuntu 18.04.2 LTS with Docker version 19.03.11. The benchmark configuration and the detailed measurement results are available at our GitHub repository [29] and on Zenodo [14].

3.1   Broker Resilience

Motivation. Many publish/subscribe protocols like MQTT or AMQP use a central message broker that manages the subscribed topics and distributes the published messages. An increasing number of clients and messages can overload the broker, resulting in higher latencies or message loss. When selecting an appropriate broker or for configuration tuning the resilience can be a crucial criterion.

Study Objectives. A first study objective that can be analyzed using ComBench is the load level at which the broker start to delay messages (SO 1.1). As brokers are implemented in different programming languages and might operate less or more performant, their resilience may differ. Identifying performance variations between the brokers at different load levels is a further objective of our study (SO 1.2).

Scenario. To compare the resilience of different brokers, we stress the broker by horizontal scaling, i.e., by increasing the number of clients and messages. We analyze the captured latency of the message transmission of all clients (ref. Section 2.1: Eq. (4)) and compare different load levels of different broker implementations. We apply a supermarket company's supply chain load profile as introduced by the SPECjms2007 benchmark [25]. Due to space limitations, we refer to the SPECjms2007 documentation for a detailed workload description. We perform measurement runs of 60 s, each with three repetitions at the scaling factors one, five, ten, 15, 20, 30, 40, 45, 50, and 55.

Conclusions. Figure 3 depcits the average latency per scaling factor for each broker . The x-axis shows the different scaling levels, while the y-axis represents the measured average latency in milliseconds. The analysis indicates no noticeable increase in latency until a scaling factor of 30 for RabbitMQ and up to 45 for Mosquitto and EMQ X (SO 1.1). All brokers perform similar until a scaling factor of 45, at which the latency of RabbitMQ increases rapidly (SO 1.1). EMQ X and Mosquitto show only a slight latency increase starting at a scaling factor of 50 (SO 1.2).

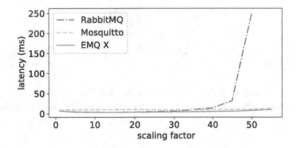

Fig. 3. Average latency for message transmission for different scaling factors and MQTT broker implementations.

3.2   Protocol Efficiency

Motivation. In addition to the message payload, the protocol headers of the protocol stack also affect the actual number of transmitted bytes at the network interface. Various publish/subscribe protocols generate different overheads due to their headers and the underlying protocols. Especially for low data rates, as often present in Wireless Sensor Networks (WSNs), low overhead is crucial. The protocol efficiency (ref. Sect. 2.2: Eq. (5)) indicates the proportion of the payload, i.e., the message content, compared to the transferred bytes observed at the network interface.

Study Objectives. A first study analyzes and compares the protocol efficiency of AMQP, MQTT, and CoAP at different payload sizes (SO 2.1). The objective is to identify which of the protocols is best suited for low-bandwidth networks due to its efficiency (SO 2.2). Furthermore, the study compares the protocol efficiency of MQTT at the three quality of service (QoS) levels, which differs through additional control messages (SO 2.3).

Scenario. To investigate protocol efficiency, we deploy a simple setup consisting of one publisher and one subscriber, without any configured restrictions related to the network conditions. The publisher sends ten messages per second over 60 s with a fixed payload via the broker to the subscriber. The measurements are repeated for payload sizes between 100 and 1000 bytes in step sizes of 100 bytes for the protocols MQTT, AMQP, and CoAP. To compare the protocol efficiency of MQTT at the QoS levels 0 (default), 1, and 2, we executed the measurement series with payload sizes between 200 and 10000 bytes in step sizes of 200 bytes.

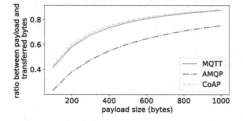

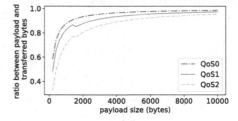

Fig. 4. Ratio between payload and transmitted bytes for different protocols.

Fig. 5. Ratio between payload and transmitted bytes for different MQTT QoS levels.

Conclusions. Figure 4 shows the protocol efficiency of the protocols MQTT, AMQP, and CoAP. The x-axis depicts the size of the payload in bytes, while the y-axis indicates the ratio between the payload and the transferred bytes observed at the network interface. The graph shows that AMQP has a lower efficiency than MQTT and CoAP, i.e., it has a higher overhead due to, e.g., protocol headers (SO 2.1). MQTT and CoAP have similar efficiency; therefore, both protocols are well suitable for low bandwidth networks from the protocol overhead perspective (SO 2.2). As expected, the efficiency increases with the payload size since the payload takes a higher proportion of the transferred bytes than the header. Figure 5 depicts the efficiency of MQTT at different QoS levels, with an identical axis interpretation to Fig. 4. This measurement complies with the expectations that with an increasing QoS level, the efficiency decreases due to the additional control traffic (SO 2.3).

3.3 Effect of packet Loss

Motivation. Packets can get lost during data transmission, especially in unstable wireless networks. A few lost packets are usually compensated by the protocols or the underlying protocol layers through re-transmissions so that messages on application layer still arrive (usually delayed). For applications running on unstable networks, a protocol with adequate compensation mechanisms should be chosen. The message loss rate can be determined by Eq. (1) (see Sect. 2.2).

Study Objectives. One objective of this study is to determine at which packet loss rate messages get lost (SO 3.1). Based on this, we test at which packet loss rate the communication is no longer possible (SO 3.2). For particularly lossy connections, a comparison of the message loss rates of protocols would support identifying particularly robust protocols (SO 3.3).

Scenario To investigate the packet loss, we again use a setup consisting of one publisher and one subscriber. The publisher sends ten messages per second with 20 repetitions. The network interface of the publisher is configured with a constant packet loss rate, increasing along with a measurement series from 0 to 100% in steps of 5%. Further traffic and network disturbances are not configured. Lost messages are determined by matching the IDs of the messages sent by the publisher and the messages received by the subscriber. This study is performed for MQTT and AMQP.

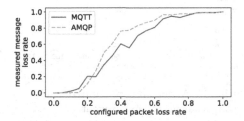 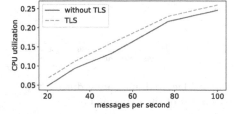

Fig. 6. Message loss rate at artificially configured packet loss.

Fig. 7. Impact of TLS on client's CPU load on an increasing number of messages.

Conclusions. Figure 6 depicts the message loss rate for different configured packet loss rates for the protocols MQTT and AMQP. The x-axis shows the measured message loss rate, as indicated by Eq. (1). The y-axis depicts the configured packet loss rate at the subscriber's network interface. The results show that the first message losses occur at a packet loss rate of 10% for MQTT and 15% for AMQP (SO 3.1). Communication is impossible, i.e., almost all messages are lost, for both protocols at a packet loss rate of approximately 85% (SO 3.2). The two considered protocols behave similarly in terms of robustness against packet loss (SO 3.3). Further measurements are necessary to identify significant differences between these protocols related to the sensitivity to packet loss.

3.4   Impact of TLS on Client CPU Utilization

Motivation. TLS enables encrypted communication for MQTT on the transport layer (Layer 4 on ISO OSI model). However, the handshake, encryption, and decryption of TLS may introduce additional load on the client's CPU. Especially for resource-constrained IoT devices, the influence of encryption mechanisms on the system load should be taken into account.

Study Objectives. The first objective of this study is to quantify the additional load on the CPU by enabling TLS (SO 4.1). Furthermore, the effect of an increased number of encrypted messages on the client's CPU utilization is analyzed (SO 4.2).

Scenario. For this scenario, we use a setup consisting of one publisher and one subscriber. The publisher sends messages with a fixed rate for 60 s with a payload of 100 bytes, which the subscriber will receive. This message rate is increased at rates between 20 and 100 messages per second within a measurement series. No encryption is used for a first measurement series, while in a second series all measurements are repeated with TLS enabled.

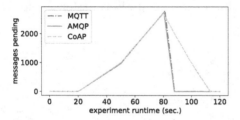
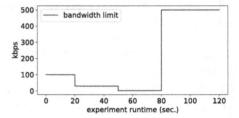

Fig. 8. Delayed messages during the execution of the experiment.

Fig. 9. Artificially configured bandwidth during the experiment.

Conclusions. For the evaluation, we use the reported CPU utilization on the publisher. Figure 7 shows the impact of TLS on the client's CPU utilization. The x-axis depicts the number of sent messages, and the y-axis shows the ratio of utilized CPU. The graph shows that TLS introduces approximately 2% more CPU utilization compared to the scenario without encryption (SO 4.1). It can moreover be seen that the influence of TLS introduces an almost constant overhead regardless of the number of messages (SO 4.2).

3.5   Influence of Fluctuating Bandwidth

Motivation. IoT applications in wireless networks often face fluctuating network quality, like limited bandwidths. Fluctuating bandwidths occur particularly often for moving devices. The communication behavior at such fluctuating bandwidths is of particular interest in mobile applications. Besides the behavior in case of a static bandwidth limitation, it is especially relevant how the protocol behaves at bandwidth variations.

Study Objectives. When considering varying bandwidths, it can be essential to know the bandwidth limit to which packets can be sent without any delay for the protocols MQTT, AMQP, and CoAP (SO 5.1). In contrast, it could further be relevant how fast queued packets are sent after bandwidth limitation (SO 5.2). It is also essential if packets were lost during this time (SO 5.3).

Scenario. To analyze the influence of fluctuating bandwidth, we use a setup with one publisher and one subscriber. The publisher sends 67 messages per second for 120 s with a payload of 1000 bytes each. The bandwidth is not set with a static value. Instead, different bandwidth limits are specified as time series, which are applied automatically by the client. The bandwidth is set to 100 kbps initially, limited to 30 kbps from second 20, and reduced to 2 kbps from second 50 to imitate a decreasing signal quality, as depicted in Fig. 9. From second 80, it is assumed that the mobile network is available again, and the bandwidth limitation is increased to 500 kbps. The experiment is performed for MQTT, AMQP, and CoAP.

Conclusion. Figure 8 shows the number of delayed messages during the experiment for MQTT, AMQP, and CoAP. The x-axis depicts the runtime of the experiment, while the y-axis indicates the number of messages that were not delivered at the respective time. The first delay for the three protocols occurs in second 20, when the bandwidth is limited to 30 kbps (ref. Figure 9), which means that transmission at the previous 100 kbps was possible without delay (SO 5.1). The graph follows the expectation that from second 50 on, more messages are delayed due to the further decreased bandwidth, which is similar for all used protocols. After pushing the bandwidth limit to 500 kbps in second 80, MQTT and AMQP send the delayed messages in about 8 s, while CoAP takes about 35 s (SO 5.2). Since no delayed messages remain at the end of the experiment, all messages were delivered without loss (SO 5.3).

4 Threats to Validity

The previous section has shown the wide variety of use cases in which ComBench can be used. Nevertheless, there are a few weaknesses that a benchmark operator should be aware of. This section discusses these vulnerabilities and provides some approaches to remedy or mitigate them.

ComBench offers several advantages in usability, verifiability, and metrics due to its unified, multi-protocol client and its instrumentation. However, the protocol adapters contained in the client create a small performance overhead that must be considered compared to a single-protocol implementation. Also, the client's instrumentation for logging the messages and recording CPU, RAM, and network utilization introduces additional load on the client. We assume that this overhead can be neglected; however, a detailed analysis of the overhead is part of our future work. If the pure client performance needs to be measured, a single-protocol client and another instrumentation should be used, which is out of this work's scope. By implementing the REST interface against our controller, alternative clients—also if necessary with alternative protocol implementations—can be integrated transparently.

Another possible vulnerability of our benchmarking framework is time synchronization. Latency measurements determine the delta between the timestamps from sending the message to receiving it. An accurate calculation requires exact time synchronization between all participants. Some previous work implements a request/reply pattern, i.e., the subscriber responds to the sender, to measure the round trip time [23]. However, we deliberately decided against this procedure because we do not consider request/reply very practical in publish/subscribe environments. Instead, we rely on the Precision-Time-Protocol (PTP), and the benchmark verifies before each run that the participants' times are synchronized. Studies show a deviation of less than $1\,\mu s$ when using PTP [21], which we assume is satisfiable in practice. Please note that in small setups, where all clients are running on a single host, the issue of possible time deviation is irrelevant as all guest systems use the host's clock.

5   Related Work

Testing the performance of communication protocols is not a new field in academia. Therefore, tools dealing with load generation and measurements related to communication systems are especially important for this paper. Section 5.1 presents some work that deals with the comparison of IoT network protocols. While Sect. 5.2 presents tools that focus on load testing and measurements, Sect. 5.3 identifies benchmarks providing predefined load scenarios in addition to load generation and measurement instrumentation. Section 5.4 summarizes the related work by listing all the works with their primary focus and characteristics.

5.1   Protocol Comparisons

A large number of papers deal with the comparison of different communication protocols, which emphasizes the relevance of a benchmarking framework. In the following, we provide a possible categorization of related comparisons of IoT application layer protocols. We have assigned the studies to their primary focus, but some can also be well suited in multiple categories.

The first category is the theoretical comparison of protocols [10,20]. Header structure, payload size, and security features are discussed and compared. Furthermore, related protocol comparisons deal with scalability and analyzing the resource consumption of clients or brokers [17,28]. Another category of related studies focuses on network performance and network load of protocols [3,15,22].

More specialized studies in this area analyze and compare the performance of protocols under constraint networks [5,6,23] and different topologies [12].

The studies provide valuable insights into the protocols, some of them also with artificially restricted network communication [5,6,23]. Although some authors make their developed test tools publicly available, they are usually targeted to the specific test and do not allow free configuration as we expect from a benchmarking framework.

5.2 Load Testing Frameworks

Tools and benchmarking frameworks for load testing are most related to our work. These artifacts are characterized by their accessibility as a tool that can both generate loads and perform measurements. In the following, we present a selection of some well-known load testing frameworks.

JMeter [13] is a Java-based open-source application designed to load test functional behavior and measure performance. While it was originally designed to test web applications, extensions add other features and communication protocols such as MQTT. *LoadRunner* [31] is a commercial testing solution supporting a wide range of technologies and protocols in the industry with focus on testing applications and measuring system behaviour and performance under load. It supports the IoT protocols MQTT and CoAP and provides a IDE for scripting and running unit tests. *Gatling* [9] is an open-source load testing framework written in Scala for analyzing and measuring the performance of different services, focusing on web applications. Community plugins can be used to add protocol support for, e.g., MQTT, and AMQP. *MZBench* [8] is a community-driven open-source benchmarking framework written in Erlang and focusing on testing software products. Among others the communication protocols XMPP, and AMQP are supported by MZBench, furthermore can be added as extension. *LOCUST* [7] is another community-driven open source load testing tool focusing on load testing of web applications. Although the common protocols for web applications (HTTP, Websocket, ...) are supported, there are currently no clients for IoT communication protocols. Although load test tools usually have a wide range of configuration options and measurement methods, to the best of our knowledge there are no tools that support comprehensive network connectivity constraints.

5.3 Benchmarks

In this section, we discuss existing IoT benchmarks using publish/subscribe protocols and show that already some research effort was made on a static analysis and comparison of publish/subscribe protocols. Benchmarks differ from load test frameworks primarily in that one or more predefined load profiles are provided in addition to the tooling.

Sachs et al. propose their *SPECjms2007* benchmark [25], focusing on evaluating the performance of message-oriented middleware (MOM), i.e., the Java message service (JMS). As a follow-up work, *Sachs et al.* proposed the *jms-2009-PS* benchmark [24], focusing on publish/subscribe patterns. Afterward, *Appel et al.* add another use case by changing the used protocol to AMQP in the jms2009-PS benchmark to analyze the MOM [2]. *Zhang et al.* presented in 2014 *PSBench*, a benchmark for content- and topic-based publish/subscribe systems [32]. Under the name *IoTBench* a research initiative pursues the vision of a generic IoT benchmark [4]. Their goal is to test and compare low-power wireless network protocols. This initiative collaborates on the vision of this benchmark and already published first steps toward a methodology in [16]. *RIoTBench* is a real-time IoT benchmark suite for distributed stream processing systems [26]. The covered performance metrics include latency, throughput, jitter, and CPU and memory utilization.

The presented benchmarks are a valuable contribution to the investigation of the performance of communication protocols including competitive workloads. However, we could not identify a fluctuating network quality in any of the benchmarks, which is an inherent issue for IoT systems.

5.4 Summary

As our review for related work shows, the evaluation of IoT communication protocols has a significant relevance in current research [10,20]. While individual

Table 1. Matrix summarizing the scope and characteristics of related work compared to our ComBench benchmarking framework.

Approach	Type	IoT protocol support	Client perf. measurement	Network perf. measurement	Influence netw. conditions	Released tool inc. reporting
Dizdarević et al. [10]	Survey	✓				
Naik et al. [20]	Survey	✓				
Talaminos-Barroso et al. [28]	Evaluation	✓	✓	✓		✓
Kayal et al. [17]	Evaluation	✓	✓			
Iglesias-Urkia et al. [15]	Evaluation	✓		✓		
Bansal et al. [3]	Evaluation	✓		✓	✓	
Pohl et al. [22]	Evaluation	✓		✓	✓	
Chen et al. [5]	Evaluation	✓		✓	✓	
Collina et al. [6]	Evaluation	✓		✓	✓	
Profanter et al. [23]	Evaluation	✓	✓	✓		
JMeter [13]	LT Tool	✓	✓	✓		✓
LoadRunner [31]	LT Tool	✓	✓	?		✓
Gatling [9]	LT Tool	✓	✓			✓
MZBench [8]	LT Tool	✓	✓			✓
LOCUST [7]	LT Tool	✓	✓			✓
SPECjms2007 [25]	Benchmark		✓	✓		✓
jms2000-PS [24]	Benchmark		✓	✓		✓
Appel et al. [2]	Evaluation	✓	✓			
PSBench [32]	Benchmark	✓	✓			
IoTBench [4,16]	Benchmark	✓				
RIoTBench [26]	Benchmark	✓	✓	✓		✓
ComBench	LT Tool	✓	✓	✓	✓	✓

evaluations usually focus on one or multiple specific aspects [3,22], load test frameworks or benchmarks typically offer a more universal applicability due to their configurable workload and included reporting [13,26]. Most evaluations and tools assume a lossless connectivity, while some work shows the demand to study the protocols also under connections with constrained network quality [5,23]. However, to the best of our knowledge, no benchmarking framework currently exists that focuses on the evaluation of IoT communication protocols under constrained network quality, such as packet loss or a temporary link failure.

Table 1 summarizes our considered related works. The first column names the approach, while the second column classifies it according to the category of its main purpose. We distinguish between *surveys* providing a theoretical comparison but no measurements, *evaluations* comparing different protocols through measurements, *load test tools* (*"LT tool"*) focusing particularly on the reuse of measurement tools, and *benchmarks* providing additional predefined competitive workload scenarios. The subsequent columns indicate by '✓' whether certain properties are met by the approach. An '?' indicates that we have been unable to determine this property. Note that we only consider the primary focus of the approach, and for tools, we only check properties that can be enabled through official plugins and settings within the tool. IoT protocol support is fulfilled if at least one of the protocols MQTT, AMQP, XMQP, CoAP, ZeroMQ, DDS, or OPC UA is supported. Client performance measurements specifies whether the approach observes client resource consumption such as CPU or RAM, while network performance measurements targets aspects such as latency or bandwidth consumption. The aspect influence network conditions is fulfilled if at least one of the configurations packet loss, transmission delay or link failure is present or examined. If an official tool is provided as an artifact including a reporting, the requirement for the last column is met.

The summary shows that so far there is no specific load test tool or benchmarking framework for the evaluation of IoT communication protocols under network constraints, which motivates us for the developing ComBench.

6   Conclusion

In this paper, we presented ComBench, our publish/subscribe benchmarking framework for IoT systems. The framework is designed to analyze and compare different application layer protocols and includes three key features. Firstly, ComBench can be useful to investigate the effects of varying network quality on communication behavior. Second, due to its multi-protocol capability, ComBench can compare different protocols and their features. At third, our benchmarking framework supports designers, developers, and operators of IoT systems, analyzing the scalability, robustness, and reliability of clients, networks, and brokers. For all these areas, ComBench offers the appropriate instrumentation for collecting and analyzing measurement data. Section 3 presented some exemplary benchmarking scenarios and pointed out some briefly answered objectives during the case study. During the development of ComBench, we paid attention to high usability, which is especially characterized by the multi-protocol client, the central benchmark controller, the deployment in a containerized environment, and the included generation of reports. ComBench is published as open-source under the Apache License 2.0 on GitHub [29] and is accessible to other researchers, system designers, software engineers, and developers.

In the future, we want to continue developing ComBench and add additional technical features. One goal is to implement additional publish/subscribe protocol adapters for, e.g., ZeroMQ, DDS, and OPC UA pub/sub. Furthermore, we plan to add request/response patterns as used in REST applications. A graphical, interactive web interface can present the results in a more comfortable and user-friendly way, especially for users from the industry. Last, we aim to add representative workloads and establish a standardized benchmark from these.

Acknowledgement. This project is funded by the Bavarian State Ministry of Science and the Arts and coordinated by the Bavarian Research Institute for Digital Transformation (bidt).

References

1. Al-Fuqaha, A., Guizani, M., Mohammadi, M., Aledhari, M., Ayyash, M.: Internet of things: a survey on enabling technologies, protocols, and applications. IEEE Commun. Surv. Tutor. **17**(4), 2347–2376 (2015). https://doi.org/10.1109/COMST.2015.2444095

2. Appel, S., Sachs, K., Buchmann, A.: Towards benchmarking of AMQP. In: Proceedings of the Fourth ACM International Conference on Distributed Event-Based Systems, pp. 99–100 (2010)

3. Bansal, S., Kumar, D.: IoT application layer protocols: performance analysis and significance in smart city. In: 2019 10th International Conference on Computing, Communication and Networking Technologies (ICCCNT), pp. 1–6. IEEE (2019)

4. Boano, C.A., et al.: IoTBench: towards a benchmark for low-power wireless networking. In: 2018 IEEE Workshop on Benchmarking Cyber-Physical Networks and Systems (CPSBench). IEEE (2018)

5. Chen, Y., Kunz, T.: Performance evaluation of IoT protocols under a constrained wireless access network. In: 2016 International Conference on Selected Topics in Mobile & Wireless Networking (MoWNeT), pp. 1–7. IEEE (2016)

6. Collina, M., Bartolucci, M., Vanelli-Coralli, A., Corazza, G.E.: Internet of Things application layer protocol analysis over error and delay prone links. In: 2014 7th Advanced Satellite Multimedia Systems Conference and the 13th Signal Processing for Space Communications Workshop (ASMS/SPSC), pp. 398–404. IEEE (2014)

7. Community: Locust Website (2021). https://locust.io/, Accessed 15 Apr 2021

8. Community: MZBench Website (2021). https://github.com/satori-com/mzbench, Accessed 13 Apr 2021

9. Corp, G.: Gatling Website (2021). https://gatling.io/, Accessed 13 Apr 2021

10. Dizdarević, J., Carpio, F., Jukan, A., Masip-Bruin, X.: A survey of communication protocols for internet of things and related challenges of fog and cloud computing integration. ACM Comput. Surv. (CSUR) **51**, 1–29 (2019)

11. Eugster, P.T., Felber, P.A., Guerraoui, R., Kermarrec, A.M.: The many faces of publish/subscribe. ACM Comput. Surv. (CSUR) **35**(2), 114–131 (2003)

12. Gündoğran, C., Kietzmann, P., Lenders, M., Petersen, H., Schmidt, T.C., Wählisch, M.: NDN, CoAP, and MQTT: a comparative measurement study in the IoT. In: Proceedings of the 5th ACM Conference on Information-Centric Networking, pp. 159–171 (2018)
13. Halili, E.: Apache JMeter. Packt Publishing, Birmingham (2008)
14. Herrnleben, S., et al.: Evaluation results of ComBench as open data. Technical report, University of Wuerzburg (2021). https://doi.org/10.5281/zenodo.4723344, Accessed 30 Apr 2021
15. Iglesias-Urkia, M., Orive, A., Barcelo, M., Moran, A., Bilbao, J., Urbieta, A.: Towards a lightweight protocol for industry 4.0: an implementation based benchmark. In: 2017 IEEE International Workshop of Electronics, Control, Measurement, Signals and their Application to Mechatronics (ECMSM) (2017)
16. Jacob, R., Boano, C.A., Raza, U., Zimmerling, M., Thiele, L.: Towards a methodology for experimental evaluation in low-power wireless networking. In: Proceedings of the 2nd Workshop on Benchmarking Cyber-Physical Systems and Internet of Things, pp. 18–23 (2019)
17. Kayal, P., Perros, H.: A comparison of IoT application layer protocols through a smart parking implementation. In: 2017 20th Conference on Innovations in Clouds, Internet and Networks (ICIN), pp. 331–336. IEEE (2017)
18. von Kistowski, J., Arnold, J.A., Huppler, K., Lange, K.D., Henning, J.L., Cao, P.: How to build a benchmark. In: Proceedings of the 6th ACM/SPEC International Conference on Performance Engineering (ICPE 2015), ICPE '15. ACM, New York (2015)
19. Kounev, S., Lange, K.D., von Kistowski, J.: Systems Benchmarking. Springer, Heidelberg (2020). https://doi.org/10.1007/978-3-030-41705-5
20. Naik, N.: Choice of effective messaging protocols for IoT systems: MQTT, CoAP, AMQP and HTTP. In: 2017 IEEE International Systems Engineering Symposium (ISSE), pp. 1–7. IEEE (2017)
21. Neagoe, T., Cristea, V., Banica, L.: NTP versus PTP in computer networks clock synchronization. In: 2006 IEEE International Symposium on Industrial Electronics, vol. 1, pp. 317–362. IEEE (2006)
22. Pohl, M., Kubela, J., Bosse, S., Turowski, K.: Performance evaluation of application layer protocols for the internet-of-things. In: 2018 Sixth International Conference on Enterprise Systems (ES), pp. 180–187. IEEE (2018)
23. Profanter, S., Tekat, A., Dorofeev, K., Rickert, M., Knoll, A.: OPC UA versus ROS, DDS, and MQTT: performance evaluation of industry 4.0 protocols. In: Proceedings of the IEEE International Conference on Industrial Technology (ICIT) (2019)
24. Sachs, K., Appel, S., Kounev, S., Buchmann, A.: Benchmarking publish/subscribe-based messaging systems. In: Yoshikawa, M., Meng, X., Yumoto, T., Ma, Q., Sun, L., Watanabe, C. (eds.) DASFAA 2010. LNCS, vol. 6193, pp. 203–214. Springer, Heidelberg (2010). https://doi.org/10.1007/978-3-642-14589-6_21
25. Sachs, K., Kounev, S., Bacon, J., Buchmann, A.: Performance evaluation of message-oriented middleware using the SPECjms2007 benchmark. Perf. Eval. 66(8), 410–434 (2009)
26. Shukla, A., Chaturvedi, S., Simmhan, Y.: RIoTBench: an IoT benchmark for distributed stream processing systems. Concurr. Comput. Pract. Exp. 29(21), e4257 (2017)
27. Statista, IHS: Internet of Things - number of connected devices worldwide 2015–2025 (2018). https://www.statista.com/statistics/471264/iot-number-of-connected-devices-worldwide/

28. Talaminos-Barroso, A., Estudillo-Valderrama, M.A., Roa, L.M., Reina-Tosina, J., Ortega-Ruiz, F.: A machine-to-machine protocol benchmark for ehealth applications - use case: respiratory rehabilitation. Comput. Methods Prog. Biomed. **129**, 1–11 (2016)
29. University of Wuerzburg, Institute of Computer Science, Germany, Chair of Software Engineering: Git repository of ComBench (2021). https://github.com/DescartesResearch/ComBench
30. Wirawan, I.M., Wahyono, I.D., Idfi, G., Kusumo, G.R.: IoT communication system using publish-subscribe. In: 2018 International Seminar on Application for Technology of Information and Communication, pp. 61–65. IEEE (2018)
31. Zhang, H.L., Zhang, S., Li, X.J., Zhang, P., Liu, S.B.: Research of load testing and result application based on LoadRunner. In: 2012 National Conference on Information Technology and Computer Science. Atlantis Press (2012)
32. Zhang, K., Rabl, T., Sun, Y.P., Kumar, R., Zen, N., Jacobsen, H.A.: PSBench: a benchmark for content-and topic-based publish/subscribe systems. In: Proceedings of the Posters & Demos Session, pp. 17–18. Association for Computing Machinery (2014)

A Hybrid Flow/Packet Level Model for Predictive Service Function Chain Selection in SDN

Frank Wetzels[1]([⊠]), Hans van den Berg[2], and Rob van der Mei[1]

[1] CWI, Amsterdam, The Netherlands
{f.p.m.wetzels,r.d.van.der.mei}@cwi.nl
[2] TNO, The Hague, The Netherlands
j.l.vandenberg@tno.nl

Abstract. This paper is motivated by recent developments in SDN and NFV whereby service functions, distributed over a centralised controlled network, are connected to form a service function chain (SFC). Upon arrival of a new service request a decision has to be made to which one of SFCs the request must be routed. This decision is based on (1) actual state information about the background traffic through the SFC nodes, and (2) a prediction of the fraction of time that the SFC is in overflow during the course of the new flow in the system. In this paper, we propose a new method for assigning an incoming flow to an SFC. For that, we propose and compare two methods: a simple flow-based algorithm and a more refined hybrid flow/packet-based algorithm. By extensive simulations, we show that the simple flow-based algorithm works particularly well if the network is not overloaded upon new flow arrival. Moreover, the results show that the flow/packet-based algorithm enhances the flow-based algorithm as it handles initial overload significantly better. We conclude that the prediction-based SFC selection is a powerful method to meet QoS requirements in a software defined network with varying background traffic.

Keywords: Software defined network · Service function chain · Predictive selection · Varying background traffic

1 Introduction

In modern networks, the decoupling of (network) functions from the underlying hardware and the decoupling of the control plane from the data plane in network devices, has drawn a lot of attention. This led to the development of the network function virtualisation infrastructure (NFVI) [1] and software defined networking (SDN) [2] respectively. SDN, the central controlled network, makes it possible to steer traffic to functions anywhere located in the network [3]. These functions can be of any type of operation applied to traffic. For example, internet traffic

© ICST Institute for Computer Sciences, Social Informatics and Telecommunications Engineering 2021
Published by Springer Nature Switzerland AG 2021. All Rights Reserved
Q. Zhao and L. Xia (Eds.): VALUETOOLS 2021, LNICST 404, pp. 93–106, 2021.
https://doi.org/10.1007/978-3-030-92511-6_6

flowing into the network that needs to be scanned for viruses, intrusion detection and spyware. Functions, to be applied sequentially to traffic that passes are called a service function chain (SFC) [4]. The flexibility of SDN and NFVI leads to SFC resource allocation (SFC-RA) identified by the following stages: Chain composition, embedding and scheduling. As per [5–7], for each stage an optimal or near optimal solution can be determined by using specific algorithms to which stage specific conditions apply to process service requests (SRs) by the network. Different types of algorithms have been proposed to optimise the concatenation of VNFs with fixed capacity to provide a chain composition. At the embedding stage most algorithms take the chain composition as input and determine the location of where the VNFs need to be running in the network given a set of requirements like CPU and memory use. Scheduling applies to a set of different operations individually used in different, possibly multiple, chains. The sequence of required operations for each chain and the availability of resources lead to an optimal allocation sequence in time for each operation per chain.

In this paper we take a different approach by focussing on the background (BG) traffic characteristics for SFC selection. Figure 1 gives an example whereby an SR enters the network, indicated by the blue straight arrow. We assume the presence of SFCs at fixed locations in the network each consisting of identical functions applied in the same sequence and one type of SR entering the network. Two SFCs are shown and a choice must be made to which SFC the SR needs to be directed to upon arrival. In particular, we assume the presence of background traffic (indicated by the red dashed arrows) that is handled by the individual SFC nodes as well, but is not part of the SR. The background traffic affects the resource availability on each individual SFC node. We will assume certain characteristics of the background traffic and use this knowledge to make a predictive selection of the SFC that affects the SR the least. The SR is then steered to the SFC of choice until it finishes. Throughout, we will call the SR a foreground (FG) flow and the other flows BG flows.

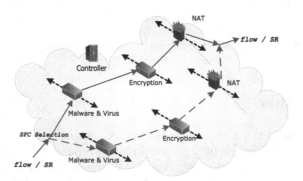

Fig. 1. Flow to be steered to one SFC based on BG traffic violating the critical level. (Color figure online)

The contribution of this paper is three-fold: We show a model and method to use characteristics of the background traffic that flows through individual SFC nodes to make a predictive SFC selection. In addition, we propose two algorithms: One that uses flow characteristics only to make a predictive SFC selection and a second algorithm that adds background packet behaviour in addition to flow characteristics to enhance the first algorithm. The second algorithm's enhancement is clearly noticed in case of high intense background traffic upon arrival of the SR. To the best of the authors' knowledge, this has not been considered before.

2   Problem Description

The problem in deciding which SFC to choose for a newly incoming FG-flow is caused by the inherent *uncertainty* about the number BG flows that travel through SFC nodes during the course of the FG-flow in the system. BG flows travelling through SFC nodes affect the resource availability dynamically, which in turn affects the processing of the FG flow travelling through the SFC itself. If the BG flows are too intense at an SFC node, available resources may be too low to process a FG flow successfully, given QoS requirements.

To effectively deal with this uncertainty, the challenge is to determine at each SFC node the expected amount of time the number of BG flows are at - or below - some level such that during the lifetime of the FG flow, the processing of the FG flow is not affected long enough by the BG flow intensity. Note that the term 'not long enough' is dictated by the requirements to which the FG flow should comply to.

To determine the expected violation duration, we need to define a critical level at each SFC node such that if BG flows do not exceed this level too long, the available resources are sufficient to process the FG flow while fulfilling FG flow requirements. As a result, we can determine what SFC affects the FG flow the least at individual SFC nodes. Secondly, at each SFC node, we will assume the BG flows are driven by a birth-death (BD) process [8]. This assumption enables us to use results of performance evaluation on BD processes in [9]. To illustrate this, Fig. 2 gives a realization of the state of the BD process (represented by X_t) over time. The red dashed horizontal line represents the critical level, $m = 6$. During the lifetime of the FG flow, the number of BG flows exceeds this threshold a certain amount of time.

In [9] a method is presented that enables us to calculate the expected amount of time a BD process is above a critical level during the lifetime of the FG flow. The expected violation duration enables us to predict which SFC affects the processing of the FG flow the least. However, the computational afford is intense. To circumvent this problem, we will use pre-calculated tables for fast decision making. In [10], a pre-calculated table is used for determining whether an SLA is met for a high-priority flow using the method in [9]. We will follow a similar method and create tables beforehand and use these for fast lookup to determine the SFC of choice. The above method ([9]) will be applied in our first algorithm. We will refer to this algorithm as a flow-based (FB) algorithm.

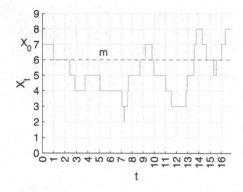
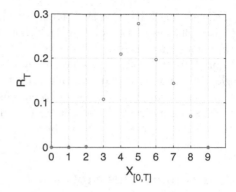

Fig. 2. Time variation in the number of BG flows. (Color figure online)

Fig. 3. Expected fractional BG flow presence duration R_T.

To enhance the FB algorithm, we will consider a second algorithm which takes into account the packet intensity per individual flow as well. This algorithm is called a flow/packet-based (FPB) algorithm. For that we will make use of the method presented in [11]. It enables us to include the transient mean sojourn time BG flow packets have when entering the SFC node.

We will combine the knowledge of BG flows above individual levels to determine the fractional presence duration for each number of BG flows. In Fig. 3 an example is given. For each number of $X_{[0,T]}$ BG flows, its expected fractional presence duration R_T during the life time $[0,T]$ of the FG flow is given. Combining the packet transient mean sojourn time together with the fractional presence duration of BG flows, a weighted transient BG packet delay can be determined. Lookup tables will be created beforehand, used for calculation and fast decision making.

3   Model and Method for Selective SFC Choice

Section 3.1 describes the network and context assumed in this paper. Next, in Sect. 3.2, additional network details, assumptions and definitions are provided. The two service chain selection algorithms, FB and FPB, will then be described in Sects. 3.3 and 3.4 respectively. Details of the analyses needed to run these algorithms are provided in Sect. 4.

3.1   Network Definition

In Fig. 4, a network is given, consisting of an entrance node A, an exit node B, an SDN controller and C parallel chains of N nodes in length. The FG flow enters the network at node A where it is directed (routed) to the chain that is expected to provide the best performance. The FG flow, undergoes the operations applied by the nodes and leaves the network at node B. All chains apply

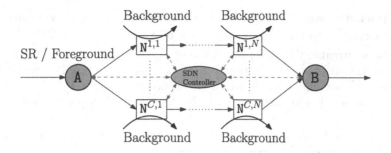

Fig. 4. C SFCs with BG flows each consisting of N-nodes.

the same functions in the same sequence. At each SFC node, BG flows enter the SFC node, undergo the same operations as the FG flow and leave the node. There is no dependency between BG flows at different SFC nodes. In this study we will focus on a single FG flow travelling through the whole chain. The number of BG flows present at a node is driven by a BD process; the BD processes for the different nodes are independent. The routing decision at A is made by the SDN controller.

3.2 Assumptions and Definitions

Numerous parameters can be varied. For example, the birth and death rates of BG flows at the various nodes, the inter-arrival time of packets per one flow of BG and FG flows, etc.

The communication between the SDN controller and the nodes, the transmission speeds, packets lengths and delay on the links are not considered. No packet loss occurs. BG flows arrive at an SFC node, leave the node after processing and will not be considered thereafter.

The following general definitions are made (note the use of the suffices 'f' and 'p' associating to flows and packets respectively):

- **Nodes** $N^{i,j}$: There are C SF chains, each consisting of N nodes. For each chain $i = 1, \ldots, C$, the nodes are connected in sequence, i.e. the j-th node in chain i, $N^{i,j}$, accepts packets belonging to the FG flow from $N^{i,j-1}$ and forwards these packets to $N^{i,j+1}$ for $j = 2, \ldots, N - 1$. Packets to $N^{i,1}$ come from A. Packets to B come from $N^{i,N}$.
- **BG BD process** $\{X_t^{i,j} | t \geq 0\}$: New BG flows arrive at node $N^{i,j}$ according to a Poisson process with rate $\lambda_f^{i,j}$, and have an exponentially distributed lifetime (duration) with parameter $1/\mu_f^{i,j}$. $X_t^{i,j}$ represents the number of BG flows present at $N^{i,j}$ at time t. The BG flow arrival and departure processes at the different nodes are independent. The state-space $S_M^{i,j}$ of the BG BD process associated to node $N^{i,j}$ may be limited by a certain maximum $M^{i,j}$.
- **Critical level** $m^{i,j}$: The critical level of BG flows at node $N^{i,j}$ (i.e. the number of BG flows at which the node gets overloaded) is represented by $m^{i,j}$.

- **BG packets rates** $\lambda_p^{i,j}$: Each BG flow consists of packets arriving at an SFC node according to a Poisson process at rate $\lambda_p^{i,j}$.
- **FG flow duration** T **and packet rates** $\hat{\lambda}_p$: We consider one FG flow of an exponential distributed duration T with parameter τ. The FG flow consists of packets arriving according to a Poisson process at rate $\hat{\lambda}_p$.
- **SFC nodes service rates** $\mu_p^{i,j}$: The Nodes $N^{i,j}$ process packets with service rate $\mu_p^{i,j}$.
- **SFC queue:** At all nodes, all packets enter an infinitely large FIFO queue. FG and BG packets are treated the same way.

3.3   Flow Based Decision Algorithm

The network is running for a while and at some point in time a FG flow arrives at node A. This is the moment $t = 0s$. The FG flow's duration is T, exponentially distributed with parameter τ. At that moment the number of BG flows arriving at all SFC nodes is recorded. That is, at node j in chain i, $X_0^{i,j}$ is set to the number of BG flows. We will leave out the indices i and j for readability hereafter.

The FB algorithm applies the method in [9] to each SFC node, whereby the BG flows are driven by BD process $\{X_t | t \geq 0\}$ with BD parameters λ_f and μ_f respectively. This gives us $E_{T,m,n}$, the fraction of the expected amount of time X_t is above critical level m at an SFC node during $t \in [0, T]$ with n BG flows at $t = 0$.

Each chain consists of a sequence of nodes. The affect of each chain to the FG flow can be determined. The chain that affects the FG flow the least will become the chain of choice to process the FG flow.

Note that the assumption on the flow characteristics may seem somewhat restrictive. The results in [9] may need to be extended to non-exponential distributions, as mentioned in Sect. 6.

3.4   Flow/Packet-Based Decision Algorithm

The same prerequisites apply as with the flow-based algorithm. However, the chain of choice is selected differently. Per SFC node, the FBP algorithm combines two parameters to create a weighted expected packet delay: The fraction of expected amount of time of the *presence* of k BG flows ($R_{T,k,n}$), for all k in the state-space S_M, and the transient mean sojourn time ($TMST_k$) of packets belonging to the presence of k BG flows.

Presence Duration of BG Flow
Determine $E_{T,k,n}$, the fraction of expected amount of time BG flows are above some level k, for all $k \in S$ during the lifetime of the FG. The number of BG flows at $t = 0$ is n. The fraction of expected amount of time of the presence of k BG flows, $R_{T,k,n}$ can be determined from $E_{T,k,n}$ and $E_{T,k+1,n}$.

Transient Mean Packet Delay
By assuming that all k BG flows are of the same type, the packet arrival rate of k BG flows equals to k times the packet rate of one single BG flow. This way we can determine the expected transient packet delay during the presence of k BG flows.

Weigthed Expected Packet Delay
By combining $R_{T,k,n}$ with $\texttt{TMST}_k$ for $k > m, k \in S$, a weighted average on the expected packet delay above m can be determined. For each individual SFC node, the weighted expected transient BG packet delay is determined. The affect of each chain to the FG flow can be determined. The chain that affects the FG flow the least will become the chain of choice to process the FG flow.

The analyses and computations needed to run the FB and FBP algorithms described above will be provided in the next section.

4   Analysis

In this section we present the theory needed to derive the quantities used in the two decision algorithms described in the previous section. We will start, in Sect. 4.1, with the analysis of the behavior of the BD process representing the number of BG flows. In particular we will use the results of [9] to predict the time that the number of BG flows at a node is above a certain critical level during transmission of the FG flow, given the number of BG flows present at the start of its transmission. This analysis is needed for both decision algorithms. Next, in Sect. 4.2, the packet level delay analysis needed for the flow/packet-based decision algorithm is performed. Finally, using these results, the two decision algorithms are further detailed and concertized in Sects. 4.3 and 4.4 respectively.

4.1   Expected Fraction of time Above Critical Level

In this section we present the theory needed for the decision algorithms described in Sect. 3. In particular, we need to determine the fraction of time that the number of BG flows at a node is above a certain critical level during the lifetime (transmission time) of the FG flow, given the number of BG flows present upon the start of the FG flow transmission. Let's therefore focus on one single node, denoted N, with the number of BG flows varying according to BD process $\{X_t | t \geq 0\}$, thus leaving out the indexing i, j in order to simplify notation. Define $E_{T,m,n}$ as the expected duration the number of ongoing BG flows is above level m during the lifetime T of the FG flow given the number of ongoing BG flows upon start of the FG flow is n, i.e.

$$E_{T,m,n} := \sum_{r=0}^{M} \mathbb{E}(U_m, X_T = r | X_0 = n).$$

In [9] a method is developed which provides a method for determining $E_{T,m,n}$ in terms of T, m, n and the parameters of the BD process driving the BG flows at the node for a finite state space of size M. This method is concerned with solving a linear system in the Laplace domain. The solutions to this system are Laplace transforms representing the amount of time a BD process is above m, starting at state n at $t = 0$ and ending at state l at $t = T$. Taking the limit to zero of all differentiated transforms of the solutions results in the expected time a BD process is above m, starting in n and ending at l. Finally, $E_{T,m,n}$ is determined by adding up the results for all end states while starting in n. The above method depends on τ, not on T.

With the expected *time* the number of BG flows is above the critical level, we can easily determine the expected *fraction* of time $E^*_{T,m,n}$ it is above m during the lifetime T of the FG flow given n BG flows at the start of its transmission:

$$E^*_{T,m,n} = \frac{1}{T} E_{T,m,n}. \tag{1}$$

Hence, the expected fraction of time that the number of BG flows $\{X_t | t \geq 0\}$ stays at or is below m during $[0, T]$ is given by $1 - E^*_{T,m,n}$. The above results (1) can be applied to each SFC node $\mathrm{N}^{i,j}$. In particular, let $E^{i,j}_{T,m^{i,j},n^{i,j}}$ be representing the expected fraction of time the number of BG flows at node $\mathrm{N}^{i,j}$ is above $m^{i,j}$ starting with $n^{i,j}$ BG flows at the start of the transmission of the FG flow. Then, the expected fraction of time the number of BG flows at *any* node in chain i exceeds the critical level of its associated node, is given by,

$$E^i_T = 1 - \prod_{j=1}^{N} \left(1 - \frac{1}{T} E^{i,j}_{T,m^{i,j},n^{i,j}}\right).$$

4.2 Weighted Expected Transient Packet Delay

For the flow/packet-based decision algorithm we are also interested in the expected fraction of time a specific number k BG flows is present during the lifetime of the FG flow. Obviously, the expected fraction of time of the presence of $k \in S_M$ BG flows at a 'general' N, denoted by $R_{T,k,n}$, is given by the expected fraction of time above $k - 1$ minus the expected fraction of time above k, i.e.

$$R_{T,k,n} = E_{T,k-1,n} - E_{T,k,n}, \qquad k = 0, 1, \dots, M. \tag{2}$$

with $E_{T,k,n} := 0$ for $k, n \geq M$ and $E_{T,-1,n} := T$. Note that $E_{T,k-1,n} \geq E_{T,k,n}$ for $k \in S_M$.

Again, consider one node. At $t = 0$s, BG packets arrive at rate $n\lambda_p$ at node N. During the FG flow presence, the number of BG flows vary. By using [11], the transient mean sojourn time TMST_k can be determined for the $k\lambda_p + 1$-th BG packet, part of k BG flows, arriving at node N.

With (2), we determine the weigthed expected transient BG packet delay, R^*, for the BG flows above m belonging to node N,

$$R^* = \sum_{k=m+1}^{M} \mathrm{TMST}_k R_{T,k,n}.$$

Apply the above to $N^{i,j}$. Then, the weighted expected transient delay for chain i, R_T^i, during the lifetime of the FG flow is approximated (see Sect. 4.4) by,

$$R_T^i = \sum_{j=1}^{N} \sum_{k=m^{i,j}+1}^{M^{i,j}} \text{TMST}_k^{i,j} R_{T,k,n^{i,j}}^{i,j}.$$

4.3   Flow-Based Decision Algorithm

The selected chain, is the one with lowest weighted expected fractional duration E,

$$E := \min_{i=1,\dots,C} E^i.$$

The chain of choice selected by the FB algorithm is given by,

$$COC_{FB} = \{i \in \{1,\dots,C\} | E_T^i = E\}. \tag{3}$$

4.4   Flow/Packet-Based Decision Algorithm

In [11] a method is presented to calculate the expected transient sojourn time of the r-th packet while w packets are in an M/M/s queue and l packets left the queue, start counting at packet 0. To simplify the calculations, we will determine the expected delay for the $n^{i,j}\lambda_p^{i,j}+1$-th packet, assuming $n^{i,j}\lambda_p^{i,j}$ packets are in the queue at $t = 0$. The '+1' assumes the next packet is the first packet of the FG flow. Define R as the minimum of expected transient delay of all chains,

$$R := \min_{i=1,\dots,C} R_T^i.$$

Then the chain of choice will be,

$$COC_{FPB} = \{i \in \{1,\dots,C\} | R_T^i = R\}. \tag{4}$$

5   Numerical Results

All simulations were conducted on a 12-core CPU system running MATLAB in combination with C-programmed code to speed up the calculation of the transient packet behaviour. To compare the delay of the selected chain against other chains, the FG packets are send to all chains immediately after chain selection. There is no cross-chain influence.

For the FPB algorithm, we combine packet behaviour and critical levels to determine the expected presence duration of the number of BG flows above the critical level. The FB and FPB algorithms have common parameters that will not be changed during simulations and other parameters will be changed during simulations. These are given in Sect. 5.1. In Sect. 5.2, the performance parameters are defined that are used to compare the simulation results in Sect. 5.3.

5.1   Simulation Parameters

The following parameters have proven to be realistic in our case. The nodes process packets at a rate of 120 pps, packets arrive at a rate of 10 pps for all flows. We run 1000 simulations per set of parameters and measure the average delay on all chains for the FG packets. Simulations are run for the FB and FPB algorithms, each with their set of parameters. We choose two chains each consisting of one node for the FB algorithm and two chains each consisting of two nodes for the FPB algorithm. The FG flow duration T is determined per single simulation and has no influence on the SFC selection.

The fixed parameters in all simulations for both algorithms are given in Sect. 5.1. In Sect. 5.1 the hierarchy of the varying parameters is shown.

Common Fixed Simulation Parameters
The common parameters for both algorithms are $(i = 1, \ldots, C, j = 1, \ldots, N)$

- The duration of the FG flow (T) is exponential distributed with parameter $\tau = \frac{1}{10}$.
- The individual BG flow mortality rate $\mu_f^{i,j}$ for all nodes is set to $\frac{1}{5}$.
- The FG packet arrival rate is set to $\hat{\lambda}_p = 10$.
- The BG packet arrival rate is set to $\lambda_p^{i,j} = 10$.
- The service rate at all nodes $\mu_p^{i,j}$ is set to 120.
- The state space of all BD processes is the same: $S_M^{i,j} = 0, 1, \ldots, M$, with $M = 26$, based on [10].

Varying Parameters for FB and FPB Algorithms
By means of Pseudocode 1 an illustration is given of the parameters that are varied and in what hierarchy during the simulations. Note that in line 1 and 3, $n^{i,j}$ and $\lambda^{i,j}$ are set for all i, j respectively. In line 4 and 5, $\lambda_f^{1,N}$ and $n^{C,N}$ are overwritten.

```
1: for all i, j : n^{i,j} = 5, ..., 13 do
2:    for all i, j : m^{i,j} = 5, ..., 13 do
3:       λ_f^{i,j} = 1.0, ∀_{i,j}
4:       λ_f^{1,N} = 2.0
5:       for n^{C,N} = 0, ..., 22 do
6:          1000 Simulations
7:       end for
8:    end for
9: end for
```
PseudoCode 1: Parameter variation and hierarchy.

5.2   Performance Metric

To compare the algorithms we need to define a performance metric that can be applied to both algorithms. A natural way to choose a performance metric is to compare the upfront selected SFC (USS) against the target selected SFC (TSS). The TSS is the SFC that led to the lowest average delay with fixed parameters, after a set of simulations. By ranging one parameter in one chain, we can detect at what values of the varying parameter the USS and TSS change. Our performance metric is given by the difference of the values at which the USS and TSS change. In our case the varying parameter is $n^{C,N}$, shown in Pseudocode 1, line 5. Although not time related, we will refer to the values of $n^{C,N}$ at which the USS and TSS change as the theoretical and target *moment* respectively.

The performance metric is the difference between target and theoretical moments, illustrating how well the algorithms behave under certain circumstances. The performance is better if its value is closer to zero, zero being ideal. A positive value, say k, means that the theoretical moment was '$+k$' moments later than the target moment, in respect to the steps in which the associated parameters varied. A negative value means that the theoretical moment was earlier then the target moment.

5.3   Main Simulation Results

The results of the simulations can be found in Sect. 5.3 for the FB algorithm with SFCs of length one and Sect. 5.3 for the FPB algorithm with SFCs of length two.

The FB Applied to Chains with 1 Node: Expected Duration of Exceeding the Critical Flow Level

Refer to Fig. 5, the results of the simulations are shown whereby parameters are varied as per Pseudocode 1. The variation of the initial values, $n^{i,j}$, (line 1 in Pseudocode 1) is shown on the vertical ax. The variation of the critical levels, $m^{i,j}$, (line 2 in Pseudocode 1) is shown on the horizontal ax. The performance

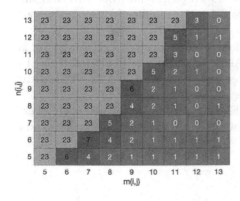

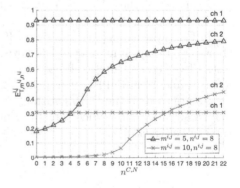

Fig. 5. Performance of FB algorithm, $N = 1$.

Fig. 6. BG flows exceed critical level, $N = 1$.

of the FB algorithm, resulting from lines 5–7 in Pseudocode 1, is given by the number in the square per $m^{i,j}$, $n^{i,j}$ combination.

Recall that the performance is given by the difference between the theoretical and target moments. The higher this number, the worse the performance. Since the range through which the $n^{C,N}$ varies is limited to $0, 1, \ldots, 22$, '+23' means that no theoretical moment exists within the $n^{C,N}$ range, i.e. the FB algorithm sticks to the same SFC during all the $n^{C,N}$ combinations. The value '+23' could be interpreted as 'off the chart'.

To elaborate more on the decision moments, refer to Fig. 6. The expected fractional time above a critical level, given by $E^{i,j}_{T,m^{i,j},n^{i,j}}$, is shown for two cases, for $n^{i,j} = 8$: $m^{i,j} = 5$ and $m^{i,j} = 10$, each for both chains. For the $m^{i,j} = 10$ case (the '×' graph), the FB algorithm selected chain 2 for $n^{C,N} < 16$ and selected chain 1 for $n^{C,N} \geq 16$, since the expected fractional time above $m^{i,j}$ for chain 2 is lower than the expected fractional time above $m^{i,j}$ for chain 1, for $n^{C,N} < 16$. For the $m^{i,j} = 10$, the theoretical moment lies at 16. From the simulations it follows that the target moment is 14. The performance for this particular case is '+2' and can be found in Fig. 5 for $m^{i,j} = 10$ and $n^{i,j} = 8$.

However, for $m^{i,j} = 5$ (the 'Δ' graph) no change will take place, as the expected fractional time above the critical level for chain 1 is greater than the expected fractional time for chain 2 above the critical level, for all $n^{C,N} = 0, 1, \ldots, 22$. As a result, the FB algorithm will stick to chain 2 and the performance will be '+23', off the chart. This result can be found in Fig. 5 for $m^{i,j} = 5$ and $n^{i,j} = 8$.

The FB algorithm does not perform well in case the initial number BG flows is (roughly) above the critical level as the performance in these circumstances results in a '+23'.

FPB Applied to Chains with Two Nodes
We have applied the FPB algorithm to SFCs consisting of two nodes each, i.e. $N = 2$. The performance results can be found in Fig. 7 whereby the parameters have been varied as per pseudocode 1. The initial values, $n^{i,j}$, are shown on the vertical ax, the critical levels, $m^{i,j}$, are shown on the horizontal ax and the performance of the FPB algorithm, is given by the number in the square per $m^{i,j}$, $n^{i,j}$ combination.

In Fig. 8 the expected duration of BG flows above a critical level (5 and 10) is shown for 13 initial number of BG flows. As per Fig. 8, for critical level 5 and 13 initial BG flows, chain 2 will not be selected for $n^{C,N} = 0, 1, \ldots, 22$. For critical level 10 and 13 initial BG flows, chain 2 is selected for $n^{C,N} \geq 21$.

As opposed to the FB algorithm, the FPB algorithm performs well in overloaded situations for which the initial number of BG flows $n^{i,j}$ are greater than the critical level $m^{i,j}$.

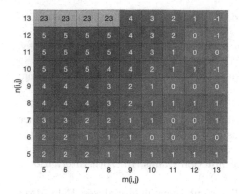

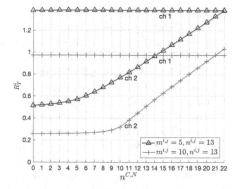

Fig. 7. Performance of FPB algorithm, $N = 2$.

Fig. 8. BG flows exceed critical level, $N = 2$.

6   Conclusion and Future Work

We created two algorithms, a Flow-based and Flow/packet-based algorithm, to select an SFC upon arrival of foreground flow relying only on the flow characteristics and the transient packet behaviour. We consider this a meaning full contribution. In particular the Flow/packet-based algorithm, as it performs well in overloaded situations. The FB algorithm does not perform well in overloaded situations. The number of SFCs and their lengths are limited in real networks and the amount of paths BG traffic can take in a network does not affect the SFC selection.

Adding **machine learning** to the decision algorithms to cope with all the different parameters and decide upfront to assign a flow to an SFC would be the next step in research in dynamic (automated) SFC-RA scheduling.

As mentioned in previous sections, our predictive results are based on exponential distributed parameters, both on a flow and a packet level. In order to work with more **realistic traffic**, results on the expected duration of exceeding a critical level, needs to be extended to other distributions as well. For example, a deterministic arrival rate of packets to simulate voice or video traffic.

References

1. ETSI Industry Specification Group (ISG). Network Functions Virtualisation (NFV); Infrastructure Overview (2015)
2. Haleplidis, E., Pentikousis, K., Denazis, S., Salim, J.H., Meyer, D., Koufopavlou, O.: Software-defined networking (SDN): layers and architecture terminology. In: IETF Network Task Force, no. rfc 7426 (2015)
3. Hantouti, H., Benamar, N., Taleb, T., Laghrissi, A.: Traffic steering for service function changing. IEEE Commun. Surv. Tutor. **21**(1), 487–507 (2018)
4. Halpern, J., Pignataro, C.: Service function chaining (SFC) architecture. In: IETF Network Task Force, no. rfc 7665 (2015)

5. Bhamare, D., Jain, R., Samaka, M., Erbad, A.: A survey on service function chaining. J. Netw. Comput. Appl **75**, 138–155 (2016)
6. Herrera, J.G., Botero, J.F.: Resource allocation in NFV: a comprehensive survey. IEEE Trans. Netw. Serv. Manag **13**(3), 518–532 (2016)
7. Xie, Y., Liu, Z., Wang, S., Wang, Y.: Service function chaining resource allocation: a survey (2016). ArXiv e-prints
8. Kleinrock, L.: Queueing Systems, Volume I: Theory. Wiley & Sons, Hoboken (1975)
9. Ellens, W., Mandjes, M., van den Berg, J., Worm, D., Blaszczuk, S.: Performance valuation using periodic system-state measurements. Perf. Eval. **93**, 27–46 (2015)
10. Wetzels, F., van den Berg, J., Bosman, J., van der Mei, R.: Dynamic priority assignment for SLA compliance in service function chains. In: 26th International Conference on Telecommunications (ICT) (2019)
11. Kaczynski, W., Leemis, L.M., Drew, J.: Transient queueuing analyses. INFORMS J. Comput. **24**(1), 10–28 (2012)

Accelerating MCMC by Rare Intermittent Resets

Vivek S. Borkar[✉][ID] and Syomantak Chaudhuri[ID]

Department of Electrical Engineering, Indian Institute of Technology Bombay,
Mumbai, India
{borkar,syomantak}@iitb.ac.in

Abstract. We propose a scheme for accelerating Markov Chain Monte Carlo by introducing random resets that become increasingly rare in a precise sense. We show that this still leads to the desired asymptotic average and establish an associated concentration bound. We show by numerical experiments that this scheme can be used to advantage in order to accelerate convergence by a judicious choice of the resetting mechanism.

Keywords: Markov Chain Monte Carlo · Time inhomogeneous
Markov chain · Rare resets · Accelerated convergence · Martingale law
of large numbers · Concentration bounds

1 Introduction

We argue that the asymptotic behavior of empirical measures of an irreducible finite state Markov chain is unaffected if we reset its state infinitely often, as long as these resets become increasingly rare in a precise sense. A judicious choice of these resets can then be leveraged to accelerate the convergence of empirical measures to the stationary distribution, a fact we validate by numerical experiments. This provides a novel method for accelerating Markov Chain Monte Carlo (MCMC) [18].

The aforementioned theorem is stated in the next section along with an interpretable concentration bound. Section 3 describes its implications to MCMC, leading to the proposed scheme. Section 4 presents some supporting numerical experiments. Section 5 presents some variants of the basic scheme. Section 6 concludes with some comments. An appendix contains some technical proofs.

While this article essentially puts forth another possible approach to speeding up MCMC which has the advantage of simplicity, along with some basic theoretical analysis, it still leaves a lot of ground yet to be covered. Section 2 also

V. S. Borkar—Work of this author was supported in part by a S. S. Bhatnagar Fellowship from the Council of Scientific and Industrial Research, Government of India.
S. Chaudhuri—Now with the Dept. of EECS, Uni. of California, Berkeley, Cory Hall, Berkeley 94720, CA.

Q. Zhao and L. Xia (Eds.): VALUETOOLS 2021, LNICST 404, pp. 107–125, 2021.
https://doi.org/10.1007/978-3-030-92511-6_7

points out a plausible approach to a more detailed analysis based on a related scheme.

This is not the first time resets have been proposed as a mechanism for speed-up. Some representative contributions are [3,4,11]. In [4], the chain, with a small probability, restarts with a uniform distribution. In [3], the restarts are to a fixed set of well chosen nodes, called a 'supernode'. In both cases, suitable mathematical relationships of the stationary expectations corresponding to the original and the modified chains are derived and used to advantage. In [11], the proposal distribution of the Metropolis-Hastings chain on $\mathcal{R}^n$ includes additional transitions. The application is limited to problems with some additional structure (specifically, stationary distributions that are mixtures of log-concave densities). In [3], another scheme that combines ideas from MCMC and reinforcement learning (RL) is presented, which has variance reduction properties (because the RL terms serve as control variates), with conditional importance sampling that facilitates restarts.

Compared to the above, the present proposal offers the following advantages.

- It is simpler to implement, since it employs predetermined deterministic times for restarts.
- A further simplicity is achieved because it works with a single running average as in the classical MCMC and does not need any extra running averages or off-line computation to map the result back to the desired average.
- It gives promising results on test problems that have a highly clustered state space.

On the flip side, a rigorous rate of convergence analysis is missing, though we try to give an intuition regarding the same.

There are several other schemes to improve upon vanilla MCMC, such as 'frontier sampling' that uses several concurrent and correlated random walks to improve mixing [21]. In [7], regenerative cycles and weighted graphs are used to propose two different modifications. See [2] and [22] for an overview of classical acceleration methods for MCMC and [9] for an overview of Markov chains with resets and their many applications.

2   A Convergence Result

This section establishes the key theoretical results of this article. The first is Theorem 1, which establishes rigorously that the rare resets do not affect the asymptotic behavior of MCMC. The second theorem and its corollary give a finite time concentration bound, which is useful for deducing sample complexity.

Consider an irreducible stochastic matrix $P = [[p(j|i)]]$ on a finite state space $S, |S| = s$, with (unique) stationary distribution π. Consider a time inhomogeneous Markov chain $\{X_n\}_{n\geq 0}$ on S with transition probabilities $q_n(j|i)$, $i, j \in S, n \geq 0$. Let $\mathcal{P}(S)$ denote the simplex of probability vectors on S. Define empirical measures $\nu_n \in \mathcal{P}(S), n \geq 0$, by:

$$\nu_n(i) = \frac{\sum_{m=1}^n I\{X_m = i\}}{n}, \ n \geq 1.$$

The following result is proved in the Appendix.

Theorem 1. *Suppose*

$$\sum_{m=1}^{n} I\{q_m \neq p\} = o(n). \tag{1}$$

Then

$$\nu_n \to \pi \ a.s. \tag{2}$$

Let $\theta, \mathbf{1}$ denote resp. the vector of all 0's and all 1's in $\mathcal{R}^s$. Let $\mu \in \mathcal{P}(S)$. Viewing π, μ as row vectors, we have

$$\pi P - \pi = \theta, \ \mu P - \mu = z$$

for some $z \in \mathbf{1}^\perp$. Thus for $I_s :=$ the $s \times s$ identity matrix,

$$z = (\pi - \mu)(I_s - P)$$
$$= (\pi - \mu)(I_s - P + \mathbf{1}\pi)$$

because $(\pi - \mu)\mathbf{1} = 0$. Hence

$$\pi - \mu = (I_s - P + \mathbf{1}\pi)^{-1} z,$$

where we use the fact that the so called 'fundamental matrix' or 'deviation matrix' $D := (I_s - P + \mathbf{1}\pi)$ is non-singular. This leads to

$$\|\mu - \pi\| \leq C\|z\| \tag{3}$$

for $C = \|D^{-1}\|$ $(:= \max_{x:\|x\|=1} \|D^{-1}x\|)$, where for vectors, $\|\cdot\|$ refers to the Euclidean norm. We next use (3) to obtain a high probability finite time bound for the 'error' $\nu_n - \pi$.

Following [20], for the time inhomogeneous Markov chain $\{X_m\}_{m=0}^{n}$ with transition probabilities $\{q_m(j|i)\}$, define the mixing time $\tau(\epsilon)$ as the minimum ℓ such that the total variation distance between $P(X_{m+\ell} = \cdot | X_m = i)$ and $P(X_{m+\ell} = \cdot | X_m = j)$ is less than ϵ for every $1 \leq m \leq n - \ell$ and $i, j \in S$. Define

$$\tau_{min} = \inf_{0 \leq \epsilon < 1} \left(\frac{2 - \epsilon}{1 - \epsilon}\right)^2 \tau(\epsilon).$$

Let $\eta(n) := \sum_{m=1}^{n} I\{\|p(\cdot|X_m) - q_m(\cdot|X_m)\| > 0\}$ for $n \geq 1$. The following result is proved in the Appendix.

Theorem 2. *For any $\delta > 0$, $\|\nu_n - \pi\| \leq \delta + C\sqrt{s}\left(\frac{\eta(n)+1}{n}\right)$ with probability at least $1 - 2se^{-\frac{\delta^2 n}{2sC^2\tau_{min}}}$.*

Let $E_s[g(X.)]$ denote the stationary expectation of $g(X_n)$ when there are no resets. Consider the problem of estimation of $E_s[g(X.)]$ for a given $g : S \mapsto \mathbb{R}$ using the empirical mean $\hat{g}^{(n)} = \frac{1}{n}\sum_{m=1}^{n} g(X_m)$. Let $\bar{g}$ denote the row vector $[g(1), \ldots, g(s)]$. The following is then immediate.

Corollary 1. *For any* $\delta > 0$,

$$P\left(|\hat{g}^{(n)} - E_s\left[g(X.)\right]| \leq \delta + \|\bar{g}\|C\sqrt{s}\left(\frac{\eta(n)+1}{n}\right)\right) \geq 1 - 2se^{-\frac{\delta^2 n}{2sC^2\|\bar{g}\|^2 \tau_{min}}}.$$

To interpret the above bounds, it is clear that the factor $\frac{\eta(n)}{n}$ captures the error due to resets. The concentration inequality, as usual, quantifies the 'high probability' bound on the error between empirical mean and the stationary expectation. What reflects the structure of the graph is the constant C. If the chain has k clusters with high conductance intra-cluster edges and low conductance inter-cluster edges, we expect k eigenvalues of P, say $\lambda_2, \cdots, \lambda_{k+1}$, satisfy $1 > |\lambda_i| \approx 1$ for $2 \leq i \leq k+1$ (see [8] for such a result in the reversible case). Then the eigenvalues $1/(1-\lambda_i), 2 \leq i \leq k+1$ of D^{-1} will be very large and the corresponding bound will be weak as expected. This insight also helps understand the role of resets in speeding up the scheme.

The concentration inequality above suggests that the sample complexity of the scheme is at worst roughly the same order as that of a classical random walk on graph without resets. What we expect, however, is that it should be much better. Unfortunately a tighter convergence rate analysis appears difficult and is left for future work. Nevertheless a comparison with a related scheme motivated by [4] is instructive. The latter uses a transition according to the random walk transition kernel with a probability $1 - \epsilon$ for some $0 < \epsilon << 1$ and a reset with uniform probability over the state space with probability $\epsilon > 0$. Denote this perturbed transition matrix as P_ϵ. The stationary average is then an $O(\epsilon)$-perturbation of the desired stationary average, but achieved much quicker. The improvement of the convergence rate is captured by the increased spectral gap which is analytically estimated in *ibid.* (see also a more general formula given by Theorem 5.1 of [16]). This clearly captures the gain in the exponential rate of convergence to stationarity. Now consider the scenario where we slowly decrease $\epsilon = \epsilon(n)$ to zero. This is in the spirit of our scheme. Then the update rule for the time n distribution $\pi(n)$, given by

$$\pi(n+1) = \pi(n)P_{\epsilon(n)},$$

and a suitable update rule for (slow) decrease of $\epsilon(n)$ may be viewed as a two time scale iteration and be analyzed as such. That is, we can treat the slowly varying resets as following a 'quasi-static' dynamics of their own and analyze the MCMC initially treating the reset mechanism as fixed. Thus the D is fixed, an approximation to the fact that in reality it is quasi-static, i.e., slowly varying. Then in the beginning when the resets are made with high probability, the corresponding chain is fast mixing (by design - it is assumed that the reset mechanism is chosen so that this is indeed so). Hence C will be moderate and the error bounds are good. As the iterate count n increases, the resets become rarer and therefore the matrix D^{-1} becomes more and more ill-conditioned, leading to increase of C. But by then the averaging of MCMC would have progressed far enough to have benefited from the high mixing of the initial stages. The tuning of C_n on the other hand aligns the stationary distribution better with the desired one. There is a clear trade-off between the two that needs to be quantified.

There is an analogy between this and the simulated annealing algorithm of [13], except that here, unlike the latter, the limiting chain is not degenerate, only slow mixing. This raises the hope of adopting the analysis of [13] and subsequent variants such as [1,15] and [23] to analyze the proposed schemes. Our scheme can be viewed as a 'derandomized' version of this chain. Our scheme is much simpler because of the deterministic schedule that spares us a randomization and update of $\epsilon(n)$ at each time, but the analysis becomes harder. Nevertheless, this connection gives additional motivation for our scheme. We have included some simulations for this alternative scheme as well, see Fig. 6.

It is worth noting that our scheme essentially averages out Markov chain runs of increasing lengths initiated at different points in the state space. The somewhat arbitrary mixing policy does not matter for the convergence, because the reset instants are increasingly rare so as to have zero 'density' in the discrete time axis. However, the choice should matter for the convergence rate. If we reset with positive frequency, e.g., periodically, then the Cesaro averages will differ from the intended ones, but only by a small amount if the resets are sufficiently infrequent, e.g., with a large period. This may not matter if the MCMC scheme is a part of an ordinal comparison or ranking exercise where only the relative orders matter, so there is certain tolerance for small errors.

3 Applications to Network Sampling

Consider a Markov chain exhibiting considerable metastability, i.e., densely connected clusters of states that are weakly connected with each other. In other words, the transitions across clusters are significantly rare compared to the transitions within clusters. This leads to behaviors such as quasistationarity [6] that significantly reduce the rate of convergence to stationarity. One can accelerate convergence to stationarity by introducing additional edges across clusters allowing for transitions that are not legitimate for the original chain. This will improve the 'conductance' of the chain and improve the rate of convergence to stationarity [19]. But this will also alter the stationary distribution. Theorem 2 suggests that making such transitions along a rare subsequence may allow us to strike a sweet spot between the two effects. The numerical results of the next section confirm this. The rest of this section describes the precise scheme we implemented.

Consider the Metropolis-Hastings (MH) MCMC algorithm on a finite state space S with stationary distribution $\pi(i)$, $i \in S$. Let $r(j|i)$ be the proposal distribution of the next state candidate j given current state i and let $a(j|i)$ be the acceptance probability thereof. The transition probability is then given by $p(j|i) = r(j|i)a(j|i)$ for $j \neq i$ and satisfies the detailed balance $p(j|i)\pi(i) = p(i|j)\pi(j)$ $\forall i, j \in S$. Our proposed algorithm modifies the MH algorithm by introducing $o(n)$ random resets in n steps, say with a prescribed transition probability $q(\cdot|\cdot)$ which need not satisfy the detailed balance. Denote this algorithm by MHRR (for Metropolis-Hastings with Random Resets). We discuss the choice of $q(\cdot|\cdot)$ later. We use the reset instants $R = \{r_k : r_k \leq n\}$ where

$$r_j - r_{j-1} = K_1 \log(K_2 + j), \quad r_0 = 10,$$

for constants $K_1, K_2 \geq 0$. It can be seen that $r_j = \Theta(j \log(j))$ so, $|R| = o(n)$.

We illustrate the proposed scheme in the case of the network sampling problem in this section. Consider the graph $G = (S, \mathcal{E})$ where $\mathcal{E} :=$ the edge set, for which we want to estimate, for a given function $g : S \to \mathbb{R}$, its average $g(S) := \frac{1}{s} \sum_{i \in S} g(i)$. For very large graphs, random walk based methods are normally preferred for this as exhaustive evaluation of $g(S)$ directly is not feasible. Usually, it is possible to query the value of $g(\cdot)$ for a particular node and also obtain its neighbors in the undirected graph. Under this setting, we describe the random walk algorithm used to estimate $g(S)$.

The simple random walk based Metropolis-Hastings algorithm in order to estimate the average node values can be obtained by taking $\pi(i) = 1/s$ and

$$r(j|i) = \begin{cases} 1/\deg(i), & \text{if } \exists(i, j) \in \mathcal{E} \\ 0, & \text{else} \end{cases}, \quad a(j|i) := \min\left\{\frac{\deg(i)}{\deg(j)}, 1\right\}.$$

If the sequence of states obtained using the MH algorithm is $\{X_i\}$, we can estimate $g(S)$ as

$$\hat{g}_{\text{MH}}^{(n)}(S) = \frac{1}{n} \sum_{i=1}^{n} g(X_i).$$

For reset probability, we propose a $q(j|i)$ that is intuitively appealing in the sense that it facilitates cross-cluster transitions in a highly clustered state space. Supposing there are c clusters (decided based on some appropriate rule, discussed later in Sect. 4) with labels chosen from $C = \{1, \ldots, c\}$ in some order. Let $N2C : S \to C$ be a mapping from nodes to its corresponding cluster and let $C2N : C \to 2^S$ be an one-to-many mapping which maps a cluster to the set of nodes present in it. If the mappings $N2C$ and $C2N$ are available, then we can choose the following transition for random resets

$$q(j|i) = \begin{cases} \frac{1}{(c-1)|C2N(N2C(j))|}, & \text{if } j \notin C2N(N2C(i)) \\ 0 & \text{else.} \end{cases}$$

In other words, choose any cluster other than the current cluster uniformly and then choose any node in that cluster uniformly. If the sequence of states obtained using the MHRR algorithm is $\{X_i\}$, we can estimate $g(S)$ as

$$\hat{g}_{\text{MHRR}}^{(n)}(S) = \frac{1}{n} \sum_{i=1}^{n} g(X_i)$$

The pseudo-code for MHRR algorithm is given in Algorithm 1.

This choice of reset probabilities is for illustrative purposes only and presupposes availability of approximate clusters without expending too much additional computation. There can be other smart choices for $q(j|i)$ in case the mappings $C2N$ and $N2C$ are not available. For example, while crawling a social network, it is quite likely that we would land in a different cluster if we randomly transition to any node/user who resides in a different country. Simply picking significant nodes (with respect to some centrality measure) that are distant from one another either geographically or in graph distance is another possibility.

For comparison purposes, we also consider a modification of the Respondent Driven Sampling (RDS) algorithm [14]. In brief, the RDS algorithm estimates the average of the node values by simple random walk where next node is randomly chosen from the set of nodes connected to the current node. The bias due to the simple random walk is removed by normalizing by the degree of the visited node. If the sequence of states obtained using the RDS algorithm is $\{X_i\}$, we can estimate $g(S)$ as

$$\hat{g}_{\mathsf{RDS}}^{(n)}(S) = \frac{\sum_{i=1}^n g(X_i)/\mathsf{deg}(X_i)}{\sum_{i=1}^n 1/\mathsf{deg}(X_i)}.$$

We also modify this algorithm to include $o(n)$ random resets and we denote this algorithm as RDSRR which is described in Algorithm 2.

4   Numerical Experiments

We compare the performance of MH, RDS, MHRR, and RDSRR algorithms on three datasets described next. Since these datasets are anonymized, we cannot

---

Algorithm 1. Metropolis-Hastings MCMC with Random Resets

1: **procedure** MHRR(K_1,K_2,B) $\triangleright$ K_1,K_2,B are constants
2:      $r_0 = 10$
3:      Generate $R = \{r_k : r_k \leq B,\ k \geq 0\}$, $r_j - r_{j-1} = K_1 \log(K_2 + j)$
4:      $\hat{f} = 0$
5:      $cv = \mathsf{Random}(\{1,\dots,s\})$ $\triangleright$ Random(A) samples from set A uniformly
6:      $\hat{f} = \hat{f} + g(cv)$
7:      **for** $i = 2,\dots,B$ **do**
8:          **if** $i \in R$ **then**
9:              $cv = \mathsf{Reset}(cv)$
10:          **else**
11:              $nv = \mathsf{Random}(\mathsf{adj}(cv))$ $\triangleright$ adj() returns the set of neighbors
12:              **if** $U() \leq \mathsf{deg}(cv)/\mathsf{deg}(nv)$ **then** $\triangleright$ U() generates sample $\sim U[0,1]$
13:                  $cv = nv$
14:          $\hat{f} = \hat{f} + g(cv)$
15:      $\hat{f} = \hat{f}/B$
16:      **Return** $\hat{f}$
17:
18: **procedure** Reset(cv)
19:      $ccom = \mathsf{N2C}(cv)$ $\triangleright$ N2C maps nodes to communities
20:      $rcom = \mathsf{Random}(C \setminus \{ccom\})$ $\triangleright$ C is the set of communities
21:      **Return** $\mathsf{Random}(\mathsf{C2N}(rcom))$ $\triangleright$ C2N lists the nodes of the community

use the methods based on the maps $N2C, C2N$ as described in Sect. 3. To overcome this problem, we first use an iterative community detection algorithm, viz., the Louvain method [5] to obtain the graph clusters. The Louvain method tries to maximize the 'modularity' of graph partitions. Modularity is a measure of goodness of partition of a community as defined below for an undirected graph $G = (S, \mathcal{E})$,

$$Q = \frac{1}{2M} \sum_{(i,j) \in S} \left(a_{ij} - \frac{k_i k_j}{2M} \right) \delta(c_i, c_j)$$

where M is the sum of all edge weights, a_{ij} is the weight of edge (i, j), k_i is the sum of weights of edges connected to node i, c_i is the community label for

---

Algorithm 2. Respondent-driven sampling with Random Resets

1: **procedure** RDSRR(K_1,K_2,B)
2:      $r_0 = 10$
3:      Generate $R = \{r_k : r_k \leq B, \ k \geq 0\}$, $r_j - r_{j-1} = K_1 \log(K_2 + j)$
4:      num = 0
5:      den = 0
6:      cv = Random($\{1, \ldots, s\}$)
7:      num = num + $g(cv)/\deg(cv)$
8:      den = den + $1/\deg(cv)$
9:      **for** $i = 2, \ldots, B$ **do**
10:         **if** $i \in R$ **then**
11:             cv = Reset(cv)
12:         **else**
13:             cv = Random(adj(cv))
14:             num = num + $g(cv)/\deg(cv)$
15:             den = den + $1/\deg(cv)$
16:     $\widehat{f}$ = num/den
17: **Return** $\widehat{f}$

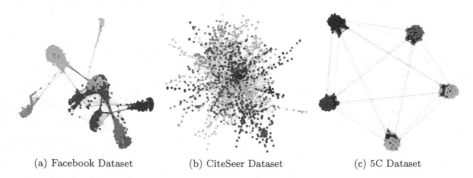

(a) Facebook Dataset   (b) CiteSeer Dataset   (c) 5C Dataset

Fig. 1. A clustered views of the datasets experimented on in this work. For each dataset, nodes from different clusters are represented by different colored dots. (Color figure online)

node i, and $\delta(\cdot, \cdot)$ is the Kronecker delta function. The following two steps are alternated on the undirected graph with the initial edge weights being set as 1 :

1. each node is mapped to a separate community label. For each node i, assign it to the community of neighbor j if doing so would result in increasing the value of modularity by maximum positive amount among all neighbors;
2. obtain a new weighted undirected graph by combining each existing community into a single node with weights between new nodes being equal to the sum of weights between the corresponding communities in the old graph. Self-loops of weight equal to the sum of weights of all intra-community edges in the old graph are to be created in the new graph;

While this is an additional preprocessing step with its own computational burden, we can expect non-anonymized social network data in practice with nodes that can be partitioned into natural regions such as geographical, and this step could be replaced by some easier and justifiable heuristic.

We conduct the experiments with the value of $K_1 = 4, K_2 = 20$ for MHRR algorithm and run for $B = 10000$ steps for both the algorithms. The value of K_1 was chosen heuristically whereas K_2 was set arbitrarily since it plays a relatively minor role in the long term behavior of the algorithm. A study on the change in performance due to variation in K_1 is described in Sect. 5. We report the Normalized Root Mean Square Error (NRMSE) value, defined as

$$\text{NRMSE} = \frac{\sqrt{\mathbb{E}[(g(S) - \hat{g}^{(n)}(S))^2]}}{g(S)}.$$

In practice, one replaces the expectation with empirical mean. In particular, we take the mean over 100 independent runs of the algorithm. The experiments were conducted with the help of the Python package Networkx [12].

For experimentation, we considered a variety of node functions whose node-average is to be estimated. They need not have any practical relevance, but suffice to examine the performance of the algorithms. Specifically, we consider the following functions:

1. $g_c(v) = I\{N2C(v) = 1\}$,
2. $g_d(v) = I\{\deg(v) > \hat{d}\}$ where $\hat{d}$ is set based on the dataset,
3. $g_p(v) = I\{v \text{ is prime}\}$, and
4. $g_r(v)$ is obtained by sampling s random values from exponential $\text{Exp}(1)$ distribution.

Before looking at the experimental results, we remark that intuitively, for functions which do not vary too much for different clusters in the graph, like g_p and g_r, one would not expect dramatic improvement in rate of convergence by intermittent resets in the MCMC algorithms. However, for functions which depend strongly on the clusters, such as g_c, the effect of random resets can be expected to be more prominent.

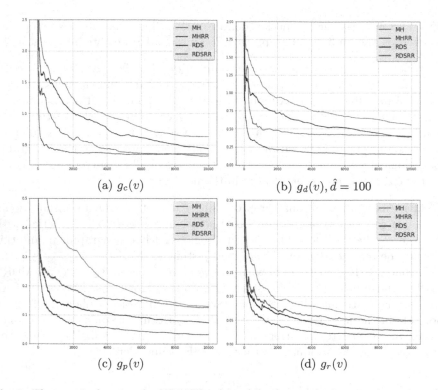

(a) $g_c(v)$

(b) $g_d(v), \hat{d} = 100$

(c) $g_p(v)$

(d) $g_r(v)$

Fig. 2. The y-axis denotes the NRMSE value while the x-axis denotes the steps taken by the algorithm (n). This figure compares the algorithms on the Facebook dataset for the described four functions.

4.1   Facebook Dataset

The Facebook dataset [17] is an undirected graph consisting of 4039 nodes and 88234 edges. As shown in Fig. 1, there is a high degree of clustering in this dataset. Note that such clustering is not surprising for graphs based on social media.

The NRMSE vs n graphs for the four algorithms - MH, MHRR, RDS, and RDSRR - are shown in Fig. 2. It can be observed that the MHRR algorithm outperforms the other three algorithms. As claimed in [3], RDS algorithm performs better than the simple MH algorithm but introducing the random resets in the MH algorithm significantly accelerates the convergence for above-mentioned functions on Facebook dataset. Adding random resets to the RDS algorithm improves the performance for functions $g_c(\cdot)$ and $g_d(\cdot)$ but it hurts the convergence of the algorithm for the functions $g_p(\cdot)$ and $g_r(\cdot)$. This can be due to the fact that the latter functions do not strongly depend on the clusters as explained earlier.

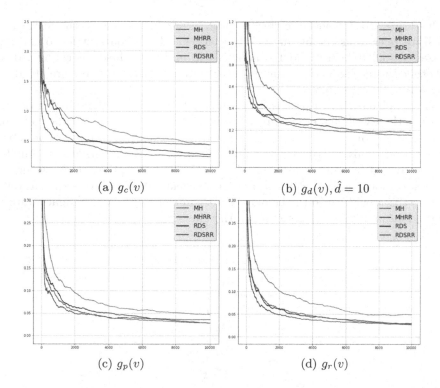

Fig. 3. The y-axis denotes the NRMSE value while the x-axis denotes the steps taken by the algorithm (n). This figure compares the algorithms on the CiteSeer dataset for the described four functions.

4.2   CiteSeer Dataset

The CiteSeer dataset [10] consists of 3313 research articles which are categorized into six classes based on the research area. It also provides 4675 directed citation-links between these articles. We treat these research articles and citations as an undirected graph consisting of 3313 nodes and 4675 edges, and then we obtain the largest connected sub-graph of 2129 nodes and 3751 edges. We ignore the research area labels available in the dataset, and stick of the community structure obtained by the Louvain method.

The NRMSE vs n graphs for the four algorithms are shown in Fig. 3. As seen visually in Fig. 1, the dataset is not well-clustered and hence, not surprisingly, random resets do not provide any visible benefits to the algorithms for functions $g_d(\cdot), g_p(\cdot)$, and $g_r(\cdot)$. For the function $g_c(\cdot)$, random resets do help to certain extent initially, but they also hurt the convergence rate later in the trajectory.

4.3   5C Dataset

It is intuitive to expect that the random reset based algorithms should perform significantly better than the other algorithms on graphs showing a high degree of clustering. To demonstrate this, we also experiment on a synthetic dataset which we refer to as the 5C dataset henceforth. The 5C dataset has 5 clusters of sizes 80, 90, 100, 110, and 120 nodes each, and each cluster is connected to any other cluster via 5 edges that were picked randomly, i.e., there are 50 edges inter-connecting the clusters.

The graph for NRMSE vs n is shown in Fig. 4. Our proposed method performs better on these well-clustered graphs, similar to the case of Facebook dataset. For the case of $g_c(\cdot)$ in both Fig. 2(a) and Fig. 4(a), a large difference in performance can be seen - this is true for the function $g_d(\cdot)$ as well. Perhaps, it can be speculated that in well-clustered graphs, the benefit of random resets would be the greatest for functions $g(\cdot)$ which has a different mean values for the clusters, because this would cause the estimates (for smaller n) to become inaccurate if the random walk tends to get stuck in the clusters.

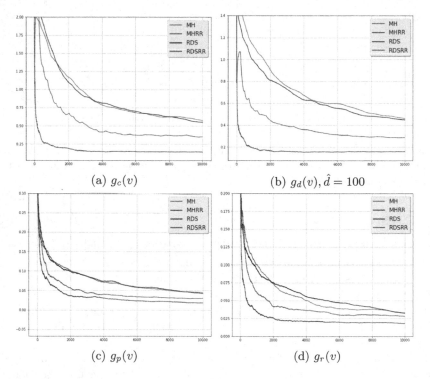

(a) $g_c(v)$ (b) $g_d(v), \hat{d} = 100$

(c) $g_p(v)$ (d) $g_r(v)$

Fig. 4. The y-axis denotes the NRMSE value while the x-axis denotes the steps taken by the algorithm (n) on the 5C dataset comprising of 5 clusters of 80, 90, 100, 110, and 120 nodes respectively.

5   Variants in the Algorithm

5.1   Variations in Transitions

We report here two variants of the basic scheme above.

In case the mappings $N2C$ and $C2N$ are not available, we can choose other heuristics for choosing the $q(j|i)$ function. For social networks, it is not uncommon to assume that we can obtain a highly-connected node in various clusters even if the entire $N2C$ mapping isn't available. For example, one can choose a popular celebrity in each country as a highly connected node. Once we obtain this set of nodes, we can randomly transition to any one of these nodes instead. Since the given data is anonymized, here we picked the highest degree nodes in the clusters obtained by the Louvain method. We conducted the same experiments for the Facebook dataset. The results with these transitions have been labeled as 'MHRR-T' and 'RDSRR-T' in Fig. 5. This pragmatic choice of transition strategy does not seem to decrease the convergence performance of the algorithms.

5.2   Variations in Resetting Instants

We also consider different the policies regarding when to reset. Since the reset strategy of transitioning to prominent nodes, described above, is more practical, we use this transitioning strategy while comparing the different policies regarding reset instants described next.

nMHRR-T : For every time step t, reset with probability $\frac{1}{1+t/K_1}$, $K_1 = 50$.
logMHRR-T : For every time step t, reset with probability $\frac{1}{1+K_1 \log(t)}$, $K_1 = 2$.

The results are shown in Fig. 6. It is difficult to draw conclusion from this experiment whether there is a clear choice among these reset policies.

5.3   Variations in Parameters

Further, we also studied the variation in performance of these algorithms by varying K_1 (see Fig. 7) for the algorithms MHRR-T, nMHRR-T, and logMHRR-T. Depending on the value of K_1, the expected number of resets varies. A judicious choice of this parameter can help strike a balance between too many resets and too few resets.

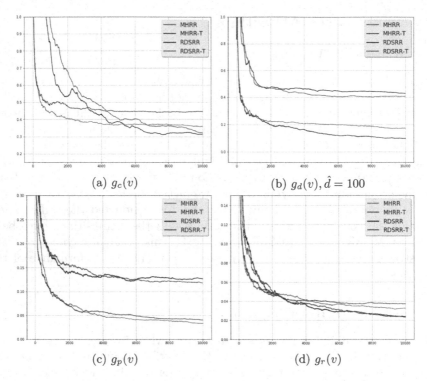

(a) $g_c(v)$ (b) $g_d(v), \hat{d} = 100$

(c) $g_p(v)$ (d) $g_r(v)$

Fig. 5. Comparison of NRMSE vs n plot for the MHRR, MHRR-T, RDSRR, and RDSRR-T. MHRR-T and RDSRR-T use a different transition probability $q(j|i)$ based on choosing a high-degree node in any different cluster uniformly.

5.4   Variations in Performance Based on Clustering in Graphs

In addition to the above variants, we study the variation in the performance of the algorithms MH and MHRR based on the degree of clustering in the graph. To this end, we consider four different graphs, each with 500 nodes.

In the first graph, labeled '10C', there are ten clusters of nodes and the cluster sizes vary from 15 nodes to 85 nodes. These clusters are sparsely inter-connected. Similarly, the other graphs have eight, four, and two nodes, and the graphs are labeled as '8C', '4C', and '2C' respectively.

Figure 8 shows the difference in convergence of the MH and MHRR algorithms on the abovementioned graphs with different degree of clustering. Although MHRR algorithm significantly improves over MH algorithm for these heavily clustered graphs, it is hard to quantify the degree of efficacy of the random resets as a function of the degree of clustering in the graphs based on Fig. 8 alone.

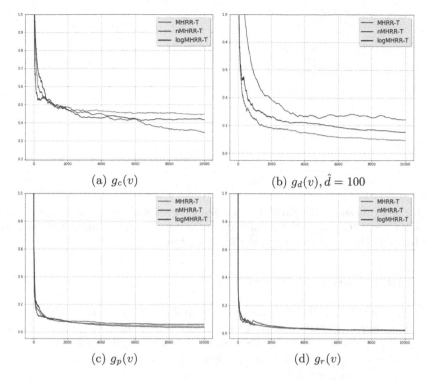

(a) $g_c(v)$ (b) $g_d(v), \hat{d} = 100$

(c) $g_p(v)$ (d) $g_r(v)$

Fig. 6. Comparison of NRMSE vs n curves for different resetting instants selection strategy.

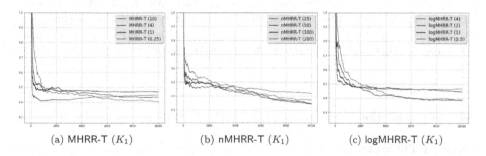

(a) MHRR-T (K_1) (b) nMHRR-T (K_1) (c) logMHRR-T (K_1)

Fig. 7. Variation of NRMSE vs n curves for the algorithms with variation in the parameter K_1 (value of K_1)

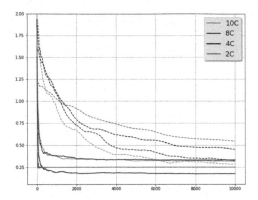

Fig. 8. The figure shows the NRMSE vs n curves for MHRR and MH algorithms for the graphs 10C, 8C, 4C, and 2C. The dashed curves represent the vanilla MH algorithm while the solid curves represent the MHRR algorithm.

6   Conclusions and future Work

We have proposed a scheme for speeding up MCMC by introducing random resets that are rare in the sense that their relative frequency $\eta(n) :=$ the fraction of times the chain was reset till time n, vanishes in the $n \uparrow \infty$ limit. We also considered two variants and provided numerical experiments for the original scheme and the variants.

We also provide an a.s. convergence result that establishes the consistency of the scheme and a finite time error bound, and interpret the expression for the latter. Nevertheless, our analysis does not quite capture the detailed nature of exactly how the resets aid the speed up. We give an intuitive interpretation for this that may serve as the basis for subsequent analysis. For this purpose, we use the variant where the reset is done at each n with a fixed reset mechanism, but with the reset probability slowly decreasing to zero.

A   Appendix

Here we provide the proofs of Theorem 1 and 2 in Sect. 2.

Proof of Theorem 1:

By the martingale law of large numbers, we have

$$\zeta_n(j) := \frac{1}{n} \sum_{m=1}^{n} (I\{X_m = j\} - \sum_i q_{m-1}(j|i)I\{X_{m-1} = i\}) \to 0$$

a.s. $\forall j \in S$. Combining this with (1), we have

$$\frac{1}{n} \sum_{m=1}^{n} (I\{X_m = j\} - \sum_i p(j|i)I\{X_{m-1} = i\}) \to 0$$

a.s. Hence any limit point ν^* of $\{\nu_n\}$ as $n \to \infty$ satisfies

$$\nu^*(j) = \sum_i \nu^*(i)p(j|i) \ \forall \ j \in S.$$

This implies (2). □

Proof of Theorem 2:

By (3), we have,

$$\|\nu_n - \pi\| \leq C\|\nu_n - \nu_n P\|. \tag{4}$$

Note that

$$\nu_n(j) - \sum_i p(j|i)\nu_n(i)$$

$$= \frac{1}{n}\sum_{m=1}^n I\{X_m = j\} - \frac{1}{n}\sum_{m=1}^n \sum_i p(j|i)I\{X_m = i\}$$

$$= \frac{1}{n}\sum_{m=1}^n I\{X_m = j\} - \frac{1}{n}\sum_{m=0}^{n-1}\sum_i q_m(j|i)I\{X_m = i\}$$

$$+ \frac{1}{n}\sum_{m=1}^n \sum_i (q_m(j|i) - p(j|i))I\{X_m = i\}$$

$$+ \frac{1}{n}\left(\sum_i q_n(j|i)I\{X_0 = i\} - \sum_i q_n(j|i)I\{X_n = i\}\right)$$

$$= \zeta_n(j) + \frac{1}{n}\sum_{m=1}^n (q_m(j|X_m) - p(j|X_m)) + \frac{1}{n}(q_0(j|X_0) - q_n(j|X_n)).$$

Bounding each term, we get

$$\left|\nu_n(j) - \sum_i p(j|i)\nu_n(i)\right| \leq |\zeta_n(j)| + \frac{1}{n} + \frac{\eta(n)}{n}.$$

Hence

$$\|\nu_n - \nu_n P\| \leq \|\zeta_n\| + \sqrt{s}\left(\frac{\eta(n)+1}{n}\right),$$

and

$$\|\nu_n - \pi\| \leq C\left[\|\zeta_n\| + \sqrt{s}\left(\frac{\eta(n)+1}{n}\right)\right].$$

For $j \in S$ and $\bar{x} := [x_0, \cdots, x_n]$, let

$$f(\bar{x}) := \frac{1}{n}\sum_{m=1}^n (I\{x_m = j\} - \sum_i p(j|i)I\{x_{m-1} = i\}).$$

Defining $\bar{y}$ analogously, note that $f(\bar{x}) - f(\bar{y}) \leq \frac{2}{n}\sum_{m=1}^{n} I\{x_i \neq y_i\}$. By Corollary 2.10 of [20], we then have, for any $\delta > 0$,

$$P\left(|\zeta_n(j)| < \frac{\delta}{C\sqrt{s}}\right) \geq 1 - 2e^{-\frac{\delta^2 n}{2sC^2\tau_{min}}}.$$

This proves the claim.   □

References

1. Anily, S., Federgruen, A.: Simulated annealing methods with general acceptance probabilities. J. Appl. Prob. **24**(3), 657–667 (1987)
2. Apers, S., Sarlette, A., Ticozzi, F.: Characterizing limits and opportunities in speeding up Markov chain mixing. Stochast. Process. Appl. **136**, 145–191 (2021)
3. Avrachenkov, K., Borkar, V.S., Kadavankandy, A., Sreedharan, J.K.: Revisiting random walk based sampling in networks: Evasion of burn-in period and frequent regenerations. Comput. Social Netw. **5**(4), 1–19 (2018)
4. Avrachenkov, K., Ribeiro, B., Towsley, D.: Improving random walk estimation accuracy with uniform restarts. In: Kumar, R., Sivakumar, D. (eds.) WAW 2010. LNCS, vol. 6516, pp. 98–109. Springer, Heidelberg (2010). https://doi.org/10.1007/978-3-642-18009-5_10
5. Blondel, V.D., Guillaume, J.L., Lambiotte, R., Lefebvre, E.: Fast unfolding of communities in large networks. J. Stat. Mech. Theory Exp. **2008**(10), P10008 (2008)
6. Collet, P., Martinéz, S., San Martin, J.: Quasi-Stationary Distributions: Markov Chains, Diffusions and Dynamical Systems. Probability and Its Applications, Springer, Heidelberg (2012). https://doi.org/10.1007/978-3-642-33131-2
7. Cooper, C., Radzik, T., Siantos, Y.: Fast low-cost estimation of network properties using random walks. Internet Math. **12**(4), 221–238 (2016)
8. Deuflhard, P., Huisinga, W., Fischer, A., Schütte, C.: Identification of almost invariant aggregates in reversible nearly uncoupled Markov chains. Linear Algebra Appl. **315**(1), 39–59 (2000)
9. Evans, M.R., Majumdar, S.N., Schehr, G.: Stochastic resetting and applications. J. Phys. A Math. Theor. **53**(19), 193001 (2020)
10. Giles, C., Bollacker, K., Lawrence, S.: Citeseer: an automatic citation indexing system. In: Proceedings of 3rd ACM Conference on Digital Libraries (2000)
11. Guan, Y., Krone, S.N.: Small-world MCMC and convergence to multi-modal distributions: from slow mixing to fast mixing. Ann. Appl. Prob. **17**(1), 284–304 (2007)
12. Hagberg, A.A., Schult, D.A., Swart, P.J.: Exploring network structure, dynamics, and function using networkx. In: Varoquaux, G., Vaught, T., Millman, J. (eds.) Proceedings of the 7th Python in Science Conference, Pasadena, CA USA, pp. 11–15 (2008)
13. Hajek, B.: Cooling schedules for optimal annealing. Math. Oper. Res. **13**(2), 311 329 (1988)
14. Heckathorn, D.D.: Respondent-driven sampling: a new approach to the study of hidden populations. Social Prob. **44**(2), 174–199 (1997)
15. Holley, R., Stroock, D.: Simulated annealing via Sobolev inequalities. Commun. Math. Phys. **115**(4), 553–569 (1988)

16. Langville, A.N., Meyer, C.D.: Deeper inside pagerank. Internet Math. **1**(3), 335–380 (2004)
17. Leskovec, J., Mcauley, J.: Learning to discover social circles in ego networks. In: Pereira, F., Burges, C.J.C., Bottou, L., Weinberger, K.Q. (eds.) Advances in Neural Information Processing Systems, vol. 25. Curran Associates, Inc. (2012)
18. Metropolis, N., Rosenbluth, A.W., Rosenbluth, M.N., Teller, A.H., Teller, E.: Equation of state calculations by fast computing machines. J. Chem. Phys. **21**(6), 1087–1092 (1953)
19. Montenegro, R., Tetali, P.: Mathematical aspects of mixing times in Markov chains. Found. Trends Theor. Comput. Sci. **1**(3), 237–354 (2006)
20. Paulin, D.: Concentration inequalities for Markov chains by Marton couplings and spectral methods. Electron. J. Prob. **20**, 1–32 (2015)
21. Ribeiro, B., Towsley, D.: Estimating and sampling graphs with multidimensional random walks. In: Proceedings of the 10th ACM SIGCOMM Conference on Internet Measurement, IMC '10, pp. 390–403. Association for Computing Machinery, New York (2010)
22. Robert, C., Elvira, V., Tawn, N., Wu, C.: Accelerating MCMC algorithms. Wiley Interdisc. Rev. Comput. Stat. **10**, e1435 (2018)
23. Tsitsiklis, J.: Markov chains with rare transitions and simulated annealing. Math. Oper. Res. **14**, 70–90 (1989)

First Frontier Monotonicity for Fluid Models of Multiclass EDF Queueing Networks

Łukasz Kruk[✉][iD]

Institute of Mathematics, Maria Curie-Skłodowska University, Lublin, Poland
lkruk@hektor.umcs.lublin.pl

Abstract. We investigate fluid models of subcritical Earliest Deadline First (EDF) multiclass queueing networks with soft deadlines. For any such model, we show that after a time proportional to the size of the initial condition, the left endpoint of the cumulative fluid mass lead time distribution, called the first frontier, is nondecreasing. Moreover, in the strictly subcritical case, the first frontier actually increases as long as there is fluid mass in the system. Stability of strictly subcritical EDF fluid models and weak stability of subcritical EDF fluid models follow from the above findings as corollaries.

Keywords: Earliest Deadline First · Fluid model · Stability

1 Introduction

In the theory of real-time queueing systems, jobs with individual timing requirements (deadlines) are considered. Several kinds of such systems, reacting differently to customer deadline misses, are known. If deadlines are hard, even a single miss results in a system failure. A firm deadline can be missed, but a late task is considered worthless and hence it is removed from the system, either by the corresponding customer (reneging), or by the system controller. Finally, systems with soft deadlines permit lateness, although they may attempt to minimize it, and use the late jobs. Applications of various real-time queueing models include telecommunication networks, voice and video transmission services, manufacturing systems, real-time vehicle control or scheduling of medical services.

A natural service discipline in the above setting is Earliest Deadline First (EDF), under which the job with the shortest lead time, defined as the difference between its deadline and the current time, is selected for service. For single server, single customer class queueing systems, EDF is known to be optimal with respect to handling real-time service requests, according to a number of performance criteria, see Liu and Layland [25], Panwar and Towsley [27,28], Moyal [26] or Kruk et al. [20].

© ICST Institute for Computer Sciences, Social Informatics and Telecommunications Engineering 2021
Published by Springer Nature Switzerland AG 2021. All Rights Reserved
Q. Zhao and L. Xia (Eds.): VALUETOOLS 2021, LNICST 404, pp. 126–143, 2021.
https://doi.org/10.1007/978-3-030-92511-6_8

To our knowledge, the first informal analysis of stochastic EDF queueing system asymptotics was due to Lehoczky [22–24] and the first mathematically rigorous paper on this topic was the heavy traffic analysis of a G/G/1 EDF queue with soft deadlines by Doytchinov, Lehoczky and Shreve [13]. The latter result was generalized to multiclass feedforward networks by Yeung and Lehoczky [32], while Kruk et al. [20] provided its counterpart for a G/G/1 EDF queue with reneging. Kruk et al. [21] investigated a possibility of a further generalization of the results of [13,32] to multiclass acyclic networks.

Recall that in classical open Jackson networks, the customer interarrival and service times are independent, exponentially distributed and each station serves one customer class (in other words, the jobs at every server are homogeneous). In generalized open Jackson networks, there are renewal arrival processes and independent, identically distributed service times which need not follow exponential distributions, but the customer population at each server is homogeneous, as in the previous case. In this sense, traditional and generalized Jackson networks are single-class models. See, e.g., Chen and Yao [9] for more details. Open multiclass queueing networks, as considered e.g., in Harrison [14], further generalize both these notions by introducing many job classes, differing in their arrival processes, service time distributions and routing through the network, with a many-to-one relation between customer classes and servers. Queueing systems of this type arise in many application areas, e.g., manufacturing and communication networks, service operations or multiprocessor computer systems. Asymptotic theory of multiclass networks is notably more difficult than its counterpart for single-class systems. In particular, it is well known that a strictly subcritical stochastic multiclass queueing network may be unstable, see Rybko and Stolyar [29] or Seidman [30].

Fluid limits, arising from the temporal and spatial scaling of the system's performance processes by the same sequence of constants, and the corresponding fluid models, i.e., formal functional law of large numbers approximations, of EDF queueing systems were investigated by several authors. Decreusefond and Moyal [12] and Atar, Biswas and Kaspi [1] derived fluid limits for nonpreemptive EDF queues with firm deadlines. Atar et al. [3] proposed a unified framework, based on the measure-valued Skorokhod map, for establishing convergence to fluid approximations for queues under a number of service disciplines, including EDF. The scope of their method was extended to a many server EDF queue by Atar, Biswas and Kaspi [2], and recently to generalized EDF Jackson networks, with soft of firm deadlines, by Atar and Shadmi [4].

Results explaining asymptotic behavior of multiclass EDF queueing networks are still in short supply. It is known that strictly subcritical networks with firm deadlines are stable under a wide range of service disciplines, including EDF [15]. In the case of soft deadlines, the following facts were established. Bramson [8] demonstrated that subsequences of suitably rescaled sample paths of the performance processes for an EDF network without preemption converge to fluid limits satisfying the First In System, First Out (FISFO) fluid model equations. Note that the FISFO protocol is a special case of EDF with all the customer initial lead times being equal (e.g., zero). Bramson [8] also showed that a class of strictly subcritical FISFO fluid models satisfying some additional technical assumptions is stable. Findings of this kind are important, because in many cases stability of fluid models implies stability of the underlying stochastic networks, see Dai [10]. In particular, the results from [8] imply stability of nonpreemptive, strictly subcritical EDF networks. The corresponding results for preemptive multiclass EDF queueing networks were provided by Kruk [15,18], who also extended Bramson's fluid model stability result to arbitrary strictly subcritical FISFO fluid models. Little is known about asymptotics of subcritical (in particular, critical) multiclass EDF, or even FISFO, fluid models. Their behavior seems to be well understood only in the feedforward case, in which such a model converges to its invariant manifold (Kruk [16,17]).

In this paper, we investigate subcritical EDF fluid models of multiclass queueing networks with soft deadlines. Our main object of study is the left endpoint of the cumulative lead time distribution corresponding to the current fluid model state, which we call the first frontier. Our main result states that after a time period proportional to the size of the initial condition, the first frontier in a subcritical EDF fluid model is nondecreasing. Moreover, in the strictly subcritical case, it actually increases at a rate bounded below by some positive constant, depending on the model primitives, as long as there is fluid mass in the system. These findings readily imply stability of strictly subcritical EDF fluid models, with the first frontier acting as the Lyapunov function. (In contrast, in previous work on convergence to equilibria for fluid models of First-In, First-Out (FIFO) networks of Kelly type [5] and head-of-the-line proportional processor sharing networks [6], suitably defined entropies were used as the corresponding Lyapunov functions.) In particular, we provide a new, simpler and perhaps more intuitive proof of the FISFO fluid model stability results from [8,18]. Let us mention, however, that although the original fluid model stability proof from [8] was notably different from ours, it was actually based on a similar observation of "the ability of the FISFO discipline to avoid idling near its oldest customer" ([8, p. 88]). Another simple consequence of our main theorem is weak stability of arbitrary subcritical EDF fluid models. The latter result, related to so-called rate stability of the underlying stochastic network, appears to be new even in the FISFO case.

Our approach was inspired by a recent article [19], in which locally edge minimal fluid models for real-time resource sharing networks were investigated. Such fluid models may be regarded, in some sense, as fluid counterparts of EDF networks with shared resources. A crucial difference between resource sharing networks and multiclass queueing networks considered in this paper is that tasks in a resource sharing network require access to all the resources along their routes at the same time, while customers of a multiclass network are being served by different stations along their routes in succession. The main result of [19] is convergence of subcritical locally edge minimal resource sharing fluid models to the corresponding invariant manifold. The main idea of its proof was to show that the frontiers of such a fluid model increase, until they stabilize at fixed levels. We hope that under suitable assumptions, this approach can be adapted to fluid models of multiclass EDF queueing networks, opening the way to a more satisfactory theory of diffusion approximations for such networks, in the spirit of Bramson [5–7] and Williams [31]. Our present paper may be considered as a step in this direction.

This paper is organized as follows. In Sect. 2, we briefly describe multiclass EDF queueing networks and we state a definition of the corresponding EDF fluid models. In Sect. 3, preliminary analysis of EDF fluid models is provided. Section 4 contains our main results and their proofs. Section 5 concludes.

2 EDF Queueing Networks and their Fluid Models

2.1 Basic Notation

The following notation will be used throughout the paper. Let $\mathbb{N} = \{1, 2, \ldots\}$ and let $\mathbb{R}$ denote the set of real numbers. The Borel σ-field on $\mathbb{R}$ will be denoted by $\mathcal{B}(\mathbb{R})$. For $a, b \in \mathbb{R}$, we write $a \vee b$ for the maximum of a and b, $a \wedge b$ for the minimum of a and b and a^+ for $a \vee 0$, respectively. The infimum taken over the empty set should be interpreted as ∞. For $n \in \mathbb{N}$, we write $\mathbb{R}^n$ to denote the n-dimensional Euclidean space. All vectors in the paper are to be interpreted as column vectors. For a vector $a = (a_1, ..., a_n) \in \mathbb{R}^n$, let $|a| \triangleq \sum_{i=1}^n |a_i|$. For $a, b \in \mathbb{R}^n$, $a = (a_1, ..., a_n)$, $b = (b_1, ..., b_n)$, the vector $(a_1 b_1, ..., a_n b_n)$ will be denoted by $a \circ b$. Vector inequalities should be interpreted componentwise, e.g., for $a \in \mathbb{R}^n$, $a = (a_1, ..., a_n)$, we write $a \geq 0$ if $a_1 \geq 0, ..., a_n \geq 0$. For a matrix A, A' denotes the transpose of A. In matrix calculations, I denotes the identity matrix. If $A = [a_{ij}]$ is a square $n \times n$ matrix, then $\|A\| = \max_{a \in \mathbb{R}^n : |a|=1} |Aa|$ is the matrix norm of A. It is easy to check that, due to our choice of the l^1 norm in $\mathbb{R}^n$, we have

$$\|A\| = \sum_{i=1}^n \max_{j=1,...,n} |a_{ij}|. \tag{1}$$

2.2 Stochastic EDF Networks

This paper investigates asymptotic properties of a family of fluid models corresponding to open multiclass queueing networks with the EDF service discipline. To motivate the introduction of these fluid models, we first provide a brief description of the corresponding queueing networks.

We consider a network consisting of J single server stations, indexed by $j = 1, ..., J$. The network is populated by K customer classes, indexed by $k = 1, ..., K$. There is a stationary external arrival process with rate α_k associated with each class k. In particular, if $\alpha_k = 0$, there are no external arrivals to class k. We put $\alpha = (\alpha_1, ..., \alpha_K)$. A customer of class k receives service at a unique station j, written $k \in \mathcal{C}(j)$ or $j = s(k)$. Let m_k be the mean service time for the class k and let $m = (m_1, ..., m_K)$. Upon being served at j, a customer of class k immediately becomes a customer of class l with probability p_{kl}, independently of its past history. Thus, the probability that a customer of class k leaves the network after completion of service equals $1 - \sum_{l=1}^{K} p_{kl}$. The routing matrix $P = (p_{kl})$ is assumed to be transient, i.e., such that the matrix

$$\Theta = (\theta_{kl}) \triangleq (I - P')^{-1} = I + P' + (P')^2 + \dots \tag{2}$$

exists. We define the *total arrival rate* vector

$$\lambda = (\lambda_1, ..., \lambda_K) = \Theta \alpha. \tag{3}$$

Without loss of generality we assume that $\lambda_k > 0$ for each k. Next, we define the *traffic intensity* at station j as

$$\rho_j = \sum_{k \in \mathcal{C}(j)} m_k \lambda_k. \tag{4}$$

When $\rho_j \leq 1$ ($\rho_j < 1$, $\rho_j = 1$) for each j, the network is called *subcritical* (*strictly subcritical*, *critical*).

Class k customers entering the network have nonnegative *initial lead times* with cumulative distribution function (c.d.f.) G_k. Note that

$$G_k(x) = 0, \qquad\qquad x < 0. \tag{5}$$

For notational convenience, we define G_k for every $k = 1, ..., K$, including classes with no external arrival streams. For k such that $\alpha_k = 0$, G_k may be chosen in an arbitrary way, subject to (5), and this choice not affect any further considerations. We put $G = (G_1, ..., G_K)$. To simplify the presentation, we assume that

$$y_k^* \triangleq \sup\{y \in \mathbb{R} : G_k(y) < 1\} < \infty, \qquad k = 1, ..., K. \tag{6}$$

To determine whether customers meet their timing requirements, one must keep track of each customer's lead time, where

$$\text{lead time} = \text{initial lead time} - \text{time elapsed since arrival}$$

for customers coming to the system after time zero and

$$\text{lead time} = \text{initial lead time} - \text{current time}$$

for *initial customers*, i.e., those who are present in the network at time zero.

Customers are served at each station according to the EDF discipline. That is, the customer with the shortest remaining lead time, regardless of class, is selected for service at each station. Late customers (customers with negative lead times) stay in the system until served to completion. Two types of EDF network protocols may be considered. In the *preemptive* case preemption occurs when a customer more urgent than the customer in service arrives (we assume preempt-resume and no set up, switch-over or other type of overhead). In EDF networks *without preemption* customer service continues until he is served to completion, even if a more urgent customer enters the station.

2.3   EDF Fluid Models

Fluid models are deterministic, continuous analogs of queueing networks, in which individual customers are replaced by a divisible commodity (fluid) of K classes, indexed by $k = 1, ..., K$, which change as the fluid moves between stations $j = 1, ..., J$ until it leaves the system. In analogy with customers of the corresponding queueing networks, class k fluid arrives exogeneously to a unique station $j = s(k)$ with rate α_k and initial lead time distribution G_k, it is processed at $s(k)$ with mean service time m_k and changes class to l with transition probability p_{kl} after service completion. As in the case of queueing networks, we say that a fluid model is *subcritical* (*strictly subcritical*, *critical*) if $\rho_j \leq 1$ ($\rho_j < 1$, $\rho_j = 1$) for each j, where ρ_j are given by (4). Fluid models are defined rigorously in terms of the appropriate *fluid model equations*.

Fluid models for EDF queueing networks consist of the six-tuples of vectors

$$\mathcal{X}(t, s) = (Z(t, s), W(t, s), A(t, s), D(t, s), T(t, s), Y(t, s)), \quad t \geq 0, \ s \in \mathbb{R}, \quad (7)$$

where the vectors $Z(t, s), W(t, s), A(t, s), D(t, s), T(t, s)$ are indexed by $k = 1, ..., K$ and the vector $Y(t, s)$ is indexed by $j = 1, ..., J$. Here $Z_k(t, s)$ denotes the amount of class k fluid with lead times less than or equal to s at time t and $W_k(t, s)$ represents the workload for station $s(k)$ associated with this fluid, i.e., the amount of time necessary for the server $s(k)$ to process it to completion (provided that the station devotes all its capacity to it, without processing any other fluids at the same time). The quantity $A_k(t, s)$ $(D_k(t, s))$ denotes the amount of fluid with lead times at time t less than or equal to s which has arrived at (departed from) class k by time t and $T_k(t, s)$ represents the amount of work devoted to this fluid by server $s(k)$ by time t. Finally, $Y_j(t, s)$ denotes the cumulative idleness by time t at station j with regard to service of fluids with lead times at time t less than or equal to s. The vectors defining $\mathcal{X}$ are the continuous analogs of the corresponding quantities in the EDF queueing network. We assume that all the components of $\mathcal{X}$ are continuous and nonnegative, with $A(\cdot, s - \cdot)$, $D(\cdot, s - \cdot)$, $T(\cdot, s - \cdot)$, $Y(\cdot, s - \cdot)$ nondecreasing in each coordinate, $A(0, s) = D(0, s) = T(0, s) = 0$ and $Y(0, s) = 0$ for $s \in \mathbb{R}$. Similarly, we assume that every coordinate of $A(t, \cdot)$, $D(t, \cdot)$, $T(t, \cdot)$, $-Y(t, \cdot)$, $Z(t, \cdot)$ and $W(t, \cdot)$ is nondecreasing for all $t \geq 0$.

The *EDF fluid model equations*, defining the model, are:

$$A(t, s) = \alpha \circ \int_0^t G(s + \eta) \, d\eta + P' D(t, s), \tag{8}$$

$$Z(t, s) = Z(0, t + s) + A(t, s) - D(t, s), \tag{9}$$

$$T(t, s) = m \circ D(t, s), \tag{10}$$

$$\sum_{k \in \mathcal{C}(j)} T_k(t, s) + Y_j(t, s) = t, \qquad j = 1, ..., J, \tag{11}$$

$$Y_j(t, s - t) \text{ can only increase in } t \text{ if } \sum_{k \in \mathcal{C}(j)} Z_k(t, s - t) = 0, \ j = 1, ..., J, \tag{12}$$

$$W(t, s) = m \circ Z(t, s), \tag{13}$$

where $t \geq 0$, $s \in \mathbb{R}$. A system (7) satisfying the Eqs. (8)–(13) will be called an *EDF fluid model*. The terms α, m, P, G and $\mathcal{C}(j)$, $j = 1, ..., J$, are the model data, given in advance. The above notions were introduced in [16], where more information, including a heuristic explanation of the form $\alpha \circ \int_0^t G(s + \eta) \, d\eta$ of the external arrival process in (8), may be found. Note that while the Eqs. (8)–(9), (11) hold regardless of the underlying scheduling protocol and the Eqs. (10), (13) are satisfied for fluid models of networks under various service disciplines, the Eq. (12) actually characterizes the EDF policy.

An important special case of the EDF fluid model equations may be obtained by putting $G_k(y) = \mathbb{I}_{[0,\infty)}(y)$ for each k, so that $y_k^* = 0$, $k = 1, ..., K$ (see (6)), and (8) takes the form

$$A(t, s) = \alpha \left(t + (s \wedge 0) \right)^+ + P'D(t, s). \tag{14}$$

The equations (9)–(14) will be referred to as the *FISFO fluid model equations*. If we change the coordinates (t, s) to $(t, \tilde{s})$, where $\tilde{s} = s - t$, in (9)–(12), (14), we obtain the FISFO fluid model equations introduced by Bramson [8]:

$$\overline{A}(t, \tilde{s}) = \alpha(t \wedge \tilde{s}) + P'\overline{D}(t, \tilde{s}), \tag{15}$$

$$\overline{Z}(t, \tilde{s}) = \overline{Z}(0, \tilde{s}) + \overline{A}(t, \tilde{s}) - \overline{D}(t, \tilde{s}), \tag{16}$$

$$\overline{D}_k(t, \tilde{s}) = \overline{T}_k(t, \tilde{s})/m_k, \qquad k = 1, ..., K, \tag{17}$$

$$\sum_{k \in \mathcal{C}(j)} \overline{T}_k(t, \tilde{s}) + \overline{Y}_j(t, \tilde{s}) = t, \quad j = 1, ..., J, \tag{18}$$

$$\overline{Y}_j(t, \tilde{s}) \text{ can only increase in } t \text{ when } \sum_{k \in \mathcal{C}(j)} \overline{Z}_k(t, \tilde{s}) = 0, \tag{19}$$

for $t, \tilde{s} \geq 0$. In (15)–(19), the coordinate $\tilde{s}$ represents the arrival times of customers (fluids) to the network, rather than their lead times.

If we take *fluid limits*, i.e., the limits of sample paths along subsequences under scaling which is linear in both time and space (called *fluid* or *hydrodynamic scaling*) obtained from a single EDF network, then the initial lead time distributions disappear in the limit, giving rise to the FISFO fluid models. This was shown in [8] and [15,18] for nonpreemptive and preemptive EDF networks, respectively. Here, following [16,17], we study somewhat more general EDF fluid models, satisfying (8)–(13) with nontrivial lead time distributions G_k. The latter setup may turn out to be useful in asymptotic analysis of a *sequence* of EDF networks in which the initial lead time distributions dilate with the same rate as the space scaling parameter. Because of such lead time scaling, used e.g., in [13,20,21,32], the customer lead times are more "realistic", i.e., they are of the same order as the queue lengths and the sojourn times.

Let

$$\mathbf{K} = \left\{ (k_1, ..., k_n) : n \in \mathbb{N}, k_1, ..., k_n \in \{1, ..., K\}, \alpha_{k_1} p_{k_1 k_2} ... p_{k_{n-1} k_n} > 0 \right\},$$

where $p_{k_1 k_2} ... p_{k_{n-1} k_n}$ should be interpreted as 1 for $n = 1$. The elements of $\mathbf{K}$ will be called *multi-indices*. They represent paths of finite length which are being followed with positive probability by customers (fluids) since their arrival to the network. For $\mathbf{k} = (k_1, ..., k_n) \in \mathbf{K}$, let $p_{\mathbf{k}} = p_{k_1 k_2} ... p_{k_{n-1} k_n}$ and $\alpha_{\mathbf{k}} = \alpha_{k_1} p_{\mathbf{k}}$. Also, for $\mathbf{k}$ as above, let $b(\mathbf{k}) = k_1$ and $e(\mathbf{k}) = k_n$ be the beginning and the end of the path $\mathbf{k}$, respectively. For $\mathbf{k} \in \mathbf{K}$, $k \in \{1, ..., K\}$ and $j \in \{1, ..., J\}$, we write $\mathbf{k} \in \tilde{C}(k)$ if $e(\mathbf{k}) = k$ and $\mathbf{k} \in \overline{C}(j)$ if $e(\mathbf{k}) \in C(j)$.

In what follows, we fix an EDF queueing network under consideration, which, for the sake of construction of the corresponding fluid models, is completely determined by α, m, P, G, together with the sets $C(j)$, $j = 1, ..., J$. We also assume a condition compatible to (6) on the initial states under consideration: there exists a constant $C_0 < \infty$ such that

$$Z(0, C_0) = \lim_{s \to \infty} Z(0, s). \tag{20}$$

In particular, the support of the measure with the distribution function $Z_k(0, \cdot)$ is bounded above by C_0 for each k. This is the case, for example, if all the mass present in the system at time zero has arrived at some prior times with initial lead time distributions G_l, $l = 1, ..., K$, upon arrival and $C_0 = \max_{k:\alpha_k > 0} y_k^*$. In most cases, without loss of generality we may additionally assume that

$$C_0 \geq \max_{k:\alpha_k > 0} y_k^*. \tag{21}$$

3   Preliminary Analysis

Consider an EDF fluid model and let $t \geq 0$, $s \in \mathbb{R}$. From (9) we have

$$D(t, s) = Z(0, t + s) - Z(t, s) + A(t, s). \tag{22}$$

Plugging this to (8), we obtain

$$(I - P')A(t, s) = \alpha \circ \int_0^t G(s + \eta) \, d\eta + P' \left(Z(0, t + s) - Z(t, s) \right),$$

which, after multiplication by Θ from the left and using (2), yields

$$A(t, s) = \Theta \left(\alpha \circ \int_0^t G(s + \eta) \, d\eta + P' \left(Z(0, t + s) - Z(t, s) \right) \right). \tag{23}$$

This, in turn, together with (22) and the identity

$$I + \Theta P' = \Theta \tag{24}$$

resulting from (2), implies

$$D(t, s) = \Theta \left(\alpha \circ \int_0^t G(s + \eta) \, d\eta + Z(0, t + s) - Z(t, s) \right). \tag{25}$$

It is not hard to see that for any EDF fluid model $\mathfrak{X}$, the function $\mathfrak{X}(t, s - t)$ is Lipschitz continuous in t for all $t \geq 0$, $s \in \mathbb{R}$. The relations (20)–(21), (25) and monotonicity of $D(t, \cdot)$, $Z(t, \cdot)$ imply that $D(t, s)$ is Lipschitz continuous in s as long as $t + s \geq C_0$. This can be proved as in [17, p. 543]. Consequently, $\mathcal{X}(t, s)$ is Lipschitz continuous in both t and s for $t + s \geq C_0$. Under the additional assumption that $Z(0, \cdot)$ is Lipschitz continuous, $\mathcal{X}(t, s)$ is Lipschitz in both variables for all $t \geq 0$, $s \in \mathbb{R}$.

We will now make an intuitively clear observation that the conditions (6) and (20)–(21) imply the absence of mass with lead times greater than C_0 in the system.

Lemma 1. *For* $k = 1, ... K$, $t \geq 0$ *and* $s \geq C_0$, *we have*

$$D_k(t, s) = D_k(t, C_0), \tag{26}$$
$$Z_k(t, s) = Z_k(t, C_0). \tag{27}$$

PROOF: Fix $k = 1, ... K$, $t \geq 0$, $s \geq C_0$. Let $l \in \{1, ..., K\}$ be such that $\theta_{kl} \alpha_l > 0$ (or, equivalently, there exists $\mathbf{k} \in \mathbf{K}$ with $b(\mathbf{k}) = l$ and $e(\mathbf{k}) = k$). By (6) and (21), we have

$$\int_0^t G_l(s + \eta) \, d\eta = \int_0^t G_l(C_0 + \eta) \, d\eta. \tag{28}$$

Now let $l \in \{1, ..., K\}$ be arbitrary. By (20), we have

$$Z_l(0, t + s) = Z_l(0, t + C_0). \tag{29}$$

The Eqs. (28)–(29), together with (25) and the fact that $D_k(t, \cdot)$ is nondecreasing, implies that $\{\Theta Z(t, s)\}_k \leq \{\Theta Z(t, C_0)\}_k$. However, $Z(t, \cdot)$ is also nondecreasing and all the entries of Θ are nonnegative by (2), so

$$Z(t, s) \geq Z(t, C_0) \tag{30}$$

and $\Theta Z(t, s) \geq \Theta Z(t, C_0)$. We have obtained

$$\{\Theta Z(t, s)\}_k = \{\Theta Z(t, C_0)\}_k. \tag{31}$$

The Eqs. (25), (28)–(29) and (31), imply (26). Finally, (30), together with (2) and (31), proves (27).   □

For $k = 1, ... K$ and $t \geq 0$, define the fluid queue lengths

$$Q_k(t) = \lim_{s \to \infty} Z_k(t, s) = Z_k(t, C_0), \tag{32}$$

where the last equality follows from Lemma 1. Let $Q(t) = (Q_1(t), ..., Q_K(t))$.

For $k = 1, ..., K$ and $y \in \mathbb{R}$, let us define $H_k(y) = \int_y^\infty (1 - G_k(\eta)) \, d\eta$. Each function H_k is finite by (6). We put $H = (H_1, ..., H_K)$. The following lemma generalizes the second part of Proposition 4.1 from [16] to EDF fluid models which are not necessarily invariant.

Lemma 2. *For* $k = 1, ...K$, $s_1 \leq s_2$ *and* $t \geq C_0 - s_1$, *the condition*

$$D_k(t, s_2) = D_k(t, s_1) \tag{33}$$

implies

$$\{\Theta\left(Z(t, s_2) - Z(t, s_1)\right)\}_k = \{\Theta\left(\alpha \circ [H(s_1) - H(s_2)]\right)\}_k. \tag{34}$$

PROOF: Fix $s_1 \leq s_2$ and let $t \geq 0$. From (25), we get

$$\Theta\left(Z(t, s_2) - Z(t, s_1)\right) = \Theta\left(Z(0, t + s_2) - Z(0, t + s_1)\right)$$
$$+ \Theta\left(\alpha \circ \left(\int_{t+s_1}^{t+s_2} G(\eta) \, d\eta - \int_{s_1}^{s_2} G(\eta) \, d\eta\right)\right)$$
$$- (D(t, s_2) - D(t, s_1)).$$

This, together with (6) and (20)–(21), implies that for $t \geq C_0 - s_1$,

$$\Theta\left(Z(t, s_2) - Z(t, s_1)\right) = \Theta\left(\alpha \circ [H(s_1) - H(s_2)]\right) - (D(t, s_2) - D(t, s_1)). \tag{35}$$

The Eq. (34) is an immediate consequence of (33) and (35). □

The following proposition is the main result of this section. It implies that after a time period proportional to the size of the initial condition, the initial fluid mass leaves the system.

Proposition 1. *Let* $m_0 = \min_{k=1,...,K} m_k$. *There exists a time* Υ *such that*

$$\Upsilon \leq C_0 + m_0(|\Theta Q(0)| + |\lambda| C_0), \tag{36}$$
$$Z(t, C_0 - t) = 0, \qquad t \geq \Upsilon. \tag{37}$$

PROOF: Let

$$\Upsilon = \inf\{t \geq C_0 : Z(t, C_0 - t) = 0\}. \tag{38}$$

We will first prove the bound (36). If $\Upsilon = C_0$, there is nothing to prove. Assume that $\Upsilon > C_0$. For every $t \in [C_0, \Upsilon)$, there exists $k \in \{1, ..., K\}$ such that $Z_k(t, C_0 - t) > 0$. By continuity of the fluid model, for some $\epsilon > 0$ we have $Z_k(\tilde{t}, C_0 - \tilde{t}) > 0$ for every $\tilde{t} \in [C_0, \Upsilon) \cap (t - \epsilon, t + \epsilon)$ and thus, by (12), for $j = s(k)$, the function $Y_j(\cdot, C_0 - \cdot)$ does not increase on $[C_0, \Upsilon) \cap (t - \epsilon, t + \epsilon)$. Consequently, for any $t \in [C_0, \Upsilon)$,

$$\sum_{j=1}^{J} (Y_j(t, C_0 - t) - Y_j(C_0, 0)) \leq (J - 1)(t - C_0),$$

and thus, by (11),

$$\sum_{k=1}^{K} (T_k(t, C_0 - t) - T_k(C_0, 0)) \geq t - C_0.$$

This, in turn, together with (10), implies that for $t \in [C_0, \Upsilon)$,

$$\sum_{k=1}^{K} D_k(t, C_0 - t) \geq \sum_{k=1}^{K} (D_k(t, C_0 - t) - D_k(C_0, 0)) \geq (t - C_0)/m_0. \quad (39)$$

For $t \geq 0$, plugging $s = C_0 - t$ into (25), we get

$$\Theta Z(t, C_0 - t) = \Theta Z(0, C_0) + \Theta\left(\alpha \circ \int_0^t G(C_0 - t + \eta)d\eta\right) - D(t, C_0 - t). \quad (40)$$

By (5), for each k and $t \in [C_0, \Upsilon)$,

$$\int_0^t G_k(C_0 - t + \eta)d\eta = \int_0^t G_k(C_0 - \eta)d\eta = \int_0^{C_0} G_k(C_0 - \eta)d\eta \leq C_0. \quad (41)$$

Moreover, by (2), all the entries of Θ are nonnegative, so $\Theta Z(t, C_0 - t) \geq 0$ and the relations (3), (32), (40)–(41) imply

$$D(t, C_0 - t) \leq \Theta Q(0) + \lambda C_0. \quad (42)$$

The relations (39) and (42) imply that $t \leq C_0 + m_0(|\Theta Q(0)| + |\lambda| C_0)$ for any $t \in [C_0, \Upsilon)$, so (36) follows.

We will now justify (37). By (38) and continuity of the fluid model, we have $Z(\Upsilon, C_0 - \Upsilon) = 0$. Let $t > \Upsilon$. Using (40) twice and observing that the equalities in (41) hold for any $t \geq C_0$, we get

$$\Theta Z(t, C_0 - t) = \Theta Z(t, C_0 - t) - \Theta Z(\Upsilon, C_0 - \Upsilon) = \Theta\left(\alpha \circ \int_0^t G(C_0 - t + \eta)d\eta\right)$$

$$-\Theta\left(\alpha \circ \int_0^\Upsilon G(C_0 - \Upsilon + \eta)d\eta\right) - (D(t, C_0 - t) - D(\Upsilon, C_0 - \Upsilon))$$

$$= -(D(t, C_0 - t) - D(\Upsilon, C_0 - \Upsilon)) \leq 0.$$

However, $\Theta Z(t, C_0 - t) \geq 0$, so actually $\Theta Z(t, C_0 - t) = 0$, yielding (37) by invertibility of Θ (see (2)). $\qquad \square$

Corollary 1. *Let Υ be as in Proposition 1. For $t \geq \Upsilon$ and $s_1 \leq s_2$, we have*

$$\Theta(Z(t, s_2) - Z(t, s_1)) \leq \Theta(\alpha \circ [H(s_1) - H(s_2)]). \quad (43)$$

PROOF: If $s_1 \geq C_0 - t$, then (43) follows from (35) and monotonicity of $D(t, \cdot)$. On the other hand, for $s \leq C_0 - t$, we have $Z(t, s) = 0$ by (37) and the function H is nonincreasing, so as long as $t \geq \Upsilon$, validity of (43) for $C_0 - t \leq s_1 \leq s_2$ implies its validity for any $s_1 \leq s_2$. $\qquad \square$

Let $t \geq \Upsilon$, $s_1 \leq s_2$. By Corollary 1, (2) and monotonicity of increments of $Z(t, \cdot)$,

$$Z(t, s_2) - Z(t, s_1) \leq \Theta(Z(t, s_2) - Z(t, s_1)) \leq \Theta(\alpha \circ [H(s_1) - H(s_2)]), \quad (44)$$

i.e., for each class k,

$$Z_k(t, s_2) - Z_k(t, s_1) \leq \sum_{\mathbf{k} \in \tilde{\mathcal{C}}(k)} \alpha_{b(\mathbf{k})} p_\mathbf{k} [H_{b(\mathbf{k})}(s_1) - H_{b(\mathbf{k})}(s_2)]$$

$$= \sum_{\mathbf{k} \in \tilde{\mathcal{C}}(k)} \alpha_\mathbf{k} [H_{b(\mathbf{k})}(s_1) - H_{b(\mathbf{k})}(s_2)]. \qquad (45)$$

In particular, since $H_k \equiv 0$ on $[y_k^*, \infty)$ by (6), we have

$$Z_k(t, y_k^{**}) = \lim_{s \to \infty} Z_k(t, s), \qquad (46)$$

where

$$y_k^{**} = \max_{\mathbf{k} \in \tilde{\mathcal{C}}(k)} y_{b(\mathbf{k})}^*. \qquad (47)$$

We will use a notion of a time-shifted EDF fluid model, introduced in [17]. For $\theta \geq 0$, we define the time-shift operator Δ_θ acting on the coordinates of a fluid model $\mathfrak{X}$ of the network under consideration as follows: for $t \geq 0$, $s \in \mathbb{R}$, $\Delta_\theta Z(t, s) = Z(t+\theta, s)$, $\Delta_\theta W(t, s) = W(t+\theta, s)$, $\Delta_\theta A(t, s) = A(t+\theta, s) - A(\theta, t+s)$, $\Delta_\theta D(t, s) = D(t+\theta, s) - D(\theta, t+s)$, $\Delta_\theta T(t, s) = T(t+\theta, s) - T(\theta, t+s)$ and $\Delta_\theta Y(t, s) = Y(t+\theta, s) - Y(\theta, t+s)$. Let

$$\Delta_\theta \mathfrak{X} = (\Delta_\theta Z, \Delta_\theta W, \Delta_\theta A, \Delta_\theta D, \Delta_\theta T, \Delta_\theta Y).$$

It is not hard to see that for any $\theta \geq 0$, $\Delta_\theta \mathfrak{X}$ satisfies the EDF fluid model equations (8)–(13).

In what follows, to simplify the presentation, we will additionally assume that (44) (and consequently (45)–(47)) holds for *every* $t \geq 0$, (rather than $t \geq \Upsilon$) and $s_1 \leq s_2$. Formally, this can be achieved by considering the fluid model $\Delta_\Upsilon \mathfrak{X}$ instead of $\mathfrak{X}$. The bound (36) assures that the length of the time interval "neglected" in this way is comparable to (and controlled by) the "size" of the initial condition. The first consequence of this additional assumption is the following refinement of Lemma 1.

Lemma 3. *For $k = 1, ...K$, $t \geq 0$ and $s \geq y_k^{**}$, we have*

$$D_k(t, s) = D_k(t, y_k^{**}), \qquad (48)$$
$$Z_k(t, s) = Z_k(t, y_k^{**}). \qquad (49)$$

Note that (49) follows immediately from (46) and monotonicity of $Z_k(t, \cdot)$. The equation (48) follows from (47) and (49) by an argument similar to the justification of (26).

4   First Frontier Monotonicity and Fluid Model Stability

For $j = 1, ..., J$, let

$$\bar{y}_j = \max_{k \in \mathcal{C}(j)} y_k^{**} = \max_{\mathbf{k} \in \tilde{\mathcal{C}}(j)} y_{b(\mathbf{k})}^*, \qquad (50)$$

and let $\bar{y}_{max} = \max_{j=1,...,J} \bar{y}_j = \max_{k:\alpha_k>0} y_k^*$. For $t \geq 0$, define

$$F_j(t) = \inf \left\{ u \in \mathbb{R} : \sum_{k \in \mathcal{C}(j)} Z_k(t, u) > 0 \right\} \wedge \bar{y}_j, \qquad j = 1, ..., J,$$

$$F^{(1)}(t) = \inf \left\{ u \in \mathbb{R} : \sum_{k=1}^{K} Z_k(t, u) > 0 \right\} \wedge \bar{y}_{max},$$

$$\mathcal{J}^{(1)}(t) = \{ j \in \{1, ..., J\} : F_j(t) = F^{(1)}(t) \}.$$

The quantity $F_j(t)$ will be called the *frontier* at station j at time t. Accordingly, $F^{(1)}(t)$ will be called the *first frontier* at time t and $\mathcal{J}^{(1)}(t)$ is the set of servers with the first frontier at this time. By (49)–(50), we have $F_j(t) = \bar{y}_j$ if and only if $\sum_{k \in \mathcal{C}(j)} Q_k(t) = 0$, while $F^{(1)}(t) = \bar{y}_{max}$ if and only if $Q(t) = 0$.

For $j = 1, ..., J$, the function F_j is upper semi-continuous, i.e., for every $t_0 \geq 0$, we have $\limsup_{t \to t_0} F_j(t) \leq F_j(t_0)$. To see this, choose any sequence $t_n \to t_0$ such that $F_j(t_n) \to a$ for some a. Then

$$0 = \sum_{k \in \mathcal{C}(j)} Z_k(t_n, F_j(t_n)) \to \sum_{k \in \mathcal{C}(j)} Z_k(t_0, a)$$

by continuity of Z, and hence $a \leq F_j(t_0)$. Similarly, the function $F^{(1)}$ is upper semi-continuous.

We will now investigate the dynamics of $F^{(1)}$. By upper semicontinuity, the function $F^{(1)}$ does not have downward jumps. Let $t \geq 0$ be such that $Q(t) \neq 0$, equivalently, $F^{(1)}(t) < \bar{y}_{max}$. Let df, dt be small positive numbers such that

$$dt \leq df \tag{51}$$

(it is convenient to think about them as infinitesimal quantities). By (23)–(24), for any s, we have

$$A(t + dt, s) - A(t, s + dt) = \Theta \left(\alpha \circ \left(\int_0^{t+dt} G(s + \eta) d\eta - \int_0^t G(s + dt + \eta) d\eta \right) \right)$$
$$- \Theta P'(Z(t + dt, s) - Z(t, s + dt))$$
$$= \Theta \left(\alpha \circ \int_0^{dt} G(s + \eta) \, d\eta \right)$$
$$- \Theta P'(Z(t + dt, s) - Z(t, s + dt))$$
$$\leq \Theta \left(\alpha \circ \int_0^{dt} G(s + \eta) \, d\eta \right) + (\Theta - I)Z(t, s + dt).$$

Consequently, the vector of fluid masses with lead times not greater than s at time $t + dt$ which can be served in the time interval $[t, t + dt]$ is bounded above as follows:

$$Z(t, s+dt) + A(t+dt, s) - A(t, s+dt) \leq \Theta \left(\alpha \circ \int_0^{dt} G(s + \eta) \, d\eta \right) + \Theta Z(t, s+dt). \tag{52}$$

Putting $s = F^{(1)}(t) + df - dt$ and using (52), the equality $Z(t, F^{(1)}(t)) = 0$, (51), (44), (3) and the definition of the functions H_k, we get

$$Z(t, F^{(1)}(t) + df) + A(t + dt, F^{(1)}(t) + df - dt) - A(t, F^{(1)}(t) + df)$$

$$\leq \Theta\left(\alpha \circ \int_0^{dt} G(F^{(1)}(t) + df - dt + \eta)\, d\eta\right) + \Theta Z(t, F^{(1)}(t) + df)$$

$$= \Theta\left(\alpha \circ \int_{F^{(1)}(t)+df-dt}^{F^{(1)}(t)+df} G(\eta)\, d\eta\right) + \Theta(Z(t, F^{(1)}(t) + df) - Z(t, F^{(1)}(t)))$$

$$\leq \Theta\left(\alpha \circ \int_{F^{(1)}(t)}^{F^{(1)}(t)+df} G(\eta)\, d\eta\right) + \Theta(\alpha \circ [H(F^{(1)}(t)) - H(F^{(1)}(t) + df)])$$

$$= \Theta(\alpha\, df) = \lambda\, df. \tag{53}$$

Let $j \in \{1, ..., J\}$. By (4), (10) and (53), the service time necessary for station j to process the fluids with lead times not greater than $F^{(1)}(t) + df - dt$ at time $t + dt$ present at the station by this time is bounded above by

$$\sum_{k \in \mathcal{C}(j)} m_k \lambda_k\, df = \rho_j\, df.$$

On the other hand, because of (12), as long as there are fluids with lead times not greater than some threshold at station j, the server will devote its full capacity to them. This leads to the main result of this paper:

Theorem 1. *The first frontier function $F^{(1)}$ corresponding to a subcritical EDF fluid model satisfying (44) for every $t \geq 0$, $s_1 \leq s_2$, is nondecreasing. Moreover, in such a model the condition $Q(t) \neq 0$ implies that*

$$F^{(1)}(t + dt) \geq F^{(1)}(t) + (1/ \max_{j=1,...,J} \rho_j - 1)\, dt. \tag{54}$$

PROOF: For $t \geq 0$ such that $Q(t) \neq 0$, letting $df = dt/ \max_{j=1,...,J} \rho_j$ in the above calculations, we see that for small dt, the time necessary for each station j to process to completion all the fluids with lead times not greater than $F^{(1)}(t) + (1/ \max_{j=1,...,J} \rho_j - 1)\, dt$ at time $t + dt$ is not greater than dt, so (54) follows. Note that in the subcritical case, even if $Q(t) = 0$, we can still repeat the above calculations with $df = dt$, getting $F^{(1)}(t + dt) \geq F^{(1)}(t) = \bar{y}_{max}$. This, by the definition of $F^{(1)}$, implies that in this case actually $F^{(1)}(t+dt) = F^{(1)}(t) = \bar{y}_{max}$, i.e., $Q(t + dt) = 0$. □

The following corollary establishes stability of strictly subcritical EDF fluid models (more precisely, a variant of the notion of fluid model stability, originally introduced by Dai [10], which is suitable in our context).

Corollary 2. *In a strictly subcritical EDF fluid model $\mathfrak{X}$ we have*

$$Q(t) = 0, \qquad t \geq \Upsilon + \frac{\bar{y}_{max} + \Upsilon - C_0}{1/ \max_{j=1,...,J} \rho_j - 1}, \tag{55}$$

where m_0 and Υ are as in Proposition 1.

PROOF: Let $\mathfrak{X}' = \triangle_{\Upsilon} \mathfrak{X}$ and let $Q'(t) = Q(t + \Upsilon)$, $F'^{(1)}(t) = F^{(1)}(t + \Upsilon)$ be the fluid queue length vector and the first frontier corresponding to $\mathfrak{X}'$, respectively. Then $\mathfrak{X}'$ satisfies (44) for every $t \geq 0$, $s_1 \leq s_2$ (see the discussion before Lemma 3) and (55) is equivalent to

$$Q'(t) = 0, \qquad t \geq \frac{\bar{y}_{max} + \Upsilon - C_0}{1/\max_{j=1,\ldots,J} \rho_j - 1}. \tag{56}$$

By (37), $Z'(0, C_0 - \Upsilon) = Z(\Upsilon, C_0 - \Upsilon) = 0$, so $F'^{(1)}(0) \geq C_0 - \Upsilon$. By Theorem 1, $F'^{(1)}(t + dt) \geq F'^{(1)}(t) + C dt$ as long as $Q(t) \neq 0$, where $C = 1/\max_{j=1,\ldots,J} \rho_j - 1 > 0$. In particular,

$$\tau = \inf\{t \geq 0 : Q'(t) = 0\} = \inf\{t \geq 0 : F'^{(1)}(t) = \bar{y}_{max}\} \leq (\bar{y}_{max} + \Upsilon - C_0)/C$$

and $F'^{(1)} \equiv \bar{y}_{max}$ on $[\tau, \infty)$ by monotonicity of $F'^{(1)}$, so (56) follows. $\quad\square$

As a special case of Corollary 2, we get stability of strictly subcritical FISFO fluid models:

Corollary 3. *In a strictly subcritical FISFO fluid model $\mathfrak{X}$ with $C_0 = 0$ in (20), we have $Q(t) = 0$ for $t \geq c|Q(0)|$, where*

$$c = \frac{m_0 \|\Theta\|}{1 - \max_{j=1,\ldots,J} \rho_j},$$

m_0 is as in Proposition 1 and $\|\Theta\| = \sum_{i=1}^{K} \max_{j=1,\ldots,K} \theta_{ij}$ (see (1)).

This result was first proved, under some additional technical assumptions, as Theorem 2 in Bramson [8], with a somewhat larger constant, namely

$$c = \frac{11(\sum_{j=1}^{J} \rho_j + 2)|e'_k M \Theta|}{1 - \max_{j=1,\ldots,J} \rho_j} = \frac{11(\sum_{j=1}^{J} \rho_j + 2) \sum_{i=1}^{K} \sum_{j=1}^{K} m_i \theta_{ij}}{1 - \max_{j=1,\ldots,J} \rho_j},$$

where $e_K = (1, \ldots, 1) \in \mathbb{R}^K$ and M is the $K \times K$ diagonal matrix with $m_1, \ldots, m_K$ on the main diagonal. Kruk [18, Theorem 3.1], extended the scope of Bramson's proof to arbitrary strictly subcritical FISFO fluid models, at a price of an additional technical argument. Our present approach, based on Theorem 1, seems to be simpler and somewhat stronger than the FISFO fluid model stability arguments from [8,18]. In particular, we can also establish some kind of EDF fluid model stability in the general subcritical case.

Definition 1 (Compare [11], Definition 6). *A fluid model is said to be weakly stable if for every fluid model solution with $Q(0) = 0$, we have $Q(t) = 0$, $t \geq 0$.*

Note that by (2) and (8)–(9), in a weakly stable EDF fluid model with $Q(0) = 0$, we have

$$A(t, s) = D(t, s) = \Theta \left(\alpha \circ \int_0^t G(s + \eta) \, d\eta \right), \qquad t \geq 0, s \in \mathbb{R}.$$

This notion is related to *rate stability of the corresponding stochastic network*, introduced by Dai and Prabhakar [11], which in our context means

$$\lim_{r \to \infty} \frac{1}{r} D(rt, rs) = \lim_{r \to \infty} \frac{1}{r} A(rt, rs) = \Theta \left(\alpha \circ \int_0^t G(s + \eta) \, d\eta \right)$$

for any $t \geq 0$, $s \in \mathbb{R}$, where A and D are the arrival and departure processes in the *stochastic network* approximated by the EDF fluid model solutions.

Corollary 4. *Any subcritical EDF fluid model is weakly stable.*

PROOF: Let $\mathfrak{X}$ be a subcritical EDF fluid model such that $Q(0) = 0$. Then (20) holds with $C_0 = 0$. With this choice, we have $\Upsilon = 0$, where Υ is defined by (38). Moreover, (37) holds by the proof of Proposition 1. However, for $C_0 = 0$, the condition (21) may fail and consequently Lemma 2 may no longer hold. Nevertheless, for $t \geq 0$ and $s_1 \leq s_2$, by (25), the equality $Q(0) = 0$ and monotonicity of $D(t, \cdot)$, we get

$$\begin{aligned}
\Theta \left(Z(t, s_2) - Z(t, s_1) \right) &= \Theta \left(Z(0, t + s_2) - Z(0, t + s_1) \right) \\
&\quad + \Theta \left(\alpha \circ \left(\int_{t+s_1}^{t+s_2} G(\eta) \, d\eta - \int_{s_1}^{s_2} G(\eta) \, d\eta \right) \right) \\
&\quad - (D(t, s_2) - D(t, s_1)) \\
&\leq \Theta \left(\alpha \circ \left(\int_{t+s_1}^{t+s_2} d\eta - \int_{s_1}^{s_2} G(\eta) \, d\eta \right) \right) \\
&\quad - (D(t, s_2) - D(t, s_1)) \\
&\leq \Theta(\alpha \circ [H(s_1) - H(s_2)]),
\end{aligned}$$

so Corollary 1 and its consequences (44)–(47) hold. Therefore, our justification of Theorem 1 is valid in the current setting. Hence, for any $t \geq 0$, $F^{(1)}(t) = F^{(1)}(0) = \bar{y}_{max}$ which, together with (47), implies that $Q(t) = 0$.   □

5   Conclusion

We studied fluid models of subcritical EDF multiclass queueing networks with soft job deadlines. We established monotonicity of the first frontier function $F^{(1)}$ corresponding to such a model. Stability of strictly subcritical EDF fluid models and weak stability of their subcritical counterparts follow easily from this finding.

Our main result implies the existence of a finite limit

$$F^{(1)}(\infty) := \lim_{t \to \infty} F^{(1)}(t) \tag{57}$$

in each subcritical EDF fluid model. In fact, Corollary 2 assures that in the strictly subcritical case, we have $F^{(1)}(\infty) = \bar{y}_{max}$ and this limit is actually attained in finite time. It would be interesting to characterize the rate of convergence in (57) for other subcritical EDF fluid models, in particular, to find sufficient conditions for this limit to be attained in finite time. More generally, we would like to find sufficient conditions for convergence of the frontier $F_j(t)$ at each station j as $t \to \infty$ for a subcritical EDF fluid model. This would imply convergence of the fluid model states to the invariant manifold (so-called asymptotic stability of the fluid model). The latter result may be very helpful in the development of diffusion limits for the corresponding stochastic networks.

References

1. Atar, R., Biswas, A., Kaspi, H.: Fluid limits of G/G/1+G queues under the non-preemptive earliest-deadline-first discipline. Math. Oper. Res. **40**, 683–702 (2015)
2. Atar, R., Biswas, A., Kaspi, H.: Law of large numbers for the many-server earliest-deadline-first queue. Stochast. Process. Appl. **128**(7), 2270–2296 (2018)
3. Atar, R., Biswas, A., Kaspi, H., Ramanan, K.: A Skorokhod map on measure-valued paths with applications to priority queues. Ann. Appl. Probab. **28**, 418–481 (2018)
4. Atar, R., Shadmi, Y.: Fluid limits for earliest-deadline-first networks. arXiv arXiv:2009.07169v1 (2020)
5. Bramson, M.: Convergence to equilibria for fluid models of FIFO queueing networks. Queueing Syst. Theor. Appl. **22**, 5–45 (1996)
6. Bramson, M.: Convergence to equilibria for fluid models of head-of-the-line proportional processor sharing queueing networks. Queueing Syst. Theor. Appl. **23**, 1–26 (1996)
7. Bramson, M.: State space collapse with application to heavy traffic limits for multiclass queueing networks. Queueing Syst. Theor. Appl. **30**, 89–148 (1998)
8. Bramson, M.: Stability of earliest-due-date, first-served queueing networks. Queueing Systems. Theor. Appl. **39**, 79–102 (2001)
9. Chen, H., Yao, D.D.: Fundamentals of Queueing Networks. Springer Science+Business Media, LLC, New York (2001). https://doi.org/10.1007/978-1-4757-5301-1
10. Dai, J.G.: On positive Harris recurrence of multiclass queueing networks: a unified approach via fluid limit models. Ann. Appl. Probab. **5**, 49–77 (1995)
11. Dai, J.G., Prabhakar, B.: The throughput of data switches with and without speedup. In: Proceedings IEEE INFOCOM 2000. Conference on Computer Communications. 19th Annual Joint Conference of the IEEE Computer and Communications Societies, vol. 2, pp. 556–564 (2000). https://doi.org/10.1109/INFCOM.2000.832229
12. Decreusefond, L., Moyal, P.: Fluid limit of a heavily loaded EDF queue with impatient customers. Markov Process. Relat. Fields **14**(1), 131–158 (2008)
13. Doytchinov, B., Lehoczky, J.P., Shreve, S.E.: Real-time queues in heavy traffic with earliest-deadline-first queue discipline. Ann. Appl. Probab. **11**, 332–379 (2001)
14. Harrison, J.M.: Brownian models of queueing networks with heterogeneous customer populations. In: Fleming, W., Lions, P.L. (eds.) Stochastic Differential Systems, Stochastic Control Theory and Applications. The IMA Volumes in Mathematics and Its Applications, vol. 10, pp. 147–186. Springer, New York (1988). https://doi.org/10.1007/978-1-4613-8762-6_11
15. Kruk, Ł: Stability of two families of real-time queueing networks. Probab. Math. Stat. **28**, 179–202 (2008)
16. Kruk, Ł: Invariant states for fluid models of EDF networks: nonlinear lifting map. Probab. Math. Stat. **30**, 289–315 (2010)

17. Kruk, Ł: An open queueing network with asymptotically stable fluid model and unconventional heavy traffic behavior. Math. Oper. Res. **36**, 538–551 (2011)
18. Kruk, Ł: Stability of preemptive EDF queueing networks. Ann. Univ. Mariae Curie-Skłodowska Math. A. **73**, 105–134 (2019)
19. Kruk, Ł.: Minimal and locally edge minimal fluid models for resource sharing networks. Math. Oper. Res. (2021). https://doi.org/10.1287/moor.2020.1110
20. Kruk, Ł, Lehoczky, J.P., Ramanan, K., Shreve, S.E.: Heavy traffic analysis for EDF queues with reneging. Ann. Appl. Probab. **21**, 484–545 (2011)
21. Kruk, Ł, Lehoczky, J.P., Shreve, S.E., Yeung, S.-N.: Earliest-deadline-first service in heavy traffic acyclic networks. Ann. Appl. Probab. **14**, 1306–1352 (2004)
22. Lehoczky, J.P.: Using real-time queueing theory to control lateness in real-time systems. Perform. Eval. Rev. **25**, 158–168 (1997)
23. Lehoczky, J.P.: Real-time queueing theory. In: Proceedings of the IEEE Real-Time Systems Symposium, pp. 186–195 (1998)
24. Lehoczky, J.P.: Scheduling communication networks carrying real-time traffic. In: Proceedings of the IEEE Real-Time Systems Symposium, pp. 470–479 (1998)
25. Liu, C.L., Layland, J.W.: Scheduling algorithms for multiprogramming in a hard real-time environment. J. Assoc. Comput. Mach. **20**(1), 40–61 (1973)
26. Moyal, P.: Convex comparison of service disciplines in real time queues. Oper. Res. Lett. **36**(4), 496–499 (2008)
27. Panwar, S.S., Towsley, D.: On the optimality of the STE rule for multiple server queues that serve customers with deadlines. Technical Report 88-81, Department of Computer and Information Science, University Massachusetts, Amherst (1988)
28. Panwar, S.S., Towsley, D.: Optimality of the stochastic earliest deadline policy for the G/M/c queue serving customers with deadlines. In: 2nd ORSA Telecommunications Conference. ORSA (Operations Research Society of America), Baltimore, MD (1992)
29. Rybko, A.N., Stolyar, A.L.: Ergodicity of stochastic processes describing the operations of open queueing networks. Probl. Inf. Transm. **28**, 199–220 (1992)
30. Seidman, T.I.: "First come, first served" can be unstable! IEEE Trans. Automat. Control **39**, 2166–2171 (1994)
31. Williams, R.J.: Diffusion approximations for open multiclass queueing networks: sufficient conditions involving state space collapse. Queueing Syst. Theor. Appl. **30**, 27–88 (1998)
32. Yeung, S.-N., Lehoczky, J.P.: Real-time queueing networks in heavy traffic with EDF and FIFO queue discipline. Working paper, Department of Statistics, Carnegie Mellon University (2001)

Graph Neural Network Based Scheduling: Improved Throughput Under a Generalized Interference Model

Ramakrishnan Sambamoorthy[1(✉)], Jaswanthi Mandalapu[2],
Subrahmanya Swamy Peruru[3], Bhavesh Jain[3], and Eitan Altman[1]

[1] INRIA Sophia Antipolis - Méditeranée, Valbonne, France
{ramakrishnan.sambamoorthy,eitan.altman}@inria.fr
[2] IIT Madras, Chennai, India
ee19d700@smail.iitm.ac.in
[3] IIT Kanpur, Kanpur, India
{swamyp,jbhavesh}@iitk.ac.in

Abstract. In this work, we propose a Graph Convolutional Neural Networks (GCN) based scheduling algorithm for adhoc networks. In particular, we consider a generalized interference model called the k-tolerant conflict graph model and design an efficient approximation for the well-known Max-Weight scheduling algorithm. A notable feature of this work is that the proposed method do not require labelled data set (NP-hard to compute) for training the neural network. Instead, we design a loss function that utilises the existing greedy approaches and trains a GCN that improves the performance of greedy approaches. Our extensive numerical experiments illustrate that using our GCN approach, we can significantly (4–20%) improve the performance of the conventional greedy approach.

Keywords: Resource allocation · Graph Convolutional Neural Networks · Adhoc networks

1 Introduction

The design of efficient scheduling algorithms is a fundamental problem in wireless networks. In each time slot, a scheduling algorithm aims to determine a subset of non-interfering links such that the system of queues in the network is stabilized. Depending on the interference model and the network topology, it is known that there exists a *'rate region'* - a maximal set of arrival rates - for which the network can be stabilized. A scheduling algorithm that can support any arrival rate in the rate region is said to be throughput optimal. A well-known algorithm called the Max-Weight scheduling algorithm [1] is said to be throughput optimal. However,

---

The authors are grateful to the OPAL infrastructure from Université Côte d'Azur for providing resources and support.

Q. Zhao and L. Xia (Eds.): VALUETOOLS 2021, LNICST 404, pp. 144–153, 2021.
https://doi.org/10.1007/978-3-030-92511-6_9

the Max-Weight scheduler is not practical for distributed implementation due to the following reasons: (i) global network state information is required, and (ii) requires the computation of maximum-weighted independent set problem in each time slot, which is an NP-hard problem.

There have been several efforts in the literature to design low-complex, distributed approximations to the Max-Weight algorithm [2,3]. Greedy approximation algorithms such as the *maximal* scheduling policies, which can support a fraction of the maximum throughput, are one such class of approximations [4]. On the other hand, we have algorithms like carrier sense multiple access (CSMA) algorithms [5,6], which are known to be near-optimal in terms of the throughput performance but known to suffer from poor delay performance.

Inspired by the success of deep-learning-based algorithms in various fields like image processing and natural language processing, recently, there has been a growing interest in their application in wireless scheduling as well [7–9]. Initial research in this direction focused on the adaption of widely used neural architectures like multi-layer perceptrons or convolutional neural networks (CNNs) [10] to solve wireless scheduling problems. However, these architectures are not well-suited for the scheduling problem because they do not explicitly consider the network graph topology. Hence, some of the recent works in wireless networks study the application of the Graph Neural Network (GNN) architectures for solving the scheduling problem [11]. For instance, a recent work [12] has proposed a GNN based algorithm, where it has been observed that the help of Graph Neural networks can improve the performance of simple greedy scheduling algorithms like Longest-Queue-First (LQF) scheduling.

However, this result is observed on a simple interference model called the conflict graph model, which captures only binary relationships between links. Nevertheless, in real wireless networks, the interference among the links is additive, and the cumulative effect of all the interfering links decides the feasibility of any transmission. Hence, it is essential to study whether the GNN based approach will improve the performance of greedy LQF scheduling under a realistic interference model like the (Signal-to-interference-plus-noise ratio) SINR model, which captures the cumulative nature of interference.

One of the challenges in conducting such a study is that the concept of graph neural networks is not readily applicable for the SINR interference model since a graph cannot represent it. Hence, we introduce a new interference model which retains the cumulative interference nature yet is amenable to a graph-based representation and conduct our study on the proposed interference model. This approach will provide insights into whether the GNN-based improvement for LQF will work for practical interference models.

To that end, in this paper, we study whether GNN based algorithms can be used for designing efficient scheduling under this general interference model. Specifically, we consider a k-tolerant conflict graph model, where a node can successfully transmit during a time slot if not more than k of its neighbors are transmitting in that time slot. Moreover, when k is set to zero, the k-tolerance model can be reduced to the standard conflict graph model, in which a node cannot transmit if any of its neighbors is transmitting. We finally tabulate our results and compare them with other GNN-based distributed scheduling algorithms under a standard conflict-graph-based interference model. In sum, our contributions are as follows:

(i) We propose a GCN-based distributed scheduling algorithm for a generalized interference model called the k-tolerant conflict graph model.
(ii) The training of the proposed GCN does not require a labeled data set (involves solving an NP-hard problem). Instead, we design a loss function that utilizes an existing greedy approach and trains a GCN that improves the performance of the greedy approach by 4 to 20%.

The remainder of the paper is organized as follows. In Sect. 2, we briefly present our network model. In Sect. 3, an optimal scheduling policy for k-tolerance conflict graph interference model, a GCN-based k-tolerant independent set solver, is presented. In Sect. 4, we conduct experiments on different data sets and show the numerical results of the GCN-based scheduling approach. Finally, the paper is concluded in Sect. 5.

Motivation: In the SINR interference model, a link can successfully transmit if the cumulative interference from all nodes within a radius is less than some fixed threshold value. The conflict graph model insists that all the neighbours should not transmit when a link is transmitting. However, in a real-world situation, a link can successfully transmit as long as the cumulative interference from all its neighbours (the links which can potentially interfere with a given link) is less than a threshold value. As a special case, in this paper, we consider a conservative SINR model called k-tolerance model in which, if i_{max} is the estimated strongest interference that a link can cause to another and let i_{th} be the cumulative threshold interference that a link can tolerate, then a conservative estimate of how many neighbouring links can be allowed to transmit without violating the threshold interference is given by $k = i_{th}/i_{max}$. In other words, k-neighbours can transmit while a given link is transmitting. It can be seen that this conservative model retains the cumulative nature of the SINR interference model. Hence a study on this model should give us insights into the applicability of GNN based solutions for realistic interference models.

2 Network Model

We model the wireless network as an undirected graph $\mathcal{G} = (V, E)$ with N nodes. Here, the set of nodes $V = \{v_i\}_{i=1}^{N}$ of the graph represents links in the wireless network i.e., a transmitter-receiver pair. We assume an edge between two nodes, if the corresponding links could potentially interfere with each other. Let E and $\mathbf{A}$ denote the set of edges and the adjacency matrix of graph $\mathcal{G}$ respectively. We denote the set of neighbors of node v by $\mathcal{N}(v)$ i.e., a node $v' \in \mathcal{N}(v)$, if the nodes v and v' share an edge between them. We say a node is k-tolerant, if it can tolerate at most k of its transmitting neighbors. In other words, a k-tolerant node can successfully transmit, if the number of neighbors transmitting at the same time is at most k. We define a k-*tolerant conflict graph* as a graph in which each node is k-tolerant, and model the wireless network as a $k-tolerant\ conflict\ graph$. Note that this is a generalization of the popular conflict graph model, where a node can tolerate none of its transmitting neighbors. The conflict graph model corresponds to 0-tolerant conflict graph ($k = 0$).

We assume that the time is slotted. In each time slot, the scheduler has to decide on the set of links to transmit in that time slot. A feasible schedule is a set of links that can successfully transmit at the same time. At any given time t, a set of links can successfully transmit, if the corresponding nodes form a k-*independent set* (defined below) in graph $\mathcal{G}$. Thus, a feasible schedule corresponds to a k-independent set in $\mathcal{G}$.

Definition 1. *(k-independent set) A subset of vertices of a graph $\mathcal{G}$ is k-independent, if it induces in $\mathcal{G}$, a sub-graph of maximum degree at most k.*

A scheduler has to choose a feasible schedule at any given time. Let $\mathcal{S}_\mathcal{G}$ denotes the collection of all possible $k-$independent sets i.e., the feasible schedules. We denote the schedule at time t by an N length vector $\sigma(t) = (\sigma_v(t),\ v \in V)$. We say $\sigma_v(t) = 1$ if at time t, node v is scheduled to transmit and $\sigma_v(t) = 0$, otherwise. Depending on the scheduling decision $\sigma(t) \in \mathcal{S}_\mathcal{G}$ taken at time t, node $v \in V$ (a link in the original wireless network) gets a rate of $\mu_v(t, \sigma)$. We assume that packets arriving at node v can be stored in an infinite buffer. At time t, let $\lambda_v(t)$ be the number of packets that arrive at node $v \in V$. We then have the following queuing dynamics at node v:

$$q_v(t + 1) = [q_v(t) + \lambda_v(t) - \mu_v(t, \sigma)]^+ . \tag{1}$$

The set of arrival rates for which there exist a scheduler that can keep the queues stable is known as the rate region of the wireless network.

2.1   Max-Weight Scheduler

A well known scheduler that stabilises the network is the Max-Weight algorithm [1]. The Max-Weight algorithm chooses a schedule $\sigma^*(t) \in \mathcal{S}_\mathcal{G}$ that maximizes the sum of queue length times the service rate, i.e.,

$$\sigma^*(t) = \arg \max_{\sigma \in \mathcal{S}_\mathcal{G}} \sum_v q_v(t)\mu_v(t, \sigma). \tag{2}$$

We state below one of the celebrated results in radio resource allocation.

Theorem 1. *[1] Let the arrival process $\lambda_v(t)$ be an ergodic process with mean λ_v. If the mean arrival rates (λ_v) are within the rate region, then the Max-Weight scheduling algorithm is throughput optimal.*

In spite of such an attractive result, the Max-Weight algorithm is seldom implemented in practice. This is because, the scheduling decision in (2) has complexity that is exponential in the number of nodes. Even with the simplistic assumption of a conflict graph model, (2) reduces to the NP-hard problem of finding the maximum weighted independent set. At the timescale of these scheduling decisions, finding the exact solution to (2) is practically infeasible. Hence, we resort to solving (2) using a Graph Neural Network (GNN) model. Before we explain our GNN based algorithm, we shall rephrase the problem in (2) for the k-tolerant conflict graph model below.

2.2   Maximum Weighted K-Independent Set

In the k-tolerant conflict graph model $\mathcal{G}$, the Max-Weight problem is equivalent to the following integer program:

$$\text{Maximize: } \sum_v \sigma_v w_v$$

$$\text{Such that: } \sigma_v \left(\sum_{v' \in \mathcal{N}(v)} \sigma_{v'} \right) \leq k \tag{3}$$

$$\sigma_v \in \{0, 1\}, \text{ for all } v \in \mathcal{V}$$

Here $\boldsymbol{w} = (w_v : v \in V)$ is the weight vector. The constraint in (3) ensures that whenever a node is transmitting, at most k of its neighbors can transmit. It can be observed that the maximum weight problem in (2) corresponds to using the weights $w_v = q_v(t)\mu_v(t, \sigma)$ in the above formulation. Henceforth, the rest of this paper is devoted to solving the maximum weighted k-independent set problem using a graph neural network.

3   Graph Neural Network Based Scheduler

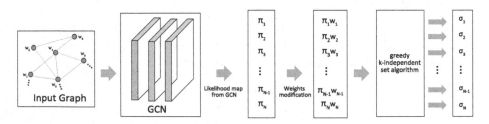

Fig. 1. The architecture of the Graph Convolutional Neural Network based maximum weighted $k-$independent set problem solver.

In this section, we present a graph neural network based solution to solve the maximum weighted k-independent set problem. We use the Graph Convolution Neural network (GCN) architecture from [13,14].

The GCN architecture is as follows: We use a GCN with L layers. The input of each layer is a feature matrix $\mathbf{Z}^l \in \mathbb{R}^{N \times C^l}$ and its output is fed as the input to the next layer. Precisely, at the $(l+1)$th layer, the feature matrix $\mathbf{Z}^{l+1}$ is computed using the following graph convolution operation:

$$\mathbf{Z}^{l+1} = \Phi(\mathbf{Z}^l \Theta_0^l + \mathcal{L} \mathbf{Z}^l \Theta_1^l), \tag{4}$$

where $\Theta_0^l, \Theta_1^l \in \mathbb{R}^{C^l \times C^{l+1}}$ are the trainable weights of the neural network, C^l denotes the number of feature channels in l-th layer, $\Phi(.)$ is a nonlinear activation function and $\mathcal{L}$ is the normalized Laplacian of the input graph $\mathcal{G}$ computed as follows: $\mathcal{L} = \mathbf{I}_N - \mathbf{D}^{-\frac{1}{2}} \mathbf{A} \mathbf{D}^{-\frac{1}{2}}$. Here, $\mathbf{I}_N$ denotes the $N \times N$ identity matrix and $\mathbf{D}$ is the diagonal matrix with entries $\mathbf{D}_{ii} = \sum_j \mathbf{A}_{ij}$.

We take the input feature matrix $\mathbf{Z}^0 \in \mathbb{R}^{N \times 1}$ as the weights $\boldsymbol{w}$ of the nodes (hence $C^0 = 1$) and $\Phi(.)$ as a ReLU activation function for all layers except for the last layer. For the last layer, we apply sigmoid activation function to get the likelihood of the nodes to be included in the k-independent set. We represent this likelihood map from the GCN network using an N length vector $\boldsymbol{\pi} = (\pi_v, v \in V) \in [0,1]^N$.

In summary, the GCN takes a graph $\mathcal{G}$ and the node weights $\boldsymbol{w}$ as input and returns a N length likelihood vector $\boldsymbol{\pi}$ (see Fig. 1). However, we need a k-independent set. In usual classification problems, such a requirement is satisfied by projecting the likelihood maps to a binary vector. Projecting the likelihood map onto the collection of k-independent sets is not straightforward, since the collection of k-independent sets are N length binary vectors that satisfy the constraints in (3). Such a projection operation by itself might be costly in terms of computation. Instead, by taking inspiration from [15], we pass the likelihood map through a greedy algorithm[1] to get a k-independent set.

The greedy algorithm requires each node to keep track of the number of its neighbours already added in k-independent set. We sort the nodes in the descending order of the product of the likelihood and the weight i.e., $\pi_v w_v$. We add the node with highest likelihood-weight product to the k-independent set, if at most k of its neighbors are already added in the k-independent set. We remove the nodes that are neighbours to a node which has already added to the set and also reached a tolerance of k. We then repeat the procedure until no further nodes are left to be added.

We use a set of node-weighted graphs to train the GCN. Since the problem at hand is NP-hard, we refrain from finding the true labels (maximum weighted k-independent set) to train the GCN. Instead, we construct penalty and reward functions using the desirable properties of the output $\boldsymbol{\pi}$. We then learn the parameters by optimizing over a weighted sum of the constructed penalties and rewards. We desire the output $\boldsymbol{\pi}$ to predict the maximum weighted k-independent set. With this in mind we construct the following rewards and penalties:

---

[1] In practice, the greedy algorithm can be replaced with a distributed greedy algorithm [16] and train the GCN model w.r.t the distributed greedy algorithm.

a) The prediction π needs to maximize the sum of the weights. So, our prediction needs to maximize $R_1 = \sum_v \pi_v w_v$.
b) The prediction π needs to satisfy the k-independent set constraints. Therefore, we add a penalty, if π violates the independent set constraints in (3), i.e., $P_1 = \sum_{v \in V} \left(\sigma_v \left(\sum_{v' \in N(v)} \sigma_{v'} - k \right) \right)^2$.
c) Recall that we use the greedy algorithm to predict the k-independent set from π. The greedy algorithm takes $(\pi_v w_v, \ v \in V)$ as the input and returns a k-independent set. We desire the total weight of the output π, i.e., $\sum_v \pi_v w_v$ to be close to the total weight of the k-independent returned by the greedy algorithm. Let W_{gcn} be the total weight of the independent set predicted by the greedy algorithm. Then, we penalise the output π if it deviates from W_{gcn}, i.e., $P_2 = |\sum_v \pi_v w_v - W_{gcn}|^2$.

We finally construct our cost function as a weighted sum of the above i.e., we want the GCN to minimize the cost function:

$$C = \beta_1 P_1 + \beta_2 P_2 - \beta_3 R_1 \tag{5}$$

where β_1, β_2 and β_3 denotes the optimization weights of the cost function defined in equation (5).

4   Experiments

We perform our experiments on a single GPU GeForce GTX 1080 Ti[2]. The data used for training, validation and testing are described in the subsection below.

4.1   Dataset

We train our GCN using randomly generated graphs. We consider two graph distributions, namely Erdos-Reyni (ER) and Barbasi-Albert (BA) models. These distributions were also used in [12]. Our choice of these graph models is to ensure fair comparison with prior work on conflict graph model [12] ($k = 0$).

In ER model with N nodes, an edge is introduced between two nodes with a fixed probability p, independent of the generation of other edges. The BA model generates a graph with N nodes (one node at a time), preferentially attaching the node to M existing nodes with probability proportional to the degree of the existing nodes.

For training purpose, we generate 5000 graphs of each of these models. For the ER model, we choose $p \in \{0.02, 0.05, 0.075, 0.10, 0.15\}$ and for the BA model we choose $M = Np$. The weights of the nodes are chosen uniformly at random from the interval $[0, 1]$. We use an additional 50 graphs for validation and 500 graphs for testing.

---

[2] Training the models took around two hours.

4.2   Choice of Hyper-Parameters

We train a GCN with 3 layers consisting i) an input layer with the weights of the nodes as input features ii) a single hidden layer with 32 features and iii) an output layer with N features (one for each node) indicating the likelihood of choosing the corresponding node in the k-independent set. This choice of using a smaller number of layers ensures that the GCN operates with a minimal number of communications with its neighbors. We fix $k = 0$, and experiment training the GCN with different choices of the optimization weights β_1, β_2 and β_3. The results obtained are tabulated in Fig. 2. Let W_{gr} denote the total weight of the plain greedy algorithm i.e., without any GCN and W_{gcn} denote the total weight of the independent set predicted by the GCN-greedy combination. We have tabulated the average ratio between the total weight of the nodes in the independent set obtained from the GCN-greedy and the total weight of the nodes in the independent set obtained from the plain greedy algorithm, i.e., W_{gcn}/W_{gr}. The average is taken over the test data set. The training was done with BA and ER models separately. We test the trained models also with test data from both models to understand if the trained models are transferable. We see that GCN trained with parameters $\beta_1 = 5$, $\beta_2 = 5$ and $\beta_3 = 10$ performs well for both ER and BA graph models. The GCN improves the total weight of the greedy

Training Data	β_1	β_2	β_3	Test Data = ER		Test Data = BA	
				Average W_{gcn}/W_{gr}	Variance $\times 10^{-3}$	Average W_{gcn}/W_{gr}	Variance $\times 10^{-3}$
BA	5	5	10	1.038	3.047	1.11	10.16
	10	10	1	1.035	3.297	1.11	10.37
	5	5	1	1.035	3.290	1.11	10.14
	1	1	1	1.034	3.253	1.10	10.23
	5	5	30	1.041	3.230	1.10	10.39
	5	5	50	1.041	3.214	1.10	10.28
	5	5	100	1.035	2.838	1.09	10.02
	30	1	1	1.031	2.401	1.07	8.25
ER	5	5	30	1.040	2.929	1.10	10.12
	5	5	10	1.039	3.145	1.11	10.71
	5	5	50	1.039	2.957	1.09	9.92
	1	1	1	1.038	3.135	1.11	10.74
	1	20	1	1.036	3.070	1.11	10.55
	10	10	1	1.034	3.428	1.11	10.34
	5	5	1	1.034	3.331	1.11	10.34
	5	5	100	1.031	2.420	1.08	8.42
Distributed scheduling using GNN [12]				1.039	3.5	1.11	11.0

Fig. 2. Table showing the average and variance of the ratio of the total weight of the nodes in the independent set ($K = 0$) obtained using GCN to that of the independent set obtained using greedy algorithm. We observe a 3% increase in the total weight for the ER model and 11% increase in the total weight for the BA model. Our performance matches with the performance of the GCN used in [12].

algorithm by 4% for the ER model and by 11% for the BA model. Also, we see that the GCN trained with ER model performs well with BA data and vice versa.

4.3 Performance for Different k

We also evaluate the performance for different tolerance values $k \in \{1, 2, 3, 4\}$. We use the parameters $\beta_1 = 5$, $\beta_2 = 5$ and $\beta_3 = 10$ in the cost function. Recall that we have come up with this choice using extensive simulations for $k = 0$. In Fig. 3, we tabulate the average ratio between the total weight of the k-independent set obtained using the GCN-greedy combo and that of the plain greedy algorithm i.e., W_{gcn}/W_{gr}. We have also included the variance from this performance. We observe that the performance for a general k is even better as compared to $k = 0$. For example, we see that for $k = 2, 3, 4$, we see 6% improvement for the ER model and close to 20% improvement for the BA model.

Training Data	k	Test Data = ER		Test Data = BA	
		Average W_{gcn}/W_{gr}	Variance $\times 10^{-3}$	Average W_{gcn}/W_{gr}	Variance $\times 10^{-3}$
BA	1	1.056	4.07	1.143	10.22
	2	1.062	5.26	1.193	10.92
	3	1.067	5.55	1.209	20.14
	4	1.063	4.53	1.241	20.57
ER	1	1.056	3.99	1.143	10.18
	2	1.064	5.12	1.187	10.81
	3	1.066	4.82	1.205	20.13
	4	1.062	4.18	1.225	20.29

Fig. 3. The table shows the average and variance of the ratio between the total weight of the k-independent set obtained using GCN-greedy combo to that of the plain greedy algorithm for $k \in \{1, 2, 3, 4\}$. We observe that the improvement is consistently above 5% for the ER model and above 14% for the BA model.

Interestingly, the GCN trained with ER graphs performs well on the BA data set as well. This indicates that the trained GCN is transferable to other models.

5 Conclusion

In this paper, we investigated the well-studied problem of link scheduling in wireless adhoc networks using the recent developments in graph neural networks. We modelled the wireless network as a k-tolerant conflict graph and demonstrated that using a GCN, we can improve the performance of existing greedy algorithms. We have shown experimentally that this GCN model improves the performance of the greedy algorithm by at least 4–6% for the ER model and 11–22% for the BA model (depending on the value of k).

In future, we would like to extend the model to a node dependent tolerance value k_v and pass the tolerance value as the node features of the GNN in addition to the weights.

References

1. Tassiulas, L., Ephremides, A.: Stability properties of constrained queueing systems and scheduling policies for maximum throughput in multihop radio networks. IEEE Trans. Autom. Control **37**(12), 1936–1948 (1992)
2. Xu, X., Tang, S., Wan, P.-J.: Maximum weighted independent set of links under physical interference model. In: Pandurangan, G., Anil Kumar, V.S., Ming, G., Liu, Y., Li, Y. (eds.) WASA 2010. LNCS, vol. 6221, pp. 68–74. Springer, Heidelberg (2010). https://doi.org/10.1007/978-3-642-14654-1_8
3. Tassiulas, L., Ephremides, A.: Dynamic server allocation to parallel queues with randomly varying connectivity. IEEE Trans. Inf. Theory **39**(2), 466–478 (1993)
4. Wan, P.-J.: Greedy approximation algorithms. In: Pardalos, P.M., Du, D.-Z., Graham, R.L. (eds.) Handbook of Combinatorial Optimization, pp. 1599–1629. Springer, New York (2013). https://doi.org/10.1007/978-1-4419-7997-1_48
5. Jiang, L., Walrand, J.: A distributed CSMA algorithm for throughput and utility maximization in wireless networks. IEEE/ACM Trans. Networking **18**(3), 960–972 (2010)
6. Swamy, P.S., Ganti, R.K., Jagannathan, K.: Adaptive CSMA under the SINR model: efficient approximation algorithms for throughput and utility maximization. IEEE/ACM Trans. Networking **25**, 1968–1981 (2017)
7. Cui, W., Shen, K., Yu, W.: Spatial deep learning for wireless scheduling. IEEE J. Sel. Areas Commun. **37**(6), 1248–1261 (2019)
8. Lecun, Y., Bottou, L., Bengio, Y., Haffner, P.: Gradient-based learning applied to document recognition. Proc. IEEE **86**(11), 2278–2324 (1998)
9. Hornik, K., Stinchcombe, M., White, H.: Multilayer feedforward networks are universal approximators. Neural Netw. **2**, 359–366 (1989)
10. Xu, D., Che, X., Wu, C., Zhang, S., Xu, S., Cao, S.: Energy-efficient subchannel and power allocation for hetnets based on convolutional neural network (2019)
11. Eisen, M., Ribeiro, A.: Optimal wireless resource allocation with random edge graph neural networks. IEEE Trans. Signal Process. **68**, 2977–2991 (2020)
12. Zhao, Z., Verma, G., Rao, C., Swami, A., Segarra, S.: Distributed scheduling using graph neural networks. In: ICASSP 2021–2021 IEEE International Conference on Acoustics, Speech and Signal Processing (ICASSP), pp. 4720–4724 (2021)
13. Kipf, T.N., Welling, M.: Semi-supervised classification with graph convolutional networks. In: International Conference on Learning Representations (ICLR) (2017)
14. Defferrard, M., Bresson, X., Vandergheynst, P.: Convolutional neural networks on graphs with fast localized spectral filtering. In: Proceedings of the 30th International Conference on Neural Information Processing Systems, NIPS 2016, (Red Hook, NY, USA), pp. 3844–3852, Curran Associates Inc., (2016)
15. Li, Z., Chen, Q., Koltun, V.: Combinatorial optimization with graph convolutional networks and guided tree search. In: Bengio, S., Wallach, H., Larochelle, H., Grauman, K., Cesa-Bianchi, N., Garnett, R. (eds.) Advances in Neural Information Processing Systems, vol. 31, Curran Associates Inc., (2018)
16. Joo, C., Lin, X., Ryu, J., Shroff, N.B.: Distributed greedy approximation to maximum weighted independent set for scheduling with fading channels. IEEE/ACM Trans. Networking **24**(3), 1476–1488 (2016)

A Novel Implementation of Q-Learning for the Whittle Index

Lachlan J. Gibson[1(✉)] ⓘ, Peter Jacko[2,3] ⓘ, and Yoni Nazarathy[1] ⓘ

[1] School of Mathematics and Physics, The University of Queensland,
Brisbane, Australia
{l.gibson1,y.nazarathy}@uq.edu.au
[2] Lancaster University, Lancaster, UK
[3] Berry Consultants, Abingdon, UK

Abstract. We develop a method for learning index rules for multi-armed bandits, restless bandits, and dynamic resource allocation where the underlying transition probabilities and reward structure of the system is not known. Our approach builds on an understanding of both stochastic optimisation (specifically, the Whittle index) and reinforcement learning (specifically, Q-learning). We propose a novel implementation of Q-learning, which exploits the structure of the problem considered, in which the algorithm maintains two sets of Q-values for each project: one for reward and one for resource consumption. Based on these ideas we design a learning algorithm and illustrate its performance by comparing it to the state-of-the-art Q-learning algorithm for the Whittle index by Avrachenkov and Borkar. Both algorithms rely on Q-learning to estimate the Whittle index policy, however the nature in which Q-learning is used in each algorithm is dramatically different. Our approach seems to be able to deliver similar or better performance and is potentially applicable to a much broader and more general set of problems.

Keywords: Multi-armed bandits · Restless bandits · Reinforcement learning · Q-learning · Markov decision processes

1 Introduction

Problems of dynamic resource allocation are ubiquitous in fields such as public health, business, communications, engineering, and agriculture as they focus on making an efficient dynamic use of limited amounts of resources. Examples include: allocation of patients to treatments in a clinical trial in order to learn about their effectiveness as quickly as possible; allocation of physicians (or beds) to patients in a hospital in order to improve their health; allocation of production lines (or manpower) in a business in order to satisfy demand of customers; allocation of shelf space to products in order to maximise revenue; allocation of online advertisements to users in order to maximise the click-through rate; allocation of webpage designs with different cognitive styles to customers in order

© ICST Institute for Computer Sciences, Social Informatics and Telecommunications Engineering 2021
Published by Springer Nature Switzerland AG 2021. All Rights Reserved
Q. Zhao and L. Xia (Eds.): VALUETOOLS 2021, LNICST 404, pp. 154–170, 2021.
https://doi.org/10.1007/978-3-030-92511-6_10

to improve their understanding; allocation of agents' time to process intelligence information in order to minimise damage caused by attacks; allocation of frequency spectrum to users in wireless networks in order to achieve a desirable quality of service; allocation of computer power/memory to simulation tasks in order to improve an existing solution. Indeed in almost every technological, societal, or logistical setting one can think of many scenarios where dynamic resource allocation plays a key role.

1.1   Problem Formulation

We consider the following formulation of the problem of dynamic resource allocation. We present it as a Markov decision process (MDP) framework, which is sufficiently versatile to allow for a variety of observability settings. A set of K heterogeneous alternative *projects* compete for a resource with the expected one-period *capacity* of M units. Each resource can be allocated to at most one project, but each project requires to be attended by a certain number of resource units. At the beginning of every time period $t \in \mathcal{T} := \{0, 1, \ldots, T-1\}$ (where T, possibly infinite, is the problem *horizon*), a decision-maker can choose which projects will be allocated to the resource units during the period. Let us denote by $\mathcal{A} := \{0, 1\}$ the *action space* of each project, where action 1 means being allocated to the resource, while action 0 means the opposite.

Each project $k \in \mathcal{K} := \{1, 2, \ldots, K\}$ is formulated as a work-reward MDP given by a tuple $(\mathcal{N}_k, (\boldsymbol{W}_k^a)_{a \in \mathcal{A}}, (\boldsymbol{R}_k^a)_{a \in \mathcal{A}}, (\boldsymbol{P}_k^a)_{a \in \mathcal{A}})$, where $\mathcal{N}_k$ is the *state space*; $\boldsymbol{W}_k^a := (W_{k,n}^a)_{n \in \mathcal{N}_k}$, where $W_{k,n}^a$ is the expected one-period capacity consumption, or *work* required by the project at state n if action a is chosen; $\boldsymbol{R}_k^a := (R_{k,n}^a)_{n \in \mathcal{N}_k}$, where $R_{k,n}^a$ is the expected one-period *reward* earned from the project at state n if action a is chosen; $\boldsymbol{P}_k^a := (p_{k,n,m}^a)_{n,m \in \mathcal{N}_k}$ is the one-period *transition probability matrix*, where $p_{k,n,m}^a$ is the probability for project k evolving from state n to state m if action a is chosen. The dynamics of project k are captured by *state process* $n_k(\cdot)$ and *action process* $a_k(\cdot)$, which correspond to state $n_k(t)$ and action $a_k(t)$ at the beginning of every time period t.

At each time period t, the choice of action $a_k(t)$ for project k in state $n_k(t)$ entails the consumption of allocated capacity (work) $W_{k,n_k(t)}^{a_k(t)}$, the gain of reward $R_{k,n_k(t)}^{a_k(t)}$, and the evolution of the state to $n_k(t+1)$. A policy's outcome is a mapping $t \mapsto \boldsymbol{a}(t)$ for all t, where $\boldsymbol{a}(t) := (a_k(t))_{k \in \mathcal{K}}$ is the vector of project actions. We are interested in the characterisation of policies with desired properties (detailed below) from the set of *admissible policies* Π, which are randomized, history-dependent, non-anticipative, and which satisfy the sample-path resource constraint $\sum_{k \in \mathcal{K}} W_{k,n_k(t)}^{a_k(t)} = M$ at every time period t. Notice that considering the resource constraint as an equality is without loss of generality, because the problem with an inequality ($\leq M$) can always be reformulated by adding a sufficient number of single-state projects, labelled, say, ℓ, with $\mathcal{N}_\ell := \{*\}, W_{\ell,*}^a := a, R_{\ell,*}^a := 0, p_{\ell,*,*}^a := 1$, to be virtually attended by the unallocated resource units.

When all the parameters of the model are known (fully observable setting) or are assumed using an observational, e.g. Bayesian model (partially observable setting), the typical objective in the field of *stochastic optimisation* is to find/characterise a policy π^{ETA} maximising the expected time-average (ETA) reward rate,

$$\mathbb{R}^{\text{ETA}}(T) := \max_{\pi \in \Pi} \lim_{\tau \to T} \mathbb{E}_0^\pi \left[\frac{1}{\tau} \sum_{t=0}^{\tau-1} \sum_{k \in \mathcal{K}} R_{k,n_k(t)}^{a_k(t)} \right], \tag{1}$$

where $\mathbb{E}_0^\pi$ denotes the expectation over state processes $n_k(\cdot)$ and action processes $a_k(\cdot)$ for all k conditioned on initial states $n_k(0)$ for all k and on policy π. Technical assumptions are required for the maximum and the limit to exist, however for simplicity we use notation as above. Variants of (1), e.g. total discounted reward, are also of interest.

When parameters are not known or it is not desirable to assume any observational model (limited or non-observable/non-parametric setting) and when an analytic solution to a problem is not available, the typical objective in the field of *reinforcement learning* is to find/characterise a policy minimising the expected cumulative regret (ECR) of the reward which captures the lost reward due to the limited knowledge of parameters or observations with respect to the expected time-average reward obtainable in the fully observable setting over an infinite horizon. Several variants of this problem can be considered, depending on the assumptions made about which parameters are known by the decision-maker and what is being observed at each period.

By allowing for a more general set of resources and more general dynamics, the above problem formulation covers the classic multi-armed bandit problem, a popular model in the field of *design of sequential experiments*, which is one of the fundamental topics in statistics and machine learning. The classic *multi-armed bandit* problem can be obtained by setting $M := 1$, $W_{k,n}^a := a$, $R_{k,n}^0 := 0$ for all k, n, a, and taking $\boldsymbol{P}_k^0$ as an identity operator/matrix (i.e. unattended projects do not change state). When unattended projects are allowed to change state (i.e., $\boldsymbol{P}_k^0$ is not necessarily an identity operator) and the resource capacity M is allowed to be any integer between 1 and $K - 1$, this more general problem is known as the *restless multi-armed bandit*.

On the other hand, our problem formulation belongs to a broader, multidisciplinary field of *decision making under uncertainty* (or, recently suggested to be called *sequential decision analytics*). The problem as formulated above thus covers an intermediate family of problems: rich enough to cover a range of important real-world problems, while the explicit structure of these problems allows to develop tractable and effective solution methods. The development of methods, algorithms, and the design of low-complexity policies with provable performance guarantees for automatically solving such problems is an *important research challenge*.

1.2   Related Work

Stochastic optimisation focusses on solving a problem by finding an exact or approximate solution of a well specified optimisation model, which captures the essential features of the problem. The objective is typically maximisation of the expected time-average or total discounted reward over a time horizon, which can be finite or infinite. Probabilistic assumptions must be made about the nature of uncertainty, which may lead to misspecified models. Nevertheless, solutions to such models are often good enough for practical purposes. Classical solution approaches based on direct use of dynamic programming occasionally allow for characterising the structure of the optimal policy. However, exact algorithms quickly become intractable because of the *curse of dimensionality*. A much more viable theoretical approach is therefore to develop and study methods that convert a problem into a simpler, tractable one, either by decomposition or by considering it in an asymptotic regime. A plethora of general methods and algorithms is available, e.g. dynamic programming, Lagrangian relaxation and decomposition, approximate dynamic programming, asymptotically optimal policies, look-ahead approximations, myopic policies, etc.

Using stochastic optimisation approaches, structural results have been obtained for certain families of resource allocation problems. In his celebrated result, Gittins established [13, 16] that a particular type of single-resource allocation problem (known as the *classic multi-armed bandit problem*, described above, with geometrically discounted rewards, and an infinite horizon) can be optimally solved by decomposing it into a collection of single-project parametric optimal-stopping problems, which in turn can be solved by assigning certain dynamic (state-dependent) quantities, now called the Gittins index values. These indices define an optimal policy, the *Gittins index rule*, which prescribes to allocate the resource at every period to the project with currently highest index value. This classic problem in Bayesian setting, under finite horizon, and with non-geometric discounting was thoroughly studied in Berry and Fristedt [6]; see also Russo and van Roy [33], Kaufmann [27].

Index rules and their generalisations became an important concept in addressing sequential resource allocation problems, which are PSPACE-hard in general [32], for their simplicity to implement, economic interpretation, and asymptotic (or near-) optimality. See, e.g. Whittle [40], Gittins [14], Weber and Weiss [39], Niño-Mora [30], Glazebrook et al. [20], Archibald et al. [2], Glazebrook et al. [21], Hauser et al. [23], Jacko [25], Ayesha et al. [4], Gittins et al. [15], Ayesta et al. [5], Villar et al. [36,37], Verloop [35], Larrañaga et al. [28] for a palette of models restricted to $W_{k,n}^a := a$ for all k, n, a. The key observation allowing to derive index rules in such more general models was made by Whittle [40], where he proposed a more general *Whittle index* definition. Subsequent work established asymptotic optimality, index characterisation, algorithms, and performance evaluation of the Whittle index rule. Further index generalisations for problems with unrestricted resource requirements $W_{k,n}^a$ were introduced in Glazebrook and Minty [17], Glazebrook et al. [18], Jacko [26], Graczová and Jacko [22], Glazebrook et al. [19], Hodge and Glazebrook [24]. It is important to note that the decomposition technique leading to index rules is not applicable when the projects are not mutually independent.

Reinforcement learning provides an alternative approach to tackling dynamic problems under uncertainty. It has gained popularity due to not requiring an observational model. Reinforcement learning focusses on solving a problem by starting from very limited or no assumptions with the intention to learn efficiently about the nature of the problem by interacting with a system to be controlled. Information may be limited either because of decision-maker's lack of historical knowledge of observations or conscious lack of model assumptions. Various methods and algorithms are available to improve decisions over time through interaction with the system, e.g., temporal-difference methods, Monte Carlo simulation, gradient-based methods, randomised greedy algorithms, etc. Some reinforcement learning methods work with an ETA objective while others work with ECR objective.

One of the earliest learning approaches is *Q-learning* [34, 38], which has become very popular as a powerful tool useful in a variety of real-life problems. Q-learning is model-free and is suited for learning an optimal policy in general MDPs with the ETA objective where the states, rewards and transition probabilities are unknown. Q-learning estimates the optimal solution by maintaining estimates of the conditional expected total reward given the current history and assuming that (what is currently estimated as) optimal actions are chosen in all future periods, known as Q-values (Q stands for 'Quantity' or more recently 'Quality'). Improvements and extensions of Q-learning are currently a very active research area, with recent advances such as deep Q-learning, which was successfully implemented in software to play games at superhuman level, including AlphaGo where a computer beat the world's best players of Go. Other techniques for learning general MDPs have been developed as well, e.g. Ortner et al. [31] proposed a general algorithm for learning certain structured MDPs, while Burnetas and Katehakis [8] developed a technique to learn an optimal policy for general MDPs where the rewards and transition probabilities are unknown.

In this context, the ECR minimization problem has been heavily studied in the literature in the i.i.d. version of the classic multi-armed bandit problem, i.e. with single resource $(M = 1)$, independent and single-state projects $(\mathcal{N}_k = \{*\}, W_{k,*}^a := a, p_{k,*,*}^a := 1)$, zero-reward unattended projects $(R_{k,*}^0 = 0)$, unknown expected one-period rewards $R_{k,*}^1$, and where observed rewards are only samples from distributions with these unknown means $R_{k,*}^1$. For the non-parametric problem setting see e.g. Burnetas and Katehakis [7], Cowan and Katehakis [10]; for parametric Gaussian problem setting see e.g. Cowan et al. [9] and the references therein. Note that problem (1) is trivial in this i.i.d. case.

Combining Optimisation and Learning. Although the two fields have common roots, they have evolved into separate theoretical research fields with an extremely limited interaction. For problems of dynamic resource allocation, there are only a few works, and typically only very special cases. The Q-learning method for the Gittins index was proposed in Duff [11]. Near-optimal regret guarantees for the finite-horizon Gittins index rule in the Gaussian problem setting are proven in Lattimore [29]. The ECR objective was studied and a learning algorithm was proposed for the setting of Whittle [40] in Ortner et al. [31].

While it is immediate that a policy for optimizing ECR is suboptimal if implemented in a fully observable setting, (1), methods for learning the policy π^{ETA} in the limited or non-observable setting are not well understood yet. For general MDPs, the Q-learning algorithm maintains $Q(n, a)$ for each state-action combination of the MDP, which is an estimate of the expected time-average (or total discounted) reward if action a is chosen in state n in the current period and the optimal policy is followed thereafter. For the problems of dynamic resource allocation, direct implementation of Q-learning would suffer from the curse of dimensionality in the same way as standard stochastic optimisation approaches.

1.3   Learning the Whittle Index Rule

Most related to our paper are works that aim at using Q-learning for learning of index rules rather than of general optimal policies for bandit problems. Specifically, we look at techniques that estimate the Whittle index policy, whereby unit values of work are assigned to the M projects with the highest estimated Whittle indices. The usual relaxation approach [40] results in the problem decomposition allowing to consider each project independently with a wage $\lambda \in \mathbb{R}$ for assigning every unit of work, thus resulting in a *profit* objective calculated as the reward less the work wage paid. Then, the quality of an action in a particular state of the kth project satisfies the dynamic programming equation

$$\beta_k(\gamma, \lambda) + Q_k(\gamma, \lambda, n, a) = R_{k,n}^a - \lambda W_{k,n}^a + \gamma \sum_{m \in \mathcal{N}_k} p_{k,n,m}^a \max_{a' \in \mathcal{A}} Q_k(\gamma, \lambda, m, a'), \quad (2)$$

where γ is the discount factor and $Q_k(\gamma, \lambda, n, a)$ is the profit quantity relative to $\beta_k(\gamma, \lambda)$ which is the optimal expected time-average profit. Under regularity conditions, the system of Eqs. (2) for fixed γ, λ and k implies that the solution $Q_k(\gamma, \lambda, n, a)$ is unique up to an additive constant [1]. Note that when $\gamma < 1$, $\beta_k(\gamma, \lambda) = 0$ and $Q_k(\gamma, \lambda, n, a)$ is the expected total discounted quantity, while when $\gamma = 1$, $Q_k(\gamma, \lambda, n, a)$ is the expected relative time-average quantity.

The policy for control on an individual project maps the set of states available to that project to either a passive (0) or active (1) action $\phi : \mathcal{N}_k \to \mathcal{A}$. An optimal policy (for an individual project) is one such mapping that chooses the action which maximises $Q_k(\gamma, \lambda, n, a)$ for all $n \in \mathcal{N}_k$. Optimal policies depend on λ. For example, choosing a sufficiently high wage will ensure that the passive policy, whereby the passive action is always chosen, becomes optimal. Conversely, choosing a sufficiently negative wage will ensure that the active policy, whereby the active action is always chosen, becomes optimal. In each extreme there exists bounds where the wage dominates the rewards and the optimal policy is fixed as either passive or active for all λ outside the bounds. Between these bounds the optimal policy can change. If the optimal policy transitions at specific values of

λ, such that the optimal policy at each state transitions exactly once, then the project is indexable and wage values at each point of transition are the Whittle index values for their corresponding state. Under this formulation, the Whittle index for an indexable project k in state n ($\lambda_{k,n}$) can be defined as the wage on the active action required to make both actions equally desirable. Namely,

$$Q_k(\gamma, \lambda_{k,n}, n, 1) - Q_k(\gamma, \lambda_{k,n}, n, 0) = 0. \qquad (3)$$

Although indexability (meaning that the project instance satisfies the law of diminishing marginal returns) in theory restricts the applicability of index rules, in practice it is rare to have non-indexable problems [30].

Learning the Whittle index values involves learning both $Q_k(\gamma, \lambda, n, a)$ and $\lambda_{k,n}$ such that (3) is satisfied. Fu et al. [12] accomplish this by storing separate tables of estimated Q values across a fixed search grid of λ values. All Q values are updated each time step using standard Q-learning updates based on a variant of Eq. (2). The Whittle indices are estimated to be the λ grid values that minimise $|Q_k(\gamma, \lambda, n, 1) - Q_k(\gamma, \lambda, n, 0)|$ in each state. A key issue with this approach is that the true Whittle index values are generally excluded from the fixed λ search grid, and the computational requirements scale with the size of the search space.

Avrachenkov and Borkar [3] resolve this issue by using a dynamic search space of $\lambda_{k,n}$ values, which represent estimates of the Whittle index of each state. At each time step after updating the Q values, the Whittle index estimates are also updated at a slower rate, such that the current estimates of $Q_k(\gamma, \lambda_{k,n}, n, 1) - Q_k(\gamma, \lambda_{k,n}, n, 0)$ move closer to 0. In this way, estimated Whittle index values slowly converge to the true Whittle index values (under certain conditions) without increasing the search space size.

We present an alternative approach which is fundamentally different from the approaches in Fu et al. [12] and Avrachenkov and Borkar [3]. These approaches aim at iteratively learning the index values, which results in continuous modification of the index value estimates. On the contrary, our approach aims at iteratively learning the index-induced policies.

We develop a method for problems of dynamic resource allocation built on an understanding of both stochastic optimisation (specifically, the Whittle index) and reinforcement learning (specifically, Q-learning). We propose a *novel implementation of Q-learning*, which exploits the structure of the problem considered, in which the algorithm maintains two sets of Q-values for each project: one for reward and one for work.

2   A New Q-Learning Method for the Whittle Index

We now present a new scheme for Q-learning the Whittle index rule that exploits the policy structure of indexable bandits to remove λ from the Q-learning updates, and significantly constrain the policy space. Our approach can be interpreted as iteratively learning the policy induced by the Whittle index values, instead of iteratively learning the Whittle index values directly which was focus of the previous literature. We believe that in certain cases, this can reduce the error introduced by incorrect estimates of the Whittle index values, thereby improving the convergence rate.

In the rest of the paper, we consider the case $\gamma = 1$, focusing on the time-average criterion, and drop the dependency on γ to simplify the notation. However, an analogous approach for the discounted case can also be formulated. Consider a policy evaluation version of the Bellman Eq. (2),

$$Q_k^\phi(\lambda, n, a) = R_{k,n}^a - \lambda W_{k,n}^a + \sum_{m \in \mathcal{N}_k} p_{k,n,m}^a Q_k^\phi(\lambda, m, \phi(m)) - \overline{Q}_k^\phi(\lambda). \quad (4)$$

There are three key differences between (2) and (4). Firstly, this is a "learning" version of (2), in which the Q values should be interpreted as estimates of the true Q values that satisfy (2). Secondly, the Q values are computed under policy ϕ, which, unlike in (2), might not be optimal. Thirdly, following Avrachenkov and Borkar [3], the $\overline{Q}_k^\phi(\lambda)$ term, where

$$\overline{Q}_k^\phi(\lambda) = \frac{1}{2|\mathcal{N}_k|} \sum_{m \in \mathcal{N}_k} \left[Q_k^\phi(\lambda, m, 1) + Q_k^\phi(\lambda, m, 0) \right], \quad (5)$$

is the time-average relative profit quantity across all states and actions for the given policy ϕ and wage λ, is included instead of $\beta_k(\lambda)$ to ensure that (4) has a unique solution in the case $\gamma = 1$ [1]. A key observation is that for a fixed policy ϕ, (4) forms a linear system. Therefore, the profit Q-value $Q_k^\phi(\lambda, n, a)$ depends linearly on λ and can be decomposed into two parts

$$Q_k^\phi(\lambda, n, a) = Q_k^\phi(n, a) - \lambda H_k^\phi(n, a), \quad (6)$$

where $Q_k^\phi(n, a) := Q_k^\phi(0, n, a)$ represents the profit Q-value when wage $\lambda = 0$, i.e. can be interpreted as the reward Q-value, and $H_k^\phi(n, a)$ can be interpreted as the work Q-value, both under policy ϕ. Policy evaluation equations for both $Q_k^\phi(n, a)$ and $H_k^\phi(n, a)$ can be derived by substituting (6) into (4) and (5) and applying separation of variables

$$Q_k^\phi(n, a) = R_{k,n}^a + \sum_{m \in \mathcal{N}_k} p_{k,n,m}^a Q_k^\phi(m, \phi(m)) - \overline{Q}_k^\phi, \quad (7)$$

$$H_k^\phi(n, a) = W_{k,n}^a + \sum_{m \in \mathcal{N}_k} p_{k,n,m}^a H_k^\phi(m, \phi(m)) - \overline{H}_k^\phi, \quad (8)$$

$$\overline{Q}_k^\phi = \frac{1}{2|\mathcal{N}_k|} \sum_{m \in \mathcal{N}_k} \left[Q_k^\phi(m, 1) + Q_k^\phi(m, 0) \right], \quad (9)$$

$$\overline{H}_k^\phi = \frac{1}{2|\mathcal{N}_k|} \sum_{m \in \mathcal{N}_k} \left[H_k^\phi(m, 1) + H_k^\phi(m, 0) \right]. \quad (10)$$

With this representation, direct dependence on λ is eliminated, and replaced by dependence on ϕ. Solving these new policy evaluation equations and combing the results with (6), yields an evaluation of the policy for all λ, even when the policy is not optimal.

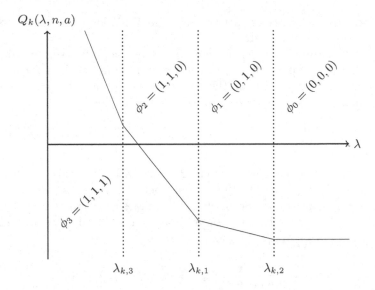

Fig. 1. A 3-state indexable bandit exemplifying the structure of $Q_k(\lambda, n, a)$. $Q_k(\lambda, n, a)$ is a piecewise linear function of λ with corners occurring at the Whittle indices. The optimal policies are independent of λ within each interval of each line segment, but change at the corners where the policy transitions and λ equals a Whittle index of the state that transitions. For example, the optimal policy changes between $(1, 1, 1)$ and $(1, 1, 0)$ on the left, so the Whittle index of the third state equals the wage at the interception point.

If the policy ϕ is optimal, then the Whittle index value for each project k in state n can be computed by combining (3) and (6)

$$\lambda_{k,n} = \frac{Q_k^\phi(n, 1) - Q_k^\phi(n, 0)}{H_k^\phi(n, 1) - H_k^\phi(n, 0)}. \tag{11}$$

Notice that during the learning process (in which policy ϕ may not be optimal), the right-hand side of (11) can be interpreted as the current estimate of the marginal productivity rate, which is the same interpretation as the Whittle index has [30]. One can therefore expect that (11) converges to the Whittle index value when ϕ is optimal, and thus might be a pragmatic choice for learning the Whittle index values and employing such an index rule in the multi-project problem. This is despite the fact that the Whittle index exists only under indexability conditions [30], and the Whittle index rule is asymptotically optimal only under certain technical assumptions [35, 39].

2.1   Algorithmic Learning Approach

Noting that $Q_k^\phi(\lambda, n, a)$ is a linear function of λ, and in cases where the space of possible policies ϕ is finite, the optimal quantity, $Q_k(\lambda, n, a)$, must be a continuous piecewise linear function of λ. If the project is indexable with a unique

Whittle index for each state, then the optimal policy is fixed within each piece. Furthermore, the set of states that the optimal policies activate in indexable projects decreases monotonically with λ, so the set of optimal policies can be represented by an ordering of the states. Figure 1 illustrates an example of this structure.

Algorithm 1: Learning Whittle index control policy via policy iteration

for $k \in \mathcal{K}$ do
    for $l \in \{0, 1 \ldots, |\mathcal{N}_k|\}$ do
        | Initialise Q_k^l and H_k^l tables;
    end
    Initialise optimal policy and Whittle index estimates;
end
for $t \in \{1, 2, \ldots, T\}$ do
    Randomly decide to explore based on an exploration schedule;
    if *Exploring* then
        | Activate M projects randomly with equal probability;
    else
        | Activate M projects with highest estimated Whittle indices (breaking
        | ties randomly);
    end
    for $k \in \mathcal{K}$ do
        for $l \in \{0, 1 \ldots, |\mathcal{N}_k|\}$ do
            | Update Q_k^l and H_k^l for observed project transition and reward via
            | equations (12) and (13);
        end
        Update policy and Whittle index estimates via algorithm 2;
    end
end

In practice, our method stores $|\mathcal{N}_k|+1$ tables for Q and H and for each project type. These correspond to the optimal policies that exist between the Whittle indices of an indexable project which occur at the policy transitions where λ is a Whittle index. We denote these policies as ϕ_l and the Q and H values as $Q_k^l(n, a) := Q_k^{\phi_l}(n, a)$ and $H_k^l(n, a) := H_k^{\phi_l}(n, a)$, where $l \in \{0, 1, \ldots, |\mathcal{N}_k|\}$ indicates the size of the set of states that the policy activates.

Our method of learning works by iteratively estimating the optimal policies ϕ_l from the $Q_k^l(n, a)$ and $H_k^l(n, a)$ estimates, and updating these estimates using the following Q updates for all l and for all observed transitions and rewards at each time step t

$$Q^\phi_{k,t+1}(n,a) = Q^\phi_{k,t}(n,a) + \alpha \left(R^a_{k,n} + Q^\phi_{k,t}(s', \phi(s')) - \overline{Q}^\phi_{k,t} - Q^\phi_{k,t}(n,a) \right),$$

(12)

$$H^\phi_{k,t+1}(n,a) = H^\phi_{k,t}(n,a) + \alpha \left(W^a_{k,n} + H^\phi_{k,t}(s', \phi(s')) - \overline{H}^\phi_{k,t} - H^\phi_{k,t}(n,a) \right),$$

(13)

where,

$$\overline{Q}^\phi_{k,t} = \frac{1}{2|\mathcal{N}_k|} \sum_{m \in \mathcal{N}_k} \left[Q^\phi_{k,t}(m,1) + Q^\phi_{k,t}(m,0) \right],$$

(14)

$$\overline{H}^\phi_{k,t} = \frac{1}{2|\mathcal{N}_k|} \sum_{m \in \mathcal{N}_k} \left[H^\phi_{k,t}(m,1) + H^\phi_{k,t}(m,0) \right],$$

(15)

where $\alpha \in (0,1)$ is a Q-learning learning rate. Algorithm 1 outlines these steps in more detail.

The policies ϕ_l, and Whittle index values are estimated using Algorithm 2. It works by beginning at $l = 0$ where ϕ_0 is the passive policy and is known to be optimal for all λ above the maximum Whittle index, and then choosing the state which maximises the marginal productivity rate computed by Eq. (11). The Whittle index of this state is estimated to be the marginal productivity rate and the next policy at $l = 1$ is estimated to be the same as the previous with the exception of this state. This process is repeated where the marginal productivity rate is maximised from the set of states available, and the subsequent optimal policy estimate ϕ_l follows

$$\phi_{l+1}(n) = \begin{cases} 1 - \phi_l(n), & n = n', \\ \phi_l(n), & \text{otherwise.} \end{cases}$$

(16)

Here n' is the available state that maximises the marginal productivity rate. Notice that estimating the optimal policies in this way constrains the space of available policy options to those of indexable projects. In doing so, we anticipate convergence to the true optimal policies could be accelerated.

The algorithm could be equivalently formulated in the reverse direction starting from $l = |\mathcal{N}_k|$ where the optimal policy $\phi_{|\mathcal{N}_k|}$ is known to be the active policy, and then choosing states that minimise the marginal productivity rate. Actually, each step of the algorithm can proceed from either the "top" or the "bottom", and we alternate between these in practice.

3   Illustrative Numerical Comparison

A comprehensive empirical comparison between control algorithms would involve controlling a broad range of restless multiarmed bandits of different sizes and types, and comparing a range of performance metrics. However, in this paper, we begin with a single test case to demonstrate that the algorithm can work in practice.

Algorithm 2: Whittle indices and policies from Q and H estimates

---

input : $Q_k^l(n,a), H_k^l(n,a) \quad \forall l \in \{0,1,\ldots,|\mathcal{N}_k|\}, \forall n \in \mathcal{N}_k, \forall a \in \mathcal{A}$

output: Estimated Whittle index ordering and values

$\Lambda_k^l(n) \leftarrow \frac{Q_k^l(n,1) - Q_k^l(n,0)}{H_k^l(n,1) - H_k^l(n,0)}$;

$(l_{\text{top}}, l_{\text{bottom}}) \leftarrow (0, |\mathcal{N}_k|)$;

$(\lambda_{\text{top}}, \lambda_{\text{bottom}}) \leftarrow (+\infty, -\infty)$;

remaining $\leftarrow \{1,\ldots,|\mathcal{N}_k|\}$;

forward order $\leftarrow$ empty tuple;

backward order $\leftarrow$ empty tuple;

while $|remaining| > 0$ **do**

    decide to iterate from the top or bottom remaining index;

    if *iterate from top* **then**

        $n' \leftarrow \arg\max_{n \in \text{remaining}} \Lambda_k^{l_{\text{top}}}(n)$;

        // clamp index between previous values

        $\lambda_{\text{top}} \leftarrow \max\left\{\lambda_{\text{bottom}}, \min\left\{\Lambda_k^{l_{\text{top}}}(n'), \lambda_{\text{top}}\right\}\right\}$;

        append n' to end of forward order;

        $\lambda_{k,n'} \leftarrow \lambda_{\text{top}}$;

        $l_{\text{top}} \leftarrow l_{\text{top}} + 1$;

        remaining $\leftarrow$ remaining $\setminus \{n'\}$;

    else

        $n' \leftarrow \arg\min_{n \in \text{remaining}} \Lambda_k^{l_{\text{bottom}}}(n)$;

        // clamp index between previous values

        $\lambda_{\text{bottom}} \leftarrow \min\left\{\lambda_{\text{top}}, \max\left\{\Lambda_k^{l_{\text{bottom}}}(n'), \lambda_{\text{bottom}}\right\}\right\}$;

        append n' to beginning of backward order;

        $\lambda_{k,n'} \leftarrow \lambda_{\text{bottom}}$;

        $l_{\text{bottom}} \leftarrow l_{\text{bottom}} - 1$;

        remaining $\leftarrow$ remaining $\setminus \{n'\}$;

    end

end

ordering $\leftarrow$ forward order concatenated with backward order;

3.1 A Birth and Death Test Case

Our restless multi-armed bandit is adapted from the Mentoring Instructions example presented by Fu *et al.* [12] and consists of $K = 100$ students (projects) and $M = 10$ mentors which can each mentor a student at each time step. Each student's state sits in one of 5 ordered study levels $\mathcal{N}_k = \{1,2,3,4,5\}$, and can increase or decrease randomly each time step according to the following birth and death transition probabilities

$$P_k^0 = \begin{bmatrix} 0.7 & 0.3 & 0 & 0 & 0 \\ 0.7 & 0 & 0.3 & 0 & 0 \\ 0 & 0.7 & 0 & 0.3 & 0 \\ 0 & 0 & 0.7 & 0 & 0.3 \\ 0 & 0 & 0 & 0.7 & 0.3 \end{bmatrix}, \qquad P_k^1 = \begin{bmatrix} 0.3 & 0.7 & 0 & 0 & 0 \\ 0.3 & 0 & 0.7 & 0 & 0 \\ 0 & 0.3 & 0 & 0.7 & 0 \\ 0 & 0 & 0.3 & 0 & 0.7 \\ 0 & 0 & 0 & 0.3 & 0.7 \end{bmatrix}, \qquad (17)$$

where the active action corresponds to the student receiving mentorship. Students in higher study levels give higher rewards each time step following $R_{k,n}^a = \sqrt{\frac{n}{5}}$. With these transition probablilities and expected rewards the Whittle indices can be computed to be

$$(\lambda_{k,n})_{n \in \mathcal{N}_k} = (0.3166\ldots, 0.5032\ldots, 0.5510\ldots, 0.5512\ldots, 0.1037\ldots). \qquad (18)$$

The Whittle index policy for controlling this system is to prioritise allocating mentors to students in the 4th study level, then 3rd, 2nd, 1st and then leaving students in the 5th study level with the lowest priority.

We compare our method to Avrachenkov and Borkar's method (denoted AB) [3] for controlling this system. In each case we use a fixed exploration rate of 10%. This means at each time step there is a 10% probability of exploring, where all mentors are allocated randomly to the students with equal probability. As outlined in Algorithm 1, when not exploring the controller assigns the mentors to the M students with highest estimated Whittle indices. Additionally we fix the learning rate in both learning algorithms to be the same at $\alpha = 0.01$, and the Whittle index learning rate in AB to 0.001. Note that the AB method requires a learning rate parameter for the Whittle index which is significantly smaller than the Q learning rate.

We ran 32 simulations for each controller for 10,000 time steps each using these parameters. The initial state of each simulation was initialised by first controlling with the Whittle index policy, including exploration, for at least 5000 time steps.

3.2   Test Case Results

We first compare how the two methods learn the Whittle index values. Figure 2 shows a typical example of how these controllers estimate the index values at each time step. There are two key differences in how the methods estimate them. Firstly, the AB method estimates of the Whittle index values vary much more smoothly than ours. However, our method appears to converge much faster before fluctuating about the correct values. The size of these fluctuations depends on the learning rate, and we anticipate that employing a learning rate schedule that tends towards zero rather than a fixed rate would allow the noise to also tend towards zero.

Despite the fluctuations, our method learns the Whittle index policy earlier than AB and consistently employs it at a higher rate. The left part of Fig. 3 plots the proportion of the 32 simulations in which the controller has learnt the correct Whittle index policy. Our method fluctuates just below 50%, about

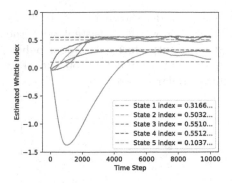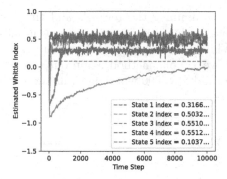

Fig. 2. Estimated Whittle index values when using AB (left) and our controller (right) for a single simulation each. Solid lines represent the estimated indices for each state at each time step and dashed lines represent the true Whittle index values. Note that the third (green) and fourth (red) Whittle index values are very close and appear as a single red dashed line at the top. AB estimates the indices more smoothly but also learns them more slowly. Our method is more noisy but could be smoothed by using a learning rate schedule that decays towards zero. (Color figure online)

double that of AB. The reason it does not converge to 100% is again due to the fixed learning rate, and also the very small difference between state 3 and state 4 Whittle indices. Learning the Whittle index policy faster and employing it more frequently is probably the reason our method achieved lower cumulative regret on average, plotted on the right part of Fig. 3.

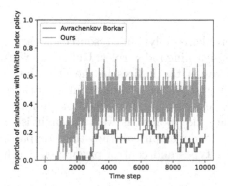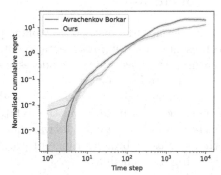

Fig. 3. Proportion of simulations that learnt the correct Whittle index state ordering (left) and normalised cumulative regret (right) averaged across 32 simulations. The cumulative regret is normalised so that the expected reward from the Whittle index policy with exploration is 1. Shaded regions represent standard errors about the sample mean. The cumulative regret when using our method is consistently lower than when using AB, probably because our method learns the Whittle index policy sooner and employs it more frequently.

Examining the rewards and time average rewards, shown in Fig. 4, shows that both methods appear to eventually produce rewards consistent with the Whittle index policy with exploration, and both outperform a random policy. However, our method again outperforms AB except around $t = 150$ and after around $t > 3000$ where both methods produce similar rewards.

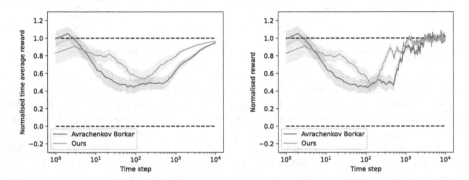

Fig. 4. Normalised time average reward (left) and normalised rewards (right) averaged across 32 simulations. Normalised so that 1 is the expected reward when following the Whittle index policy with the same exploration rate, and 0 is the expected reward following a random policy. The right is smoothed using a moving average of 100 time steps. Shaded regions represent standard errors about the sample mean.

4   Conclusion

We have developed a method controlling problems of dynamic resource allocation and presented a novel implementation of Q-learning index rules. Unlike previous Q-learning schemes which include estimates of the indices in the Q updates, our method updates based on policy estimates which are constrained to those available in indexable systems. In a single test system our method outperforms previous methods in learning the Whittle index policy as well as maximising expected rewards. These results warrant a broader empirical study that compares methods across a broader range of project types and control parameters, such as learning rate schedules.

Acknowledgements. L. J. Gibson's and Y. Nazarathy's research is supported by ARC grant DP180101602. L. J. Gibson is also supported by the ARC Centre of Excellence for Mathematical and Statistical Frontiers (ACEMS).

References

1. Abounadi, J., Bertsekas, D., Borkar, V.S.: Learning algorithms for Markov decision processes with average cost. SIAM J. Control Optim. **40**(3), 681–698 (2001)

2. Archibald, T.W., Black, D.P., Glazebrook, K.D.: Indexability and index heuristics for a simple class of inventory routing problems. Oper. Res. **57**(2), 314–326 (2009)
3. Avrachenkov, K.E., Borkar, V.S.: Whittle index based Q-learning for restless bandits with average reward. arXiv:2004.14427 (2021)
4. Ayesta, U., Erausquin, M., Jacko, P.: A modeling framework for optimizing the flow-level scheduling with time-varying channels. Perform. Eval. **67**, 1014–1029 (2010)
5. Ayesta, U., Jacko, P., Novak, V.: Scheduling of multi-class multi-server queueing systems with abandonments. J. Sched. **20**(2), 129–145 (2015). https://doi.org/10.1007/s10951-015-0456-7
6. Berry, D.A., Fristedt, B.: Bandit Problems: Sequential Allocation of Experiments. Springer, Netherlands (1985). https://doi.org/10.1007/978-94-015-3711-7
7. Burnetas, A.N., Katehakis, M.N.: Optimal adaptive policies for sequential allocation problems. Adv. Appl. Math. **17**, 122–142 (1996)
8. Burnetas, A.N., Katehakis, M.N.: Optimal adaptive policies for Markov decision processes. Math. Oper. Res. **22**(1), 222–255 (1997)
9. Cowan, W., Honda, J., Katehakis, M.N.: Normal bandits of unknown means and variances. J. Mach. Learn. Res. **18**, 154 (2017)
10. Cowan, W., Katehakis, M.N.: Minimal-exploration allocation policies: Asymptotic, almost sure, arbitrarily slow growing regret. arXiv:1505.02865v2 (2015)
11. Duff, M.O.: Q-Learning for bandit problems. In Proceedings of the Twelfth International Conference on Machine Learning, 32 p. CMPSCI Technical Report 95–26 (1995)
12. Fu, J., Nazarathy, Y., Moka, S., Taylor, P.G.: Towards q-learning the whittle index for restless bandits. In: 2019 Australian New Zealand Control Conference (ANZCC), pp. 249–254 (2019)
13. Gittins, J.C.: Bandit processes and dynamic allocation indices. J. Roy. Stat. Soc. Ser. B **41**(2), 148–177 (1979)
14. Gittins, J.C.: Multi-Armed Bandit Allocation Indices. Wiley, New York (1989)
15. Gittins, J.C., Glazebrook, K., Weber, R.: Multi-Armed Bandit Allocation Indices. Wiley, New York (2011)
16. Gittins, J.C., Jones, D.M.: A dynamic allocation index for the sequential design of experiments. In: Gani, J. (ed.) Progress in Statistics, pp. 241–266. North-Holland, Amsterdam (1974)
17. Glazebrook, K., Minty, R.J.: A generalized Gittins index for a class of multiarmed bandits with general resource requirements. Math. Oper. Res. **34**(1), 26–44 (2009)
18. Glazebrook, K.D., Hodge, D.J., Kirkbride, C.: General notions of indexability for queueing control and asset management. Ann. Appl. Probab. **21**, 876–907 (2011)
19. Glazebrook, K.D., Hodge, D.J., Kirkbride, C., Minty, R.J.: Stochastic scheduling: a short history of index policies and new approaches to index generation for dynamic resource allocation. J. Sched. **17**(5), 407–425 (2013). https://doi.org/10.1007/s10951-013-0325-1
20. Glazebrook, K.D., Kirkbride, C., Mitchell, H.M., Gaver, D.P., Jacobs, P.A.: Index policies for shooting problems. Oper. Res. **55**(4), 769–781 (2007)
21. Glazebrook, K.D., Kirkbride, C., Ouenniche, J.: Index policies for the admission control and routing of impatient customers to heterogeneous service stations. Oper. Res. **57**(4), 975–989 (2009)
22. Graczová, D., Jacko, P.: Generalized restless bandits and the knapsack problem for perishable inventories. INFORMS Oper. Res. **62**(3), 696–711 (2014)
23. Hauser, J.R., Urban, G.L., Liberali, G., Braun, M.: Website morphing. Market. Sci. **28**(2), 202–223 (2009)

24. Hodge, D.J., Glazebrook, K.D.: On the asymptotic optimality of greedy index heuristics for multi-action restless bandits. Adv. Appl. Probab. **47**(3), 652–667 (2015)
25. Jacko, P.: Dynamic Priority Allocation in Restless Bandit Models. Lambert Academic Publishing, Sunnyvale (2010)
26. Jacko, P.: Resource capacity allocation to stochastic dynamic competitors: knapsack problem for perishable items and index-knapsack heuristic. Ann. Oper. Res. **241**(1), 83–107 (2016)
27. Kaufmann, E.: On Bayesian index policies for sequential resource allocation. Ann. Stat. **46**(2), 842–865 (2018)
28. Larrañaga, M., Ayesta, U., Verloop, I.M.: Dynamic control of birth-and-death restless bandits: application to resource-allocation problems. IEEE/ACM Trans. Networking **24**(6), 3812–3825 (2016)
29. Lattimore, T.: Regret analysis of the finite-horizon Gittins index strategy for multi-armed bandits. J. Mach. Learn. Res. **49**, 1–32 (2016)
30. Niño-Mora, J.: Dynamic priority allocation via restless bandit marginal productivity indices. TOP **15**(2), 161–198 (2007)
31. Ortner, R., Ryabko, D., Auer, P., Munos, R.: Regret bounds for restless Markov bandits. Theor. Comput. Sci. **558**, 62–76 (2014)
32. Papadimitriou, C.H., Tsitsiklis, J.N.: The complexity of optimal queueing network. Math. Oper. Res. **24**(2), 293–305 (1999)
33. Russo, D., van Roy, B.: Learning to optimize via posterior sampling. Math. Oper. Res. **39**(4), 1221–1243 (2014)
34. Sutton, R.S., Barto, A.G.: Reinforcement Learning: An Introduction. MIT Press, Cambridge (1998)
35. Verloop, I.M.: Asymptotically optimal priority policies for indexable and nonindexable restless bandits. Ann. Appl. Probab. **26**(4), 1947–1995 (2016)
36. Villar, S.S., Bowden, J., Wason, J.: Multi-armed bandit models for the optimal design of clinical trials: benefits and challenges. Stat. Sci. **30**(2), 199–215 (2015)
37. Villar, S.S., Wason, J., Bowden, J.: Response-adaptive randomization for multi-arm clinical trials using the forward looking Gittins index rule. Biometrics **71**, 969–978 (2015)
38. Watkins, C.J.C.H.: Learning From Delayed Rewards. PhD thesis, Cambridge University, UK (1989)
39. Weber, R., Weiss, G.: On an index policy for restless bandits. J. Appl. Probab. **27**(3), 637–648 (1990)
40. Whittle, P.: Restless bandits: activity allocation in a changing world. In: Gani, J. (ed.), A Celebration of Applied Probability, Journal of Applied Probability, vol. 25, Issue A, pp. 287–298 (1988)

Dynamic Routing Problems with Delayed Information

Esa Hyytiä[1](✉) and Rhonda Righter[2]

[1] Department of Computer Science, University of Iceland, Reykjavik, Iceland
esa@hi.is
[2] Department of Industrial Engineering and Operations Research,
University of California Berkeley, Berkeley, CA, USA
rrighter@berkeley.edu

Abstract. The problem of routing jobs to parallel servers is known as the dispatching problem. A typical objective is to minimize the mean response time, which according to Little's result is equivalent to minimizing the mean number in the system. Dynamic dispatching policies are based on information about the state of each server. In large or real-time systems, up-to-date and accurate system state may not be available to dispatcher. We consider, cases where state information at some time in the past is available, or completed jobs are acknowledged after some propagation delay, and give efficient dispatching policies based on the incomplete state information. The dynamic dispatching policies tailored to this setting are evaluated numerically.

Keywords: Job dispatching · Delay-aware policy · JSQ · SED

1 Introduction

We consider parallel server systems known as dispatching systems under a setting of imperfect state information. Dispatching systems comprise a dispatcher and a pool of parallel servers. New jobs arrive to the dispatchers, which then routes them to different servers so as to balance the load or to minimize the mean response time. Dispatching policies such as join-the-shortest-queue (JSQ) expect the exact and current number in queue from all servers [6], which we refer to as perfect state information. With perfect state information, JSQ is often the optimal policy with respect to mean response time, e.g., with homogeneous exponential service times [16].

In contrast, we study a scenario where the dispatcher has imperfect information about the current state of each server. For example, in large or real-time systems, it may not be feasible to query the state of each server prior too each routing decision. In our setting, servers inform the dispatcher about state changes. However, such reports are not immediately available to the dispatcher, but are delayed [7]. Note that applying the basic JSQ without taking the delays

© ICST Institute for Computer Sciences, Social Informatics and Telecommunications Engineering 2021
Published by Springer Nature Switzerland AG 2021. All Rights Reserved
Q. Zhao and L. Xia (Eds.): VALUETOOLS 2021, LNICST 404, pp. 171–184, 2021.
https://doi.org/10.1007/978-3-030-92511-6_11

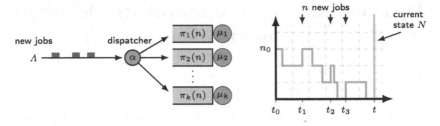

Fig. 1. Dispatching system with imperfect information (left) and evolution of the number in the server since the last status update at time t_0 (right).

into account can even lead to oscillations in the number at servers [12]. Finding optimal policies is intractable for the general model. Artiges studies properties of an optimal policy for two servers using MDP's [3]. Litvak and Yechiali [8] consider the trade-off between waiting for information before routing and routing without information, and Altman et al. [2] study a discrete-time version with deterministic service times. Most recent work on routing with delayed information has focused on asymptotic regimes [11,13,15]. We focus on analyzing heuristics.

In general, the dispatching decision can be based on different amounts of information:

1. **No information.** In this case, the server must implement a static Bernoulli split (RND) policy that may be based on the arrival rate Λ and service rates μ. In particular, in the heterogeneous case where $\mu_i \neq \mu_j$ for some i, j, one can, e.g., balance the load by routing a job to server i with probability $p_i = \mu_i / \sum_j \mu_j$. Optimal splitting probabilities can also be determined [4].
2. **Routing history** refers to the case where the dispatcher records past routing decisions, but gets no feedback from the servers. In this scenario one often implements the Round-Robin (RR) policy, or a weighted variant in the case of heterogeneous servers. RR is often the optimal policy in this type of setting [5, 9,10]. Consequently, whenever the state information is unclear, e.g., because the feedback loop is slow or acknowledgements are simply unreliable, RR can be expected to be near-optimal.
3. **Perfect information** means that the dispatcher knows the number at each server exactly. For example, the well-known JSQ and SED (Shortest Expected Delay) policies generally assume this (see Sect. 3.1).
4. **Delayed information** means that the dispatcher knows the routing history and the number at each server at some time instant in the past. Consequently, it can determine the distribution for the state at each server. This scenario is depicted in Fig. 1.

Our focus in this paper is the last scenario 4). In scenarios 1) and 2), less information is available, and thus the corresponding policies can be readily applied also in our setting. In contrast, dispatching policies based on perfect information, as assumed in scenario 3), serve as lower bounds.

2   Server State Known in the Past

Let us first consider a single server and its state distribution at some time t. Suppose that the available information is the exact number of jobs at the server at time t_0, $t_0 \le t$, after which n jobs have been routed to the server, at time instants $t_1, \ldots, t_n$, such that $t_0 < t_1 < \ldots < t_n \le t$. The unknown current number at the server is denoted by N. The situation is depicted in Fig. 1. The thick yellow curve depicts *one* possible sample path. Clearly, $0 \le N \le n_0 + n$.

To summarize, we assume that the dispatcher knows the following:

1. At time t_0 server had exactly n_0 active jobs. This is the most recent status update. Note that, because of the exponential services and Poisson arrivals, earlier updates can be ignored.
2. Since then, n new jobs have been routed to the server at time instants
   $t_1, \ldots, t_n$ (routing history is known)
3. Service times are i.i.d., $X \sim \mathrm{Exp}(\mu)$ (μ is assumed to be known/learned)

First we make an observation regarding the exponential service times.

Remark 1: *The G/M/1 queue can be seen as a system where the server runs an exponential timer that triggers at rate μ. Every time the timer goes off one job is released if present. If the system was empty, the server "shoots a blank".*

The information about the state at time t_0 could also be a distribution, but for now we assume that status updates have no errors. The first task is to determine the state probability distribution at time $t = t_{n+1}$ based on this information. Let $\Delta_i = t_i - t_{i-1}$ and $D_i \sim \mathrm{Poisson}(\mu\Delta_i)$, where D_i denotes the number of potential departures during the time interval Δ_i. Let N_1 denote the number in the system immediately after time t_1, for which it holds that

$$N_1 = 1 + (n_0 - D_1)^+, \quad \text{where } 0 < N_1 \le n_0 + 1,$$

and $(x)^+ := \max\{0, x\}$. Similarly, immediately after time t_j, $j = 2, \ldots, n$,

$$N_j = 1 + (N_{j-1} - D_j)^+, \quad \text{where } 0 < N_j \le n_0 + j.$$

No job arrives after the final time interval (t_n, t), and thus the number in the system at time t is

$$N = (N_n - D_{n+1})^+, \quad \text{where } 0 \le N \le n_0 + n.$$

Note that N depends on the service rate μ, but not on parameters of the arrival process because we know the actual arrival pattern during (t_0, t). That is, we assume exponential service times, but the arrival process can be arbitrary, i.e., the so-called G/M/1 model. Putting these together yields Algorithm 1 that computes the current state distribution $\boldsymbol{\pi} = (\pi_0, \ldots, \pi_{n_0+n})$ of the random variable N, where $\pi_i = \mathbb{P}\{N = i\}$.

---

procedure FINDN$(n_0, t_0, \ldots, t_n, t)$
  $\pi = (0, \ldots, 0, 1)$ ▷ $\pi_{n_0} = 1$
  for $i = 1, \ldots, n$ **do**
    $\pi \leftarrow$ UPDATE$(\pi, t_i - t_{i-1}, 1)$
  end for
  $\pi \leftarrow$ UPDATE$(\pi, t - t_n, 0)$
  return π
end procedure

Algorithm 1:   Computation of the state distribution based on the routing history and the state of the server at the start.

---

procedure UPDATE(π, Δ, a)
  $k \leftarrow |\pi|$ ▷ initially at most $k - 1$ jobs
  $\pi^* \leftarrow (0, \ldots, 0)$ ▷ $|\pi^*| = k - 1 + a$
  for $i = 0, \ldots, k - 1$ **do**   ▷ condition on i jobs initially
    $s \leftarrow 1$ ▷ find $s = \mathrm{P}\{$empty system$\}$
    for $j = 0, \ldots, i - 1$ **do**   ▷ condition on j jobs departing
      $p \leftarrow (\mu\Delta)^j / j! \cdot e^{-\mu\Delta}$ ▷ pr. that j jobs completed
      $\pi^*_{i-j+a} \leftarrow \pi^*_{i-j+a} + p\pi_i$
      $s \leftarrow s - p$
    end for
    $\pi^*_a \leftarrow \pi^*_a + s\pi_i$ ▷ a jobs to an empty system
  end for
  return π^*
end procedure

Update of the state distribution after time Δ.

Example 1. Let us consider the arrival pattern depicted in Fig. 1. At time $t_0 = 0$, the system has $n_0 = 3$ jobs (latest status update received). New jobs arrive at time instants $\{1.6, 3.3, 4.2\}$, and the service rate is $\mu = 1$. Figure 2 illustrates how our belief on the number N_t behaves as a function of time t. After the last arrival, at time $t = 4.2$, both the mean $\mathbb{E}[N_t]$ and the standard deviation σ_t gradually decrease to zero. Eventually we can be fairly sure that the system is empty even though no status report has been received since time $t_0 = 0$. A naïve assumption that the service time of all jobs is exactly the mean $1/\mu = 1$ would imply that the server is idle already at time $t = 6$. This is far from the reality, as we can see from the figure!

3   Dispatching with Known Past State

Suppose next that a dispatcher is routing jobs to k servers. The dispatcher keeps record of jobs it has routed to each server (say a job id and time stamp) that allows it to deduce which jobs have not been acknowledged yet by each server.

If we assume (i) first-come-first-served (FCFS) scheduling, and (ii) that there is no reordering (jobs arrive "immediately" to the server), a message about completion of job i acknowledges also all earlier jobs. Consequently, the dispatcher

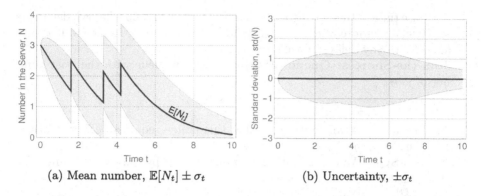

(a) Mean number, $\mathbb{E}[N_t] \pm \sigma_t$ (b) Uncertainty, $\pm \sigma_t$

Fig. 2. Evolution of the mean number in the system with confidence intervals.

knows the exact number at the server at time t_0 when the acknowledgement was generated. We note that this resembles how (basic) TCP works: cumulative acknowledgments inform the sender that all octets up to a given point have reached the destination.

In general, we can assume any work-conserving scheduling discipline because (i) each departing job is (eventually) acknowledged, (ii) our performance metric is the mean response time (cf. Little's formula), (iii) service is exponential. Hence, our results hold also for processor sharing (PS) and last-come-first-served (LCFS).

In summary, our model for the dispatching system is as follows:

1. Jobs arrive to the dispatcher at inter-arrival times[1] A_i.
2. Job sizes are independent and exponentially distributed.
3. The server pool comprises k servers with service rates $\mu_1, \ldots, \mu_k$. Thus the service time at server j is $X_j \sim \mathrm{Exp}(\mu_j)$.
4. The dispatcher routes jobs immediately upon arrival to the servers. The routing decision is irrevocable.
5. The dispatcher is aware of (e.g., it has learned) the server-specific service rates μ_j. Moreover, it is aware of the state of each server at some time in the past, $(n_0^{(j)}, t_0^{(j)})$, where $(n_0^{(j)}, t_0^{(j)})$ gets updated by acknowledgements as the process continues.

3.1 Dispatching Policies

Let us next introduce the dispatching policies considered in this paper.

P1 **Round-Robin (RR)** assigns jobs sequentially to all servers, $1, 2, \ldots,$ $k, 1, 2, \ldots$. We note that RR is the optimal policy with respect to the mean

---

[1] The results for the distribution of N, and therefore also the corresponding policies, hold for any arrival pattern, even non-renewal, though in our numerical examples we consider only Poisson arrival processes.

response time in this setting with identical servers and no acknowledgements are available. Moreover, given RR ignores any acknowledgements, it is also agnostic to any delays in them.

P2 **Join-the-Shortest-Queue (JSQ)** chooses the queue with the least number of jobs,

$$\alpha_{JSQ} := \operatorname*{argmin}_{j} N_j.$$

Ties are resolved randomly. Note that this policy requires perfect information, i.e., that the N_j are known.

P3 **Naïve JSQ$_0$:** We can adapt JSQ to our setting by falsely assuming that the available information describes the current state accurately, i.e., that $N_j \approx n_0^{(j)} + n^{(j)}$, and

$$\alpha_{JSQ_0} := \operatorname*{argmin}_{j} n_0^{(j)} + n^{(j)}.$$

In other words, we apply JSQ without being aware of the delays in acknowledgements. Note that $n_0^{(j)} + n^{(j)}$ corresponds to the maximum number of jobs server j may have at time t. This policy will be equivalent to RR if we start empty and servers are homogeneous.

P4 **Time-aware JSQ$_e$:** Given we can compute the distribution of N_j for all j, JSQ can be generalized to the case of delayed information by choosing the queue which is expected to have the least number of jobs by defining[2]

$$\alpha_{JSQ_e} := \operatorname*{argmin}_{j} \mathbb{E}[N_j].$$

3.2 Numerical Example 1

The first example system consists of 4 identical servers with service rates $\mu_i = 1$ for all i. Jobs arrive according to a Poisson process with rate $\lambda = 3$ to a dispatcher. Initially, at time $t = 0$, the servers have $\mathbf{n_0} = (1, 2, 3, 4)$ jobs, which is the available information to dispatching policies JSQ$_0$ and JSQ$_e$ throughout the time horizon (so there is no feedback from the servers). We assume RR will assign the next job to server 1. (This would be consistent with partial utilization of the available information about the initial state.) In contrast, RND is static, whereas JSQ utilizes the exact state information for every action.

The numerical results with $10,000$ simulation runs are depicted in Fig. 3. On the x-axis is the time t, and the y-axis corresponds to the average costs incurred during $(0, t)$,

---

[2] In principle, it is possible that no job has departed from a busy server since the start. That is, the number of possible states increases without bound as the process continues, and evaluating the distribution and its mean eventually becomes cumbersome. As a workaround, our implementation of JSQ$_e$ truncates the state-space to $k_{max} = 20$ jobs (per server), and updates the state probabilities accordingly whenever a new job arrives. Given the load is reasonable, the effect of this modification will be negligible.

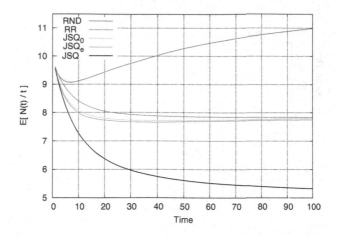

Fig. 3. The evolution of the mean cost from the known initial state $\mathbf{n_0} = (1, 2, 3, 4)$ as a function of time with four identical servers.

acknowledgements

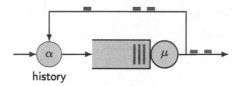

history

Fig. 4. Propagation delay influences the optimal dispatching.

$$C(t) := \frac{1}{t} \int_0^t \mathbb{E}[N_{\text{sys}}(h)] \, dh,$$

where $N_{\text{sys}}(h)$ denotes the total number of jobs in the whole system at time h. Obviously $C(t) \to \mathbb{E}[N_{\text{sys}}]$ as $t \to \infty$, and we can observe that the chosen dispatching policies fall into three groups: RND has the worst performance, $\{\text{RR}, \text{JSQ}_0, \text{JSQ}_e\}$, all reduce to RR as the time increases and the pure JSQ is the best (thanks to the exact state information). A closer look at the middle group shows that RR is initially worse than JSQ_0 and JSQ_e, of which the latter is marginally better. That is, JSQ_e utilizes the available information best.

3.3 Numerical Example 2

Let us now consider a scenario where the dispatcher receives acknowledgements from job completions after a fixed delay of d (see Fig. 4), and where t_0 is set to be the time the last acknowledgement was received.

Figure 5 depicts the mean response time with RR, JSQ, JSQ_0 and JSQ_e, as a function of the ACK delay d that is varied from zero to 4. JSQ_0 behaves as if d were zero, i.e., the decision is based on estimating the number at the server

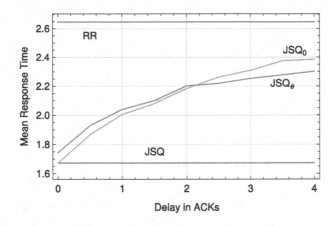

Fig. 5. Simulation results with 4 identical servers as a function of ACK delay d.

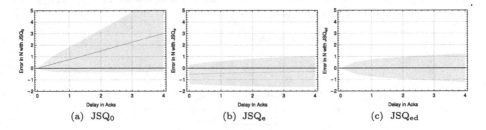

Fig. 6. Mean and standard deviation of the bias in the server state estimates, Q, with JSQ_0, JSQ_e and JSQ_{ed} as a function of ACK delay d.

simply by $\hat{N} = n_0 + n$ (which is a strict upper bound). On the other hand, JSQ_e is based on the mean of the distribution computed using Algorithm 1 with t_0, the time of the most recent acknowledgement. Initially, when $d = 0$, JSQ_0 reduces to JSQ and it is thus optimal in our setting. In contrast, the performance with JSQ_e is significantly worse. However, when d increases JSQ_e eventually becomes better than the elementary JSQ_0, and $d = \infty$ corresponds to example 1.

While RR, JSQ_0 and JSQ_e are all realistic choices in the given scenario, the pure JSQ cannot be implemented because it requires the knowledge about the exact number at each server. However, the gap between it and the realistic policies corresponds to the *price of information delay* quantifying the increase in the mean response time due to the delay d. As RR is agnostic and does not even try to utilize the incomplete information about the servers' states, it serves as an upper bound for the price of information delay in our setting.

Next we take a closer look at the quantities JSQ_0 and JSQ_e policies are based on, i.e., the expected number in each server upon dispatching, denoted by $\hat{N}_j$. We can omit the subscript j as this quantity is statistically the same for each server. Figure 6(a) and (b) depict the difference between the state estimates of

JSQ$_0$ and JSQ$_e$, and the actual number in the server upon dispatching,

$$Q := \hat{N} - N,$$

We observe that JSQ$_0$ overestimates the situation, which is explained by the fact that it is a strict upper bound for N because it includes also all the unacknowledged completed jobs into the estimate. Interestingly, JSQ$_e$ underestimates the situation, and thus both get it quite wrong!

4   Delayed Completion Acknowledgement

In the last example, we made an interesting observation. Even though JSQ$_e$ was supposed to have a better understanding about the current state of each server, this was not necessarily the case assuming delayed acknowledgements of service completions. Moreover, its performance can be worse than that of JSQ$_0$, which is easier to implement. This is somewhat unexpected and calls for a closer look.

Note that our implementation of JSQ$_e$ is based on the assumption that the absence of more recent ACKs conveys no information. Similarly, JSQ$_0$ assumes that all unacknowledged jobs are still being processed. In our example scenario with a fixed delay on acknowledgements, both assumptions are wrong! We do know that job completions that we are not yet aware of may occur only during the time interval $I_d = (t - d, t)$. Thus instead of considering time intervals $I_i = (t_i, t_{i-1})$, we need to consider their intersection with I_d, i.e. the effective time interval for unknown departures is $I_i \cap I_d$.

That is, when the constant delay parameter d is known (or its distribution), we can utilize it to obtain a better understanding of N in each server. Otherwise the procedure works similarly as before.

Let e_i denote the length of the subset of time interval (t_{i-1}, t_i) during which jobs may have departed without our knowledge, $e_i := |I_i \cap I_d|$. We have three cases for $I_i \cap I_d$,

$$\begin{cases} (t_{i-1}, t_i) & \text{when } t - d \leq t_{i-1}, \\ (t - d, t_i) & \text{when } t_{i-1} < t - d \leq t_i, \\ \emptyset, & \text{when } t_i < t - d. \end{cases} \tag{1}$$

Now it is easy to deduce that

$$e_i = \min\{t_i - t_{i-1}, \max\{0, t_i - t + d\}\}.$$

The maximum number of departures, denoted by D_i, during time interval (t_{i-1}, t_i) has a Poisson distribution with parameter μe_i, and the corresponding update rule for determining N_i is

$$N_i = (N_{i-1} - D_i)^+ + a_i,$$

where $N_0 = n_0$, and $a_i = 1$ if a job arrives at time t_i and otherwise $a_i = 0$.

Including the knowledge about the fixed delay on ACKs thus leads to a minor modification to Algorithm 1. In FindN, the calls to Update must be modified so that the second parameter is e_i.

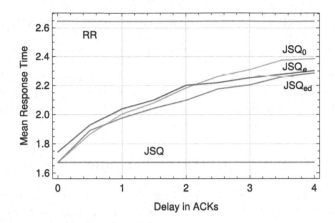

Fig. 7. Simulation results with 4 identical servers as a function of ACK delay d.

P5 **Delay-aware JSQ$_{ed}$:** Taking into account the known fixed delay d when determining the server-specific state distribution, and then computing the expected number in servers leads to delay-aware JSQ.

4.1 Numerical Example 3

Let us next consider the same setting as in the last example, i.e., four identical servers and the constant delay d of the ACKs is varied from zero to 4. Figure 7 depicts the numerical results, including for the new policy JSQ$_{ed}$. We can observe that JSQ$_{ed}$ yields the shortest mean response time among all policies that can be realized (i.e., excluding the basic JSQ that violates our assumptions on the available information). It thus gives the best estimate for to the price of information delay in our scenario.

Figure 6(c) depicts the error in JSQ$_{ed}$'s estimate on the server state. We can see that it is based on the unbiased estimate (the mean is correct), however, the standard deviation is non-negligible.

4.2 Numerical Example 4: Dependency on Load

Next we study how the offered load ρ affects the performance. We fix the delay in acknowledgements to $d = 2$ and vary the arrival rate λ. Otherwise the example system is kept the same, i.e., we have 4 identical servers with $\mu_i = 1$.

Numerical results are depicted in Fig. 8. The x-axis corresponds to the offered load, $\rho = \lambda / \sum_j \mu_j$, and the y-axis to the mean response time scaled by $(1 - \rho)$. Hence, with a single fast server with $\mu' = \sum_j \mu_j$, one would obtain the mean response time of $\mathbb{E}[T'] = 0.25$. This is depicted with the dashed constant line in the figure.

It is interesting to note that RR can be clearly better than the (naïve) JSQ$_0$. In our example case, this happens when the offered load is moderate.

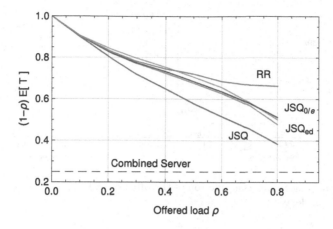

Fig. 8. Mean response time as a function of the offered load ρ with four identical servers and ACK delay of $d = 2$.

5 Heterogeneous Servers

Often servers are not identical. For example, the server pool may comprise two types of hardware due to an upgrade. In this section, we generalize the heuristic policies to the case of heterogenenous service rates.

We adapt the previously discussed dispatching policies accordingly. Instead of RR, we can use the so-called generalized Round-Robin (GRR) [1]. With GRR, faster servers appear more frequently in the number sequence defining the assignment pattern.

Similarly, instead of JSQ we consider the *shortest-expected-delay* (SED) [14],

$$\alpha_{\text{SED}} := \underset{j}{\text{argmin}}\ \frac{n_j + 1}{\mu_j}.$$

where μ_j is the service rate of server j, and n_j is the current number in server j. Given the exact state information is not available, we resort to estimating the mean $\mathbb{E}[N_j]$ the same way as before. Similarly as with JSQ, the subscript indicates how $\mathbb{E}[N_j]$ is estimated (based on delayed acknowledgements).

5.1 Numerical Example 5

As an example scenario, suppose we have three servers with service rates $\mu = \{2, 1, 1\}$, so that server 1 is twice as fast as the other two servers. As in our earlier numerical examples, jobs arrive according to Poisson process. The propagation delay of the acknowledgements is constant, either small $d = 0.5$, or large $d = 3$.

As mentioned, the standard round robin sequence would be ill-fitted given one server is significantly faster than other, and we will resort to the generalized Round-Robin (GRR). In our example case the optimal sequence is trivially $1, 2, 1, 3, 1, 2, \dots$.

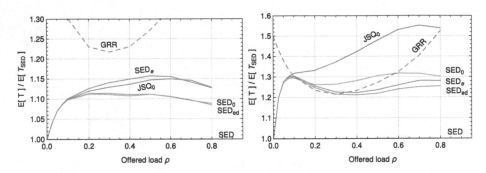

Fig. 9. Mean response time as a function of the offered load ρ with three heterogeneous servers and ACK delay of $d = 0.5$ and $d = 3$.

Numerical results are depicted in Fig. 9. The y-axis corresponds to the mean response time relative to SED with perfect state information. We also included the (naïve) JSQ, resolving ties in favor of the fast server, for comparison.

When the propagation delay of ACKs is small (left figure), we can see that both SED_0 and SED_{ed} work well, whereas SED_e and JSQ_0 have similar performance. GRR utilizes the slower servers unnecessarily, especially at low loads, and it is no match to these dynamic policies.

However, when the propagation delay increases to $d = 3$ things change dramatically. First, SED_{ed} yields the shortest mean response time, followed by SED_e, as expected. These are the policies that take into account the delay in acknowledgements, and its role is now sufficiently large that even SED_e does a good job. JSQ_0 shows significantly worse performance than SED policies. At the same time, GRR starts to move up in the ranks. This is expected as GRR is insensitive to delay in acknowledgements. In fact, it is also insensitive to delay from dispatcher to server, should there be such.

To summarize, our numerical experiments suggest RR and its generalized version (GRR) are good robust options when the state information available to the dispatcher is vague and/or noisy, especially when load is relatively high. Moreover, false assumptions on the process describing the delivery of acknowledgements can severely degrade the performance of the system when a JSQ- or SED-type of policy is applied. With accurate modelling under the right assumptions, JSQ- and SED-policies can work well even under uncertainties about the server states. However, when delays are small, the standard JSQ and SED remain near-optimal and should be favored for their simplicity.

6   Conclusions

In this paper, we considered elementary yet tractable models for dispatching systems subject to varying information and (random) delays in servers' status updates. Such settings arise, e.g., in large or real-time systems, where it is not practical to query the state of each server for every job.

The main contributions of this paper are:

1. We give computationally straightforward algorithms to determine distributions for the server states based on incomplete information. The arrival process can be arbitrary, but exponential service times are assumed, which is a reasonable assumption, e.g., with a large number of servers.
2. The (naïve) JSQ, referred to as JSQ_0, turned out to be worse than RR in some cases. This is due to the fact that its estimate on server states has bias. In other words, before applying JSQ, one should ensure that the state information is timely and accurate [11].
3. Moreover, we observed that it is also important to take into account the delay pattern appropriately. Our first trial, JSQ_e turned out to have defiencies especially when completed jobs are acknowledged and ACK delays are short.
4. Our delay-aware policy, JSQ_{ed}, takes the delay structure appropriately into account. In the example cases, its performance was better than with any other comparable policy based on the same amount of information.
5. With heterogeneous servers, JSQ is often replaced by SED, which takes into account the service rates. Similarly as with JSQ, SED can also be adapted to the setting of uncertain state information. The numerical results suggest similar observations; with a short delay the standard SED works well, but as the delay increases it must be taken into account. SED_{ed} turned out to be superior across different example scenarios.
6. Generalized round robin (GRR) is well-suited for heterogenerous servers. It is a robust policy that works well when state information is vague or contains unknown errors (e.g. wrong assumptions about the ACK propagation delay processs). However, with sufficiently accurate state information even the basic JSQ_0 was better than GRR.

The dynamic decision making scenario studied in this paper can be approached in the framework of partially observable MDPs. Similarly, one can develop "black-box" machine learning models that can be expected to perform well after a sufficiently long training period. However, these approaches are less insightful on how the system works, and the learning phase in real operation can be disruptive when the operational parameters change. In contrast, our approach can be *immediately* adapted to a new scenario, e.g., if the arrival rate Λ or the size of the server pool changes. This also allows dynamic dimensioning of the server pool without the long experiments that machine learning approaches would require.

References

1. Altman, E., Gaujal, B., Hordijk, A.: Balanced sequences and optimal routing. J. ACM **47**(4), 752–775 (2000)
2. Altman, E., Kofman, D., Yechiali, U.: Discrete time queues with delayed information. Queueing Syst. **19**, 361–376 (1995)
3. Artiges, D.: Optimal routing into two heterogeneous information. In: Proceedings of the 32nd Conference on Decision and Control, San Antonio, Texas (1993)

4. Buzen, J.P., Chen, P.P.: Optimal load balancing in memory hierarchies. In: Proceedings of the 6th IFIP Congress, pp. 271–275, Stockholm, Sweden, August 1974
5. Ephremides, A., Varaiya, P., Walrand, J.: A simple dynamic routing problem. IEEE Trans. Automatic Control **25**(4), 690–693 (1980)
6. Haight, F.A.: Two queues in parallel. Biometrika **45**(3–4), 401–410 (1958)
7. Lipschutz, D.: Open problem-load balancing using delayed information. Stochas. Syst. **9**(3), 305–306 (2019)
8. Litvak, N., Yechiali, U.: Routing in queues with delayed information. Queueing Syst. **43**, 147–165 (2003)
9. Liu, Z., Righter, R.: Optimal load balancing on distributed homogeneous unreliable processors. Oper. Res. **46**(4), 563–573 (1998)
10. Liu, Z., Towsley, D.: Optimality of the round-robin routing policy. J. Appl. Probab. **31**(2), 466–475 (1994)
11. Mitzenmacher, M.: How useful is old information? IEEE Trans. Parallel Distrib. Syst. **11**(1), 6–20 (2000)
12. Novitzky, S., Pender, J., Rand, R.H., Wesson, E.: Nonlinear dynamics in queueing theory: determining the size of oscillations in queues with delay. SIAM J. Appl. Dyn. Syst. **18**(1), 279–311 (2019)
13. Pender, J., Rand, R., Wesson, E.: A stochastic analysis of queues with customer choice and delayed information. Math. Oper. Res. **45**(3), 1104–1126 (2020)
14. Selen, J., Adan, I., Kapodistria, S., van Leeuwaarden, J.: Steady-state analysis of shortest expected delay routing. Queueing Syst. **84**(3), 309–354 (2016)
15. Whitt, W.: On the many-server fluid limit for a service system with routing based on delayed information. Oper. Res. Lett. **49**, 316–319 (2021)
16. Winston, W.: Optimality of the shortest line discipline. J. Appl. Probab. **14**, 181–189 (1977)

Evolutionary Vaccination Games with Premature Vaccines to Combat Ongoing Deadly Pandemic

Vartika Singh$^{(\boxtimes)}$, Khushboo Agarwal, Shubham, and Veeraruna Kavitha

IEOR, IIT Bombay, Mumbai, India
{vsvartika,agarwal.khushboo,191190009,vkavitha}@iitb.ac.in

Abstract. We consider a vaccination game that results with the introduction of premature and possibly scarce vaccines introduced in a desperate bid to combat the otherwise ravaging deadly pandemic. The response of unsure agents amid many uncertainties makes this game completely different from the previous studies. We construct a framework that combines SIS epidemic model with a variety of dynamic behavioral vaccination responses and demographic aspects. The response of each agent is influenced by the vaccination hesitancy and urgency, which arise due to their personal belief about efficacy and side-effects of the vaccine, disease characteristics, and relevant reported information (e.g., side-effects, disease statistics etc.). Based on such aspects, we identify the responses that are stable against static mutations. By analysing the attractors of the resulting ODEs, we observe interesting patterns in the limiting state of the system under evolutionary stable (ES) strategies, as a function of various defining parameters. There are responses for which the disease is eradicated completely (at limiting state), but none are stable against mutations. Also, vaccination abundance results in higher infected fractions at ES limiting state, irrespective of the disease death rate.

Keywords: Vaccination games · ESS · Epidemic · Stochastic approximation · ODEs

1 Introduction

The impact of pandemic in today's world is unquestionable and so is the need to analyse various related aspects. There have been many disease outbreaks in the past and the most recent Covid-19 pandemic is still going on. Vaccination is known to be of great help; however, the effectiveness depends on the responses of the population ([1] and references therein). The vaccination process gets more challenging when vaccines have to be introduced prematurely, without much

The work of first and second author is partially supported by Prime Minister's Research Fellowship (PMRF), India.

Q. Zhao and L. Xia (Eds.): VALUETOOLS 2021, LNICST 404, pp. 185–206, 2021.
https://doi.org/10.1007/978-3-030-92511-6_12

information about their efficacy or side effects, to combat the ravaging on-going pandemic. The scarcity of vaccines makes it all the more challenging.

There is a vast literature developed during the current pandemic that majorly focuses on exhaustive experiments. Recently authors in [2] discuss the importance of game theoretic and social network models for better understanding the pandemic. Our paper exactly aims to achieve this purpose. We aim to develop a mathematical framework that mimics the ongoing pandemic as closely as possible. *We consider vaccination insufficiency, hesitancy, impact of individual vaccination responses, possibility of excess deaths, lack of information (e.g., possible end of the disease, vaccine details) etc. Our model brings together the well known epidemic SIS model [3, 4], evolutionary game theoretic framework [11], dynamic behavioural patterns of the individuals and demographic aspects.*

Majority of the literature on vaccination games assumes some knowledge which aids in vaccination plans; either they consider seasonal variations, or the time duration and or the time of occurrence of the disease is known etc. For example, in [5, 6] authors consider a replicator dynamics-based vaccination game; each round occurs in two phases, vaccination and the disease phase. Our paper deals with agents, that continually operate under two contrasting fears, vaccination fear and the deadly pandemic fear. They choose between the two fears, by estimating the perceived cost of the two aspects. In [7] authors consider perceived vaccination costs as in our model and study the 'wait and see' equilibrium. Here the agents choose one among 53 weeks to get vaccinated. In contrast, *we do not have any such information*, and the aim is to eradicate the deadly disease.

In particular, we consider a population facing an epidemic with limited availability of vaccines. Any individual can either be infected, susceptible or vaccinated. *It is evident nowadays (with respect to Covid-19) that recovery does not always result in immunization. By drawing parallels, the recovered individuals become susceptible again.* We believe the individuals in this current pandemic are divided in opinion due to vaccination hesitancy and vaccination urgency. A vaccination urgency can be observed depending upon the reported information about the sufferings and deaths due to disease, lack of hospital services etc. On the other hand, the vaccination hesitancy could be due to individuals' belief in the efficacy and the emerging fears due to the reported side effects.

To capture above factors, we model a variety of possible vaccination responses. We first consider the individuals who exhibit the *follow-the-crowd (FC)* behavior, i.e., their confidence (and hence inclination) for vaccination increases as more individuals get vaccinated. In other variant, the interest of an individual in vaccination reduces as vaccinated proportion further grows. Basically such individuals attempt to enjoy the resultant benefits of not choosing vaccination and still being prevented from the disease; we refer them as *free-riding (FR) agents*. Lastly, we consider individuals who make more informed decisions based also on the infected proportion; these agents exhibit increased urge towards vaccination as the infected population grows. We name such agents as *vigilant agents*.

In the era of pre-mature vaccine introduction and minimal information, the agents depend heavily on the perceived cost of the two contrasting (vaccination/infection cost) factors. The perceived cost of infection can be large leading to a vaccination urgency, when there is vaccine scarcity. In case there is abundance, one may perceive a smaller risk of infection and procrastinate the vaccination till the next available opportunity. *Our results interestingly indicate a larger infected proportion at ES equilibrium as vaccine availability improves.*

We intend to investigate the resultant of the vaccination response of the population and the nature of the disease. Basically, we want to understand if the disease can be overpowered by given type of vaccination participants. Our analysis is layered: firstly, *we use stochastic approximation techniques to derive time-asymptotic proportions and identify all possible equilibrium states, for any given vaccination response and disease characteristics.* Secondly, we derive the vaccination responses that are *stable against mutations*. We slightly modify the definition of classical evolutionary stable strategy, which we refer as *evolutionary stable strategy against static mutations (ESS-AS)*[1]. We study the equilibrium/limiting states reached by the system under ES vaccination responses.

Under various ES (evolutionary stable) vaccination responses, the dynamic behaviour at the beginning could be different, but *after reaching equilibrium, individuals either vaccinate with probability one or zero*. Some interesting patterns are observed in ES limiting proportions as a function of important defining parameters. For example, the ES limit infected proportions are concave functions of birth rate . *With increased excess deaths, we have smaller infected proportions at ES limiting state.* Further, *there are many vaccination responses which eradicate the disease completely at equilibrium; but none of them are evolutionary stable, unless the disease can be eradicated without vaccination.* At last, we corroborate our theoretical results by performing exhaustive Monte-Carlo simulations.

2 Problem Description and Background

We consider a population facing an epidemic, where at time t, the state of the system is $(N(t), S(t), V(t), I(t))$. These respectively represent total, susceptible, vaccinated and infected population. Observe $N(t) = S(t) + V(t) + I(t)$.

At any time t, any susceptible individual can contact anyone among the infected population according to exponential distribution with parameter $\lambda/N(t)$, $\lambda > 0$ (as is usually done in epidemic models, e.g., [8]). In particular, a contact between a susceptible and an infected individual may result in spread of the disease, which is captured by λ. Any infected individual may recover after an exponentially distributed time with parameter r to become susceptible again. A susceptible individual can think of vaccination after exponentially distributed time with parameter ν. At that epoch, the final decision (to get vaccinated) depends on the information available to the individual. We refer to the probability of a typical individual getting vaccinated as q; more details will follow

---

[1] Agents use dynamic policies, while mutants use static variants (refer Sect. 2).

below. *It is important to observe here that ν will also be governed by vaccination availability; the individual decision rate is upper bounded by availability rate.* Further, there can be a birth after exponentially distributed time with parameter $b N(t)$. A death is possible in any compartment after exponentially distributed time with parameter $d I(t)$ (or $d S(t)$ or $d V(t)$), and excess death among infected population with parameter $d_e I(t)$. Furthermore, we assume $b > d + d_e$.

One of the objectives is to analyse the (time) asymptotic proportions of the infected, vaccinated and susceptible population depending upon the disease characteristics and vaccination responses. To this end, we consider the following fractions:

$$\theta(t) := \frac{I(t)}{N(t)}, \ \psi(t) := \frac{V(t)}{N(t)}, \ \text{and} \ \phi(t) := \frac{S(t)}{N(t)}. \tag{1}$$

2.1   Responses Towards Vaccination

In reality, the response towards available vaccines depends upon the cost of vaccination, the severity of infection and related recovery and death rates. However, with lack of information, the vaccination decision of any susceptible depends heavily on the available data, $(\theta(t), \psi(t))$, the fractions of infected and vaccinated population at that time. Further, it could depend on the willingness of the individuals towards vaccination. The final probability of getting vaccinated at the decision/availability epoch is captured by probability $q = q(\theta, \psi, \beta)$; this probability is also influenced by the parameter β which is a characteristic of the given population. Now, we proceed to describe different possible behaviours exhibited by agents/individuals in the population:

Follow the Crowd (FC) Agents: These type of agents make their decision to get vaccinated by considering only the vaccinated proportion of the population; they usually ignore other statistics. They believe vaccination in general is good, but are hesitant because of the possible side effects. The hesitation reduces as the proportion of vaccinated population increases, and thus accentuating the likelihood of vaccination for any individual. In this case, the probability that any individual will decide to vaccinate is given by $q := \beta\psi(t)$.

Free-Riding (FR) Agents: These agents exhibit a more evolved behaviour than FC individuals, in particular free-riding behaviour. In order to avoid any risks related to vaccination or paying for its cost, they observe the crowd behaviour more closely. Their willingness towards vaccination also improves with $\psi(t)$, however they tend to become free-riders beyond a limit. In other words, as $\psi(t)$ further grows, their tendency to vaccinate starts to diminish. We capture such a behaviour by modelling the probability as $q := \beta\psi(t)(1 - \psi(t))$.

Vigilant Agents: Many a times individuals are more vigilant, and might consider vaccination only if the infected proportion is above a certain threshold, in which case we set $q = \beta\psi(t)\mathbb{1}_{\{\theta(t)>\Gamma\}}$ (Γ is an appropriate constant). This dependency could also be continuous, we then model the probability as $q = \beta\theta(t)\psi(t)$. We refer to them as Vigilant Follow-the-Crowd (VFC) agents.

Vaccination Policies: We refer the vaccination responses of the agents as vaccination policy, which we represent by $\pi(\beta)$. When the vaccination response policy is $\pi(\beta)$, the agents get vaccinated with probability $q_{\pi(\beta)}(\theta, \psi)$ at the vaccination decision epoch, based on the system state (θ, ψ) at that epoch. For ease of notation, we avoid π and β while representing q.

As already explained, we consider the following vaccination policies: a) when $\pi(\beta) = F_C(\beta)$, i.e., with follow-the-crowd policy, the agents choose to vaccinate with probability $q(\theta, \psi) = \min\{1, \tilde{q}(\theta, \psi)\}$, $\tilde{q}(\theta, \psi) = \beta\psi$ at vaccination decision epoch; b) when $\pi(\beta) = F_R(\beta)$, the free-riding policy, $\tilde{q}(\theta, \psi) = \beta\psi(1 - \psi)$; and (c) the policy $\pi(\beta) = V_{FC}^1(\beta)$ represents vigilant (w.r.t. θ) follow the crowd (w.r.t. ψ) policy and $\tilde{q}(\theta, \psi) = \beta\theta\psi$. We define $\Pi := \{F_C, F_R, V_{FC}^1\}$ to be the set of all these policies. We discuss the fourth type of policy V_{FC}^2 with $\tilde{q}(\theta, \psi) = \beta\theta\mathbb{1}_{\{\theta > \Gamma\}}$ separately, as the system responds very differently to these agents. Other behaviour patterns are for future study.

2.2 Evolutionary Behaviour

The aim of this study is three fold: (i) to study the dynamics and understand the equilibrium states (settling points) of the disease depending upon the agents' behavior and the availability of the vaccine, (ii) to compare and understand the differences in the equilibrium states depending on agents' response towards vaccine, and (iii) to investigate if these equilibrium states are stable against mutations, using the well-known concepts of evolutionary game theory. Say a mutant population of small size invades such a system in equilibrium. We are interested to investigate if the agents using the original (vaccination response) are still better than the mutants. Basically we use the concept of evolutionary stable strategy (ESS), which in generality is defined as below (e.g., [11]):

By definition, for $\pi(\beta)$ to be an ESS, it should satisfy two conditions, i) $\pi(\beta) \in \arg\min_{\pi \in \widetilde{\Pi}} u(\pi; \pi(\beta))$, where $\widetilde{\Pi}$ is the set of the policies, $u(\pi, \pi')$ is the utility/cost of the (can be mutant) user that adapts policy π, while the rest of the population uses policy π'; and ii) it should be stable against mutations, i.e. there exist an $\bar{\epsilon}$, such that for all $\epsilon \leq \bar{\epsilon}(\pi)$,

$$u(\pi, \pi_\epsilon(\beta, \pi)) > u(\pi(\beta), \pi_\epsilon(\beta, \pi)) \text{ for any } \pi \neq \pi(\beta),$$

where $\pi_\epsilon(\beta, \pi) := \epsilon\pi + (1 - \epsilon)\pi(\beta)$, represents the policy in which ϵ fraction of agents (mutants) use strategy π and the other fraction uses $\pi(\beta)$.

In the current paper, we restrict our definition of ESS to cater for mutants that use static policies, Π^D. Under any static policy $\pi \equiv q$, the agent gets vaccinated with constant probability q at any decision epoch, irrespective of the system state. We now *define the Evolutionary Stable Strategies, stable Against Static mutations* and the exact definition is given below.

Definition 1. [ESS-AS] *A policy $\pi(\beta)$ is said to be ESS-AS, i) if $\{q^*_{\pi(\beta)}\} = \mathcal{B}(\pi(\beta))$, where the static-best response set*

$$\mathcal{B}(\pi(\beta)) := \arg\min_{q \in [0,1]} u(q, \pi(\beta)),$$

and $q^*_{\pi(\beta)} = q(\theta^*, \psi^*)$ is the probability with which the agents get vaccinated after the system reaches equilibrium under strategy $\pi(\beta)$; and ii) there exists an $\bar{\epsilon}$ such that $\{q^*_{\pi(\beta)}\} = \mathcal{B}(\pi_\epsilon(\beta, q))$, for any $\epsilon \le \bar{\epsilon}(q)$ and any q.

Anticipated Utility/Cost of a User: Once the population settles to an equilibrium (call it $\hat{\theta}, \hat{\psi}$) under a certain policy $\pi(\hat{\beta})$, the users (assume to) incur a cost that depends upon various factors. To be more specific, any user estimates[2] its overall cost of vaccination considering the pros and cons as below to make a judgement about vaccination.

The cost of infection (as perceived by the user) can be summarized by $p_I(\hat{\theta})(c_{I_1} + c_{I_2} d_e \hat{\theta})$, where $p_I(\hat{\theta})$ equals the probability that the user gets infected before its next decision epoch (which depends upon the fraction of infected population $\hat{\theta}$ and availability/decision rate ν), c_{I_1} is the cost of infection without death (accounts for the sufferings due to disease, can depend on r), while $c_{I_2} d_e \hat{\theta}$ is the perceived chance of death after infection. Observe here that $d_e \hat{\theta}$ is the fraction of excess deaths among infected population which aids in this perception.

On the contrary, the cost of vaccination is summarized by $c_{v_1} + \min\{\bar{c}_{v_2}, c_{v_2}/\hat{\psi}\}$, where c_{v_1} is the actual cost of vaccine. Depending upon the fraction $\hat{\psi}$ of the population vaccinated and their experiences, the users anticipate additional cost of vaccination (caused due to side-effects) as captured by the second term $c_{v_2}/\hat{\psi}$. Inherently *we assume here that the side effects are not significant,* and hence in a system with a bigger vaccinated fraction, the vaccination hesitancy is lesser. Here $\bar{c}_{v_2}$ accounts for maximum hesitancy. In all, the expected anticipated cost of vaccination by a user in a system at equilibrium (reached under $\pi(\hat{\beta})$) equals: the probability of vaccination (say q) times the anticipated cost of vaccination, plus $(1 - q)$ times the anticipated cost of infection. Thus we define:

Definition 2. [User utility at equilibrium]
When the population is using policy $\pi(\hat{\beta})$ and an agent attempts to get itself vaccinated with probability q, then, the user utility function is given by:

$$u(q; \pi(\hat{\beta})) := q\left(c_{v_1} + \min\left\{\bar{c}_{v_2}, \frac{c_{v_2}}{\hat{\psi}}\right\}\right) + (1 - q)p_I(\hat{\theta})(c_{I_1} + c_{I_2} d_e \hat{\theta})$$

$$= qh(\pi(\hat{\beta})) + p_I(\hat{\theta})(c_{I_1} + c_{I_2} d_e \hat{\theta}), \quad where \tag{2}$$

$$h(\pi(\hat{\beta})) = h(\hat{\theta}, \hat{\psi}) := c_{v_1} + \min\left\{\bar{c}_{v_2}, \frac{c_{v_2}}{\hat{\psi}}\right\} - p_I(\hat{\theta})(c_{I_1} + c_{I_2} d_e \hat{\theta}).$$

In the next section, we begin with ODE approximation of the system, which facilitates in deriving the limiting behaviour of the system. Once the limiting behaviour is understood, we proceed towards evolutionary stable strategies.

---

[2] We assume mutants are more rational, estimate various rates using reported data.

3   Dynamics and ODE Approximation

Our aim in this section is to understand the limiting behaviour of the given system. The system is transient with $b > d + d_e$; it is evident that the population would not settle to a stable distribution (it would explode as time progresses with high probability). However the fraction of people in various compartments (given by (1)) can possibly reach some equilibrium and we look out for this equilibrium or limiting proportions (as is usually considered in literature [9]).

To study the limit proportions, it is sufficient to analyse the process at transition epochs. Let τ_k be the kth transition epoch, and infected population immediately after τ_k equals $I_k := I(\tau_k^+) = \lim_{t \downarrow \tau_k} I(t)$; similarly, define N_k, S_k and V_k. Observe here that $T_{k+1} := \tau_{k+1} - \tau_k$ is exponentially distributed with a parameter that depends upon previous system state (N_k, S_k, I_k, V_k).

Transitions: Our aim is to derive the (time) asymptotic fractions of (1). Towards this, we define the same fractions at transition epochs,

$$\theta_k := \frac{I_k}{N_k}, \psi_k := \frac{V_k}{N_k}, \text{ and } \phi_k := \frac{S_k}{N_k}.$$

Observe that $\theta_k + \psi_k + \phi_k = 1$. To facilitate our analysis, we also define a slightly different fraction, $\eta_k := N_k/k$ for $k > 1$ and $\eta_0 := N(0)$, $\eta_1 := N(1)$. As described in previous section, the size of the infected population evolves between two transition epochs according to:

$$I_{k+1} = I_k + G_{I,k+1}, \text{ with } G_{I,k+1} := \mathbb{I}_{k+1} - \mathbb{R}_{k+1} - \mathbb{D}_{I,k+1}, \qquad (3)$$

where $\mathbb{I}_{k+1}$ is the indicator that the current epoch is due to a new infection, $\mathbb{R}_{k+1}$ is the indicator of a recovery and $\mathbb{D}_{I,k+1}$ is the indicator that the current epoch is due to a death among the infected population. Let $\mathcal{F}_k := \sigma(I_j, S_j, V_j, N_j, j \leq k)$ represent the sigma algebra generated by the history until the observation epoch k and let $E_k[\cdot]$ represent the corresponding conditional expectation. By conditioning on $\mathcal{F}_k$, using the memory-less property of exponential random variables,

$$E_k[\mathbb{I}_{k+1}] = \frac{\frac{\lambda I_k S_k}{N_k}}{N_k b + N_k d + I_k d_e + \frac{\lambda I_k S_k}{N_k} + S_k \nu + I_k r} = \frac{\lambda \theta_k \phi_k}{\varrho_k}, \text{ with,}$$

$$\varrho_k := b + d + d_e \theta_k + \lambda \theta_k \phi_k + \nu \phi_k + r \theta_k,$$

$$E_k[\mathbb{R}_{k+1}] = \frac{r\theta_k}{\varrho_k}, \text{ and, } E_k[\mathbb{D}_{I,k+1}] = \frac{\theta_k(d + d_e)}{\varrho_k}. \qquad (4)$$

In similar lines the remaining types of the population evolve according to the following, where $\mathbb{V}_{k+1}, \mathbb{B}_{k+1}, \mathbb{D}_{V,k+1}$ and $\mathbb{D}_{S,k+1}$ are respectively the indicators of vaccination, birth and corresponding deaths,

$$V_{k+1} = V_k + G_{V,k+1}, \quad G_{V,k+1} := \mathbb{V}_{k+1} - \mathbb{D}_{V,k+1}, \qquad (5)$$

$$N_{k+1} = N_k + G_{N,k+1}, \quad G_{N,k+1} := \mathbb{B}_{k+1} - \mathbb{D}_{I,k+1} - \mathbb{D}_{V,k+1} - \mathbb{D}_{S,k+1}, \text{ and,}$$

$$S_{k+1} = N_{k+1} - I_{k+1} - V_{k+1}.$$

As before,

$$E_k[\mathbb{V}_{k+1}] = \frac{\nu q(\theta_k, \psi_k)\phi_k}{\varrho_k}, \quad E_k[\mathbb{D}_{V,k+1}] = \frac{\psi_k d}{\varrho_k},$$

$$E_k[\mathbb{D}_{S,k+1}] = \frac{\phi_k d}{\varrho_k}, \text{ and, } E_k[\mathbb{B}_{k+1}] = \frac{b}{\varrho_k}. \tag{6}$$

Let $\Upsilon_k := [\theta_k, \psi_k, \eta_k]^T$. The evolution of Υ_k can be studied by a three dimensional system, described in following paragraphs. To facilitate tractable mathematical analysis we consider a slightly modified system that freezes once η_k reaches below a fixed small constant $\delta > 0$. The rationale and the justification behind this modification is two fold: a) once the population reaches below a significantly small threshold, it is very unlikely that it explodes and the limit proportions in such paths are no more interesting; b) the initial population $N(0)$ is usually large, let $\delta = 2/(N(0) - 1)$ and then with $b > d + d_e$, it is easy to verify that the probability, $P\Big(\eta_k < \delta \text{ for some } k \Big| N(0)\Big) \to 0$ as $N(0) \to \infty$. From (3),

$$\frac{I_{k+1}}{N_{k+1}} = \frac{I_k}{N_k} + \frac{I_{k+1}}{N_{k+1}} - \frac{I_k}{N_k} = \frac{I_k}{N_k} + \frac{1}{k+1}\frac{k+1}{N_{k+1}}\left[I_{k+1} - \frac{I_k N_{k+1}}{N_k}\right],$$

$$= \frac{I_k}{N_k} + \frac{1}{k+1}\frac{k+1}{N_{k+1}}\left[G_{I,k+1} - \frac{N_{k+1}-N_k}{N_k}I_k\right], \text{ and thus including } \mathbb{1}_{\{\eta_k > \delta\}},$$

$$\theta_{k+1} = \theta_k + \epsilon_k \frac{\mathbb{1}_{\{\eta_k > \delta\}}}{\eta_{k+1}}[G_{I,k+1} - (N_{k+1}-N_k)\theta_k], \quad \epsilon_k := \frac{1}{k+1}. \tag{7}$$

We included the indicator $\mathbb{1}_{\{\eta_k > \delta\}}$, as none of the population types change (nor there is any evolution) once the population gets almost extinct. Similarly, from equation (5),

$$\psi_{k+1} = \psi_k + \epsilon_k \frac{\mathbb{1}_{\{\eta_k > \delta\}}}{\eta_{k+1}}[G_{V,k+1} - (N_{k+1}-N_k)\psi_k], \tag{8}$$

$$\eta_{k+1} = \eta_k + \mathbb{1}_{\{\eta_k > \delta\}}\epsilon_k[G_{N,k+1} - \eta_k].$$

We analyse this system using the results and techniques of [10]. In particular, we prove equicontinuity in extended sense for our non-smooth functions (e.g., $q(\theta, \psi)$ may only be measurable), and then use [10, Chapter 5, Theorem 2.2]. Define $L_{k+1} := [L_{k+1}^\theta, L_{k+1}^\psi, L_{k+1}^\eta]^T$, with

$$L_{k+1}^\theta = \frac{\mathbb{1}_{\{\eta_k > \delta\}}}{\eta_{k+1}}[G_{I,k+1} - (N_{k+1}-N_k)\theta_k], \tag{9}$$

$$L_{k+1}^\psi = \frac{\mathbb{1}_{\{\eta_k > \delta\}}}{\eta_{k+1}}[G_{V,k+1} - (N_{k+1}-N_k)\psi_k], \text{ and, } L_{k+1}^\eta = \mathbb{1}_{\{\eta_k > \delta\}}(G_{N,k+1} - \eta_k).$$

Thus (7)–(8) can be rewritten as, $\Upsilon_{k+1} = \Upsilon_k + \epsilon_k L_{k+1}$. Conditioning as in (4):

$$E_k[L_{k+1}^\theta] = \frac{\theta_k \mathbb{1}_{\{\eta_k > \delta\}}}{\eta_k \varrho_k}[\phi_k \lambda - r - d_e - (b - d_e\theta_k)] + \alpha_k^\theta \tag{10}$$

$$=: g^\theta(\Upsilon_k) + \alpha_k^\theta, \text{ where, } \alpha_k^\theta = E_k\left[L_{k+1}^\theta - \frac{\eta_{k+1}}{\eta_k}L_{k+1}^\theta\right].$$

In exactly similar lines, we define $g(\Upsilon_k) = [g^\theta(\Upsilon_k), g^\psi(\Upsilon_k), g^\eta(\Upsilon_k)]^T$ (details just below) and α_k^ψ such that $E_k[L_{k+1}^\psi] = g^\psi(\Upsilon_k) + \alpha_k^\psi$ and $E_k[L_{k+1}^\eta] = g^\eta(\Upsilon_k)$. Our claim is that the error terms would converge to zero (shown by Lemma 3 in Appendix) and ODE $\dot{\Upsilon} = g(\Upsilon)$ approximates the system dynamics, where,

$$g^\theta(\Upsilon) = \frac{\theta \mathbb{1}_{\{\eta > \delta\}}}{\eta \varrho} \left[\phi\lambda - r - d_e - (b - d_e\theta)\right], \quad \phi = 1 - \theta - \psi$$

$$g^\psi(\Upsilon) = \frac{\mathbb{1}_{\{\eta > \delta\}}}{\eta \varrho} \left[q(\theta, \psi)\phi\nu - (b - d_e\theta)\psi\right], \quad \text{and,} \tag{11}$$

$$g^\eta(\Upsilon) = \mathbb{1}_{\{\eta > \delta\}} \left(\frac{b - d - d_e\theta}{\varrho} - \eta\right), \quad \varrho = b + d + d_e\theta + \lambda\theta\phi + \nu\phi + r\theta.$$

We now state our first main result (with proof in Appendix A).

A. Let the set A be locally asymptotically stable in the sense of Liapunov for the ODE (11). Assume that $\{\Upsilon_n\}$ visits a compact set, S_A, in the domain of attraction, D_A, of A infinitely often (i.o.) with probability $\rho > 0$.

Theorem 1. *Under assumption* **A.**, *i) the sequence converges,* $\Upsilon_n \to A$ *as* $n \to \infty$ *with probability at least* ρ; *and ii) for every* $T > 0$, *almost surely there exists a sub-sequence* (k_m) *such that:* $(t_k := \sum_{i=1}^k \epsilon_i)$

$$\sup_{k:t_k \in [t_{k_m}, t_{k_m}+T]} d(\Upsilon_k, \Upsilon_*(t_k - t_{k_m})) \to 0, \quad \text{as } m \to \infty, \text{ where,}$$

$\Upsilon_*(\cdot)$ *is the solution of ODE* (11) *with initial condition* $\Upsilon_*(0) = \lim_{k_m} \Upsilon_{k_m}$. ∎

Using above Theorem, one can derive the limiting state of the system using that of the ODE (in non-extinction sample paths, i.e., when $\eta_k > \delta$ for all k). Further, for any finite time window, there exists a sub-sequence along which the disease dynamics are approximated by the solution of the ODE. The ODE should initiate at the limit of the system along such sub-sequence.

4   Limit Proportions and ODE Attractors

So far, we have proved that the embedded process of the system can be approximated by the solutions of the ODEs (11) (see Theorem (1)). We will now analyse the ODEs and look for equilibrium states for a given vaccination policy $\pi(\hat{\beta})$. The following notations are used throughout: *we represent the parameter by* $\hat{\beta}$ *and the corresponding equilibrium states by* $(\hat{\theta}, \hat{\psi})$. Let $\hat{q} := q(\hat{\theta}, \hat{\psi})$ and $\hat{\varrho} := \varrho(\hat{\theta}, \hat{\psi})$. In the next section, we identify the evolutionary stable (ES) equilibrium states (θ^*, ψ^*), among these equilibrium states, and the corresponding vaccination policies $\pi(\beta^*)$.

We now identify the attractors of the ODEs (11) that are locally asymptotically stable in the sense of Lyapunov (*referred to as attractors*), which is a requirement of the assumption **A.** However, we are yet to identify the domains of attraction, which will be attempted in future. Not all infectious diseases lead to deaths. One can either have: (i) *non-deadly disease*, where only natural deaths occur, $d_e = 0$, or (ii) *deadly disease* where in addition, we have excess deaths due to disease, $d_e > 0$. We begin with the non-deadly case and FC agents.

The equilibrium states for FC agents ($\hat{\beta} \geq 0$) are (proof in Appendix B):

Theorem 2. [FC agents] *Define* $\rho := \lambda/(r + b + d_e)$, $\mu := b/\nu$. *When* $d_e = 0$ *and* $\tilde{q}(\theta, \psi) = \hat{\beta}\psi$, *at the attractor we have* $\hat{\eta} = (b-d)/\hat{\varrho}$. *The remaining details of the attractors are in Table 1. The interior attractors (when* $(\hat{\theta}, \hat{\psi}) \in (0,1) \times (0,1)$) *are the zeros of the right hand side (RHS) of ODE (11).* ∎

Table 1. Attractors for FC agents, $(\theta_E, \psi_E) := \left(1 - \frac{1}{\rho} - \frac{1}{\mu\rho}, \frac{1}{\mu\rho}\right)$

Nature	Parameters		$(\hat{\theta}, \hat{\psi})$
	$\hat{\beta} < \mu\rho$		$\left(1 - \frac{1}{\rho}, 0\right)$
Endemic, $\rho > 1$	$\hat{\beta} > \mu\rho$, $\tilde{q}(0, 1 - \frac{\mu}{\hat{\beta}}) < 1$ implies $\hat{\beta} < \mu + 1$*		$\left(0, 1 - \frac{\mu}{\hat{\beta}}\right)$
	$\hat{\beta} > \mu\rho$, $\tilde{q}(0, 1 - \frac{\mu}{\hat{\beta}}) > 1$ implies $\hat{\beta} > \mu + 1$	$\mu\rho < \mu + 1$	$\left(0, \frac{1}{\mu+1}\right)$
		$\mu\rho > \mu + 1$	(θ_E, ψ_E)
SE, $\rho < 1$	$\mu > \hat{\beta}$		$(0, 0)$

In all the tables of this section, the * entries are also valid when $\rho < 1$. When $\hat{\beta} = \mu\rho$, the ODE (and hence the system) is not stable; such notions are well understood in the literature and we avoid such marginal cases. We now consider the FR agents (proof again in Appendix B):

Theorem 3. [FR agents] *When* $\tilde{q}(\theta, \psi) = \hat{\beta}\psi(1 - \psi)$ *and* $d_e = 0$, *then the attractors for ODE (11) are* $(\hat{\theta}, \hat{\psi}, (b-d)/\hat{\varrho})$, *which are provided in Table 2. The interior attractors are the zeros of the RHS of ODE (11).* ∎

Table 2. Attractors for FR agents, $(\theta_E, \psi_E) := \left(1 - \frac{1}{\rho} - \frac{1}{\mu\rho}, \frac{1}{\mu\rho}\right)$

Nature	Parameters		$(\hat{\theta}, \hat{\psi})$
	$\hat{\beta} < \mu\rho$		$\left(1 - \frac{1}{\rho}, 0\right)$
	$\hat{\beta} > \mu\rho$, $\tilde{q}(\hat{\theta}, \hat{\psi}) < 1$	$\hat{\beta} > \rho^2\mu$*	$\left(0, 1 - \sqrt{\frac{\mu}{\hat{\beta}}}\right)$
		$\hat{\beta} < \rho^2\mu$	$\left(\frac{\mu\rho}{\hat{\beta}} - \frac{1}{\rho}, 1 - \frac{\mu\rho}{\hat{\beta}}\right)$
Endemic, $\rho > 1$	$\hat{\beta} > \rho^2\mu$, $\tilde{q}\left(0, 1 - \sqrt{\frac{\mu}{\hat{\beta}}}\right) > 1$	$\mu + 1 > \mu\rho$	$\left(0, \frac{1}{\mu+1}\right)$
		$\mu + 1 < \mu\rho$	(θ_E, ψ_E)
	$\mu\rho < \hat{\beta} < \rho^2\mu, \tilde{q}\left(\frac{\mu\rho}{\hat{\beta}} - \frac{1}{\rho}, 1 - \frac{\mu\rho}{\hat{\beta}}\right) > 1$ $\implies \mu\rho > (\mu + 1)$		(θ_E, ψ_E)
SE, $\rho < 1$	$\mu > \hat{\beta}$		$(0, 0)$

As seen from the two theorems, we have two types of attractors: (i) interior attractors (for e.g., third row in Tables 1 and 2) in which $(\hat{\theta}, \hat{\psi}) \in (0,1) \times (0,1)$, and (ii) boundary attractors, where at least one of the components is 0. In the latter case, either the disease is eradicated with the help of vaccination ($\hat{\theta} = 0$, $\hat{\psi} > 0$) or the disease gets cured without the help of vaccination ($\hat{\theta} = \hat{\psi} = 0$) or no one gets vaccinated ($\hat{\psi} = 0$). In the last case, the fraction of infected population reaches maximum possible level for the given system, which *we refer to as non-vaccinated disease fraction (NVDF)*, ($\hat{\theta}^N = 1 - 1/\rho$) *(first row in Tables 1 and 2)*. Further, $\hat{\eta}$ is always in $(0,1)$. Furthermore, important characteristics of the attractors depend upon quantitative parameters describing the nature and the spread of the disease, and the vaccination responses:

- *Endemic disease:* The disease is not self-controllable with $\rho > 1$ and the eventual impact of the disease is governed by the attitude of agents towards vaccination (all rows other than the last in Tables 1 and 2).
- *Self-eradicating (SE) Disease:* The disease is not highly infectious ($\rho < 1$) and can be eradicated without exogenous aid (vaccine).

The above characterisation interestingly draws parallels from queuing theory. One can view λ as the arrival of infection and $r + b + d_e$ as its departure. Then $\rho = \lambda/(r+b+d_e)$, resembles the load factor. It is well known that queuing systems are stable when $\rho < 1$, similarly in our case, the disease gets self-eradicating with $\rho < 1$. The attractors for VFC1 agents (proof in Appendix B):

Theorem 4. [VFC1 agents] *When $\hat{\beta} \leq 2\mu\rho^2$ or when $\hat{q} = 1$ with $\tilde{q}(\hat{\theta}, \hat{\psi}) \neq 1$, the attractors for ODE (11) are $(\hat{\theta}, \hat{\psi}, (b-d)/\hat{\theta})$, and are provided in Table 3. The interior attractors are the zeroes of the RHS of ODE (11).* ∎

Table 3. Attractors for VFC1 agents: Disease never gets eradicated with $\rho > 1$

Nature	Parameters		$(\hat{\theta}, \hat{\psi})$
Endemic, $\rho > 1$	$\hat{\beta} < \mu \left(\frac{\rho^2}{\rho - 1} \right)$		$\left(1 - \frac{1}{\rho}, 0 \right)$
	$\hat{\beta} > \mu \left(\frac{\rho^2}{\rho - 1} \right)$, $\tilde{q}(\hat{\theta}, \hat{\psi}) < 1$		$\left(\frac{\mu\rho}{\hat{\beta}}, 1 - \frac{1}{\rho} - \frac{\mu\rho}{\hat{\beta}} \right)$
	$\tilde{q}\left(\frac{\mu\rho}{\hat{\beta}}, 1 - \frac{1}{\rho} - \frac{\mu\rho}{\hat{\beta}} \right) > 1 \implies \rho\mu > \mu + 1$ and $\hat{\beta} > \frac{(\rho\mu)^2}{\rho\mu - \mu - 1}$		(θ_E, ψ_E)
SE, $\rho < 1$			$(0, 0)$

Key observations and comparisons of the various equilibrium states:

- When the disease is self-eradicating, agents need not get vaccinated to eradicate the disease. For all the type of agents, $\hat{\theta} = 0$ (and so is $\hat{\psi} = 0$) for all $\hat{\beta} < \mu$.

- With endemic disease, *it is possible to eradicate the disease only if the agents get vaccinated aggressively.* The FC agents with $\hat{\beta} > \rho\mu$ and FR agents with a bigger $\hat{\beta} > \rho^2\mu$ can completely eradicate the disease; this is possible only when $\mu\rho < \mu + 1$. However interestingly, *these parameters can't drive the system to an equilibrium that is stable against mutations, as will be seen in the next section.*
- With a lot more aggressive FR/FC agents, the system reaches disease free state ($\hat{\theta} = 0$ for all bigger $\hat{\beta}$), however with bigger vaccinated fractions ($\leq$ $1/(\mu+1)$).
- Interestingly *such an eradicating equilibrium state is not observed with vigilant agents (Table 3).* This is probably analogous to the well known fact that the rational agents often pay high price of anarchy.
- For certain behavioural parameters, system reaches an equilibrium state at which vaccinated and infected population co-exist. Interestingly for all three types of agents some of the co-existing equilibrium are exactly the same (e.g., (θ_E, ψ_E) in Tables and left plot of Fig. 1). In fact, such equilibrium are stable against mutations, as will be seen in the next section.

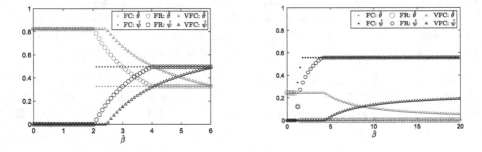

Fig. 1. Attractors for FC, FR, VFC1 agents versus β

A numerical example is presented in Fig. 1 that depicts many of the above observations. The parameters in respective plots are ($r = 1.188, \nu = 0.904$, $\lambda = 8.549$) and ($r = 1.0002, \nu = 0.404, \lambda = 1.749$), with $b = 0.322$. From the left plot, it can be seen that for all agents, $(\hat{\theta}, \hat{\psi})$ equals NVDF for smaller values of $\hat{\beta}$. As $\hat{\beta}$ increases, proportions for FC agents directly reach (θ_E, ψ_E). However, with FR and VFC1 agents, the proportions gradually shift from another interior attractor to finally settle at (θ_E, ψ_E). Further, for the right plot ($\mu\rho < \mu+1$) the disease is eradicated with FC (when $\hat{\beta} > \rho\mu$), FR (when $\hat{\beta} > \rho^2\mu$), i.e., $(\hat{\theta}, \hat{\psi})$ settles to $(0, 1/\mu+1)$, which is not a possibility in VFC1 agents. For the latter type, the proportions traverse through an array of co-existence equilibrium, and would approach $(0, 1 - 1/\rho)$ as $\hat{\beta}$ increases (see row 2 in Table 3).

Table 4. Attractors for FC agents (deadly disease)

Nature	Parameters		$(\hat{\theta}, \hat{\psi})$
Endemic, $\rho > 1$	$\hat{\beta} > \rho\mu$ *		$\left(0, 1 - \frac{\mu}{\hat{\beta}}\right)$
	$\hat{\beta} < \rho\mu$	$\rho_e\mu_e > 1$ or $\hat{\beta}\nu < b - d_e$	$\left(1 - \frac{1}{\rho_e}, 0\right)$
		$\rho_e\mu_e < 1, \hat{\beta}\nu > b - d_e$	$\left(\theta^*, 1 - \theta^*\left(1 - \frac{d_e}{\lambda}\right) - \frac{1}{\rho}\right),$ $\theta^* = \frac{\hat{\beta}\nu\left(\frac{\mu}{\hat{\beta}} - \frac{1}{\rho_e}\right)}{d_e\left(1 - \frac{\hat{\beta}\nu}{\lambda}\right)}$
SE, $\rho < 1$	$\hat{\beta} < \mu$		$(0, 0)$

Deadly Disease: We now consider the deadly disease scenario ($d_e > 0$) with FC and FR agents. Let $\rho_e := \frac{\lambda - d_e}{r + b}, \mu_e := \frac{b - d_e}{\hat{\beta}\nu - d_e}$. We conjecture the attractors with $\hat{\beta} \leq 1$ in Tables 4, 5 respectively, with $A := d_e\hat{\beta}\nu$, $B := -[(r + b + d_e)\hat{\beta}\nu + d_e\lambda]$, and $C := b\lambda + d_e(r + d_e)$. Further, the candidate attractors with $\tilde{q}(\hat{\theta}, \hat{\beta}) > 1$ are provided in the next section, which is of interest to ESS-AS. We expect that the proofs can be extended analogously and will be a part of future work, along with identifying and proving the attractors for other cases.

Table 5. Attractors for FR agents (deadly disease)

Nature	Parameters			$(\hat{\theta}, \hat{\psi})$
Endemic, $\rho > 1$	$\hat{\beta} > \rho\mu$	$\hat{\beta} > \rho^2\mu$ *		$\left(0, 1 - \sqrt{\frac{\mu}{\hat{\beta}}}\right)$
		$\hat{\beta} < \rho^2\mu$		$\left(1 - \frac{1}{\rho_e} - \frac{\lambda\hat{\psi}}{\lambda - d_e}, \hat{\psi}\right),$ $\hat{\psi} = 1 - \frac{-B - \sqrt{B^2 - 4AC}}{2A}$
	$\hat{\beta} < \rho\mu$	$\rho_e\mu_e > 1$ or $\hat{\beta}\nu < b - d_e$		$\left(1 - \frac{1}{\rho_e}, 0\right)$
		$\rho_e\mu_e < 1, \hat{\beta}\nu > b - d_e$		$\left(1 - \frac{1}{\rho_e} - \frac{\lambda\hat{\psi}}{\lambda - d_e}, \hat{\psi}\right),$ $\hat{\psi} = 1 - \frac{-B - \sqrt{B^2 - 4AC}}{2A}$
SE, $\rho < 1$	$\hat{\beta} < \mu$			$(0, 0)$

5 Evolutionary Stable Vaccination Responses

Previously, for a given user behaviour, we showed that the system reaches an equilibrium state, and identified the corresponding equilibrium states in terms of limit proportions. If such a system is invaded by mutants that use a different vaccination response, the system can get perturbed, and there is a possibility that the system drifts away. We now identify those equilibrium states, which are evolutionary stable against static mutations. Using standard tools of evolutionary game theory, we will show that the mutants do not benefit from deviating under certain subset of policies. That is, we identify the ESS-AS policies defined at the end of Sect. 2.

We begin our analysis with $d_e = 0$, the case with no excess deaths (due to disease). From Tables 1, 2 and 3, with $\rho > 1$, for all small values of $\hat{\beta}$ (including $\hat{\beta} = 0$), irrespective of the type of the policy, the equilibrium state remains the same (at NVDF), $(\hat{\theta}, \hat{\psi}) = (1 - 1/\rho, 0)$. Thus, the value of $h(\cdot)$ in user utility function (2) for all such small $\hat{\beta}$ equals the same value and is given by:

$$h_m := h(\pi(0)) = h\left(1 - \frac{1}{\rho}, 0\right) = c_{v_1} + \bar{c}_{v_2} - p_I(\theta)c_{I_1}, \text{ with, } p_I(\theta) := \frac{\lambda\theta}{\lambda\theta + \nu}. \quad (12)$$

In the above, $p_I(\cdot)$ is the probability that the individual gets infected before the next vaccination epoch. The quantity h_m is instrumental in deriving the following result with $\rho > 1$. When $\rho < 1$, the equilibrium state for all the policies (and all $\hat{\beta}$) is $(0, 0)$ leading to the following (proof in [12]):

Lemma 1. *If $\rho < 1$, or if $\rho > 1$ with $h_m > 0$, then $\pi(0)$ is an ESS-AS, for any $\pi \in \Pi$.*   ∎

When the disease is self-eradicating ($\rho < 1$), the system converges to $(0, 0)$, an infection free state on it's own without the aid of vaccination. Thus we have the above ESS-AS. When $\rho > 1$, if the inconvenience caused by the disease captured by $-h_m$ is not compelling enough (as $-h_m < 0$), the ES equilibrium state again results at $\hat{\beta} = 0$. *In other words, policy to never vaccinate is evolutionary stable in both the cases.* Observe this is a static policy irrespective of agent behaviour (i.e., for any $\pi \in \Pi$), as with $\hat{\beta} = 0$ the agents never get vaccinated irrespective of the system state.

From tables of the previous section, there exists $\bar{\beta}(\pi)$ such that the equilibrium state remains at NVDF for all $\hat{\beta} < \bar{\beta}$ (including $\hat{\beta} = 0$) and for all $\pi \in \Pi$. For such $\hat{\beta}$, we have $h(\hat{\theta}, \hat{\psi}) < 0$ if $h_m < 0$. Thus from the user utility function (2) and ESS-AS definition, the static best response set $\mathcal{B}(\pi(\hat{\beta})) = \{1\}$ for all $\pi \in \Pi$ and all $\hat{\beta} < \bar{\beta}$. Thus with $\rho > 1$ and $h_m < 0$, any policy $\pi(\hat{\beta})$ such that $\hat{\beta} < \bar{\beta}$ is not an ESS-AS. This leads to the following (proof available in [12]):

Theorem 5. [Vaccinating-ESS-AS] *When $\rho > 1$ and $h_m < 0$, there exists an ESS-AS among a $\pi \in \Pi$ if and only if the following two conditions hold:*

(i) there exist a $\beta^* > 0$ such that $q(\Upsilon^*) = 1$ and $\tilde{q}(\Upsilon^*) \neq 1$ under policy $\pi(\beta^*)$,
(ii) the equilibrium state is $(\theta^*, \psi^*) = (\theta_E, \psi_E)$, with $\mu\rho > \mu + 1$ and the corresponding user utility component, $h(\pi(\beta^*)) = h(\theta_E, \psi_E) < 0$.   ∎

Remarks: Thus when $\rho > 1$ and $h_m < 0$, there is no ESS-AS for any $\pi \in \Pi$ if $h(\theta_E, \psi_E) \geq 0$; observe that (θ_E, ψ_E) is equilibrium state with $\hat{q} = 1$ and hence from ODE (11), does not depend upon π. On the contrary, if $h(\theta_E, \psi_E) < 0$, $\pi(\beta^*)$ is ESS-AS for any $\pi \in \Pi$, with $\beta^* > \mu\rho$, $\beta^* > (\mu\rho)^2/(\mu\rho - 1)$, and $\beta^* > (\mu\rho)^2/(\mu\rho - 1 - \mu)$ respectively for FC, FR and VFC policies (using Tables 1, 2 and 3).

Thus interestingly, evolutionary stable behaviour is either possible in all, or in none. However, the three types of dynamic agents require different set of parameters to arrive at ES equilibrium. *An ES equilibrium with vaccination is possible only when* $\mu\rho > \mu + 1$ and interestingly, the infected and vaccinated fractions at this equilibrium are indifferent of the agent's behaviour. *In conclusion the initial dynamics could be different under the three different agent behaviours, however, the limiting proportions corresponding to any ESS-AS are the same.*

Numerical Examples: We study the variations in vaccinating ESS (θ_E, ψ_E), along with others, with respect to different parameters. In these examples, we set the costs of vaccination and infection as $c_{v_1} = 2.88, c_{v_2} = 0.65, \bar{c}_{v_2} = 1.91, c_{I_1} = 4.32/r$. Other parameters are in the respective figures; black curves are for $d_e = 0$. In Fig. 2, we plot the ESS-AS for different values of birth-rate. Initially, $\rho > 1$, and vaccinating ESS-AS exists for all $b \leq 0.54$; here $h(\theta_E, \psi_E) < 0$ as given by Theorem 5. As seen from the plot, θ_E is decreasing and approaches zero at $b \approx 0.54$. Beyond this point there is no ESS because $\mu\rho$ reduces below $\mu + 1$. With further increase in b, non-vaccinating ESS emerges as ρ becomes less than one. Interestingly, a much larger fraction of people get vaccinated at ESS for smaller birth-rates. This probably could be because of higher infection rate per birth. In fact from the definition of θ_E, the infected fractions at ES equilibrium are concave functions of birth rate. When infection rate per birth is sufficiently high, *it appears people pro-actively vaccinate themselves, and bring infected fraction (at ES equilibrium) lower than those at smaller ratios of infection rate per birth.*

In Fig. 3 we plot ESS-AS for different values of ν. For all $\nu \leq 0.31$, the vaccinating ESS-AS exists. Beyond this, there is no ESS because $\mu\rho$ reduces below $\mu + 1$. As ν further increases, h_m becomes[3] positive, leading to $NVDF$ as ESS. One would expect a smaller infected proportion at ES equilibrium with increased availability rate, however we observe the converse; this is because the users' perception about infection cost changes with abundance of vaccines.

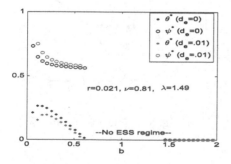

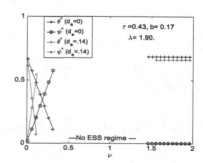

Fig. 2. ESS versus birth-rate (Color figure online)

Fig. 3. ESS versus vaccine availability (Color figure online)

---

[3] This is because the chances of infection before the next vaccination epoch decrease with increase in the availability rate ν.

With Excess Deaths $(d_{e>0})$: The analysis will follow in exactly similar lines as above. In this case, we have identified the equilibrium points of the ODE (11) but are yet to prove that they are indeed attractors. We are hoping that proof of attractors will go through similar to case with $d_e = 0$, but we have omitted it due to lack of time and space. Once we prove that the equilibrium points are attractors, the analysis of ESS would be similar to the previous case. When $\rho < 1$ or if $\rho > 1$ and $h_m^{d_e} := h(\pi(0)) = h\left(1 - \frac{1}{\rho_e}, 0\right) > 0$ (see (2), Table 4 and 5), then $\pi(0)$ is an ESS-AS, as in Lemma 1. Now, we are only left with case when $\rho > 1$ and $h_m^{d_e} < 0$ and one can proceed as in Theorem 5. In this case the only candidate for ESS-AS is $\pi(\hat{\beta})$ with $\hat{\beta}$ such that $q = 1$ and $\tilde{q} \neq 1$ and $h(\hat{\theta}, \hat{\psi}) < 0$. So, we will only compute the equilibrium points for $\hat{\beta}$ such that $q = 1$ and $\tilde{q} \neq 1$. From ODE (11), such an equilibrium point is given by $(B := \lambda b + d_e(r + d_e - \lambda - \nu))$:

$$\theta_E^{d_e} = 1 - \frac{1}{\rho_e} - \frac{\lambda \psi_E^{d_e}}{\lambda - d_e} \text{ and } \psi_E^{d_e} = \frac{-B + \sqrt{B^2 + 4\lambda d_e \nu (r + b)}}{2\lambda d_e}.$$

One can approximate this root for small d_e (by neglecting second order term $\lambda d_e \approx 0$), the corresponding ES equilibrium state $(\theta_E^{d_e}, \psi_E^{d_e})$ (again from (11)):

$$\psi_E^{d_e} \approx \frac{(r + b)\nu}{\lambda b + d_e(r + d_e - \lambda - \nu)}, \text{ that is,}$$

$$(\theta_E^{d_e}, \psi_E^{d_e}) \approx \left(1 - \frac{1}{\rho_e} - \frac{o^{d_e}}{\mu \rho_e}, \frac{o^{d_e}}{\mu \rho_e} \frac{\lambda - d_e}{\lambda}\right) \text{ with } o^{d_e} := \frac{1}{1 + \frac{d_e(r + d_e - \lambda - \nu)}{\mu \lambda \nu}}.$$

As before, there is no ESS if $\mu + o^{d_e} \geq \mu \rho_e$ (for larger d_e, when $\theta_E^{d_e} < 0$). From Figs. 2, 3 (red curves), the ES equilibrium state in deadly case has higher vaccinated fraction and lower infected fraction as compared to the corresponding non-deadly case (all parameters same, except for d_e). More interestingly the variations with respect to the other parameters remain the same as before.

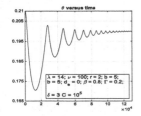
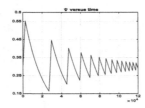
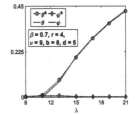

Fig. 4. VFC2 agents: Limit behaviour   **Fig. 5.** FC agents, against λ

6   Numerical Experiments

We performed Monte-Carlo simulations to reinforce our ODE approximation theory. We plotted attractors of the ODE (11) represented by (θ, ψ), and the corresponding infected and vaccinated fractions (θ^s, ψ^s) obtained via simulations for different values of λ, ν and β. Our Monte-Carlo simulation based dynamics mimic the model described in Sect. 3. In all these examples we set $N(0) = 40000$. The remaining parameters are described in the respective figures. We have several plots in Figs. 5, 6, 7 and 8, which illustrate that the ODE attractors well approximate the system limits, for different sets of parameters.

VFC2 Agents: These agents attempt to vaccinate themselves only when the disease is above a certain threshold Γ, basically $\tilde{q} = \hat{\beta}\psi \mathbb{1}_{\{\theta > \Gamma\}}$. As one may anticipate, the behaviour of such agents is drastically different from the other type of agents. Theorem 1 is applicable even for these agents (approximation in finite windows will be required here). However with a close glance at the ODE, one can identify that the ODE does not have a limit point or attractor, but rather would have a limiting set. From the RHS of the ODE (11), one can observe that the ψ derivative fluctuates between positive and negative values, and hence ψ goes through increase-decrease phases if at-all $\theta(t)$ reaches near Γ. This indeed happens, the fact is supported by a numerical example of Fig. 4. Thus interestingly with such a vaccine response behaviour, the individuals begin to vaccinate the moment the infection is above Γ, which leads to a reduced infection, and when it reaches below Γ, individuals stop vaccinating themselves. This continues forever, and one can observe such behaviour even in real world.

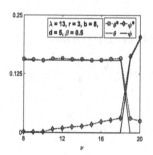

Fig. 6. FC agents vs ν

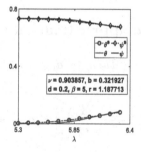

Fig. 7. FR agents vs λ

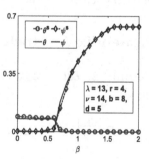

Fig. 8. FC agents vs β

7   Conclusions

With the ongoing pandemic in mind, we consider a scenario where the vaccines are being prematurely introduced. Further, due to lack of information about the side-effects and efficacy of the vaccine, individuals exhibit vaccination hesitancy. This chaos is further amplified sometimes due to reported disease statistics, unavailability of the vaccine, leading to vaccination urgency. We developed an epidemic SIS model to capture such aspects, where the system changes due to births, deaths, infections and recoveries, while influenced by the dynamic vaccination decisions of the population. As observed in reality, a variety of behavioral patterns are considered, in particular follow-the-crowd, free-riding and vigilant agents. Using stochastic approximation techniques, we derived the time asymptotic proportions of the infected and vaccinated population for a given vaccination response. Additionally, we considered the conflict of stability for dynamic policies against static mutations, identified the strategies which are stable against static mutations and studied the corresponding equilibrium states.

Interestingly, the agents exhibit different behaviors and lead to different equilibrium states, however, at ESS-AS, all of the agents reach same limit state, where they choose vaccination either with probability 1 or 0, only based on system parameters. Also, by analysing the corresponding ODEs, we obtained many responses under which disease can be eradicated completely, but none of those are stable against mutations. Ironically, this is a resultant of the rationality exhibited by agents, which prevents them from reaching the disease-free state.

We observed certain surprising patterns at evolutionary stable equilibrium: (a) no one gets vaccinated with abundant vaccines; scarcity makes them rush; (b) the limit infected proportions are concave functions of birth rates.

Lastly, the excess deaths did not change the patterns of ES equilibrium states versus parameters, however, the limit vaccination fractions are much higher. So, in all it appears individuals rush for vaccine and we have smaller infected fractions at ES equilibrium, when there is a significant scare (of either deaths, or of scarcity of vaccines, or high infection rates etc.). Ironically, the disease can be better curbed with excess deaths.

Appendix A: Stochastic approximation related proofs

Lemma 2. *Let $\delta = 2/(N(0)-1)$. Then for any k, $\eta_k \geq \bar{\delta} := \frac{N(0)-3}{(N(0)-1)^2}$ a.s. And thus,*

$$E\left[\frac{1}{\eta_k^2}\right] \leq \frac{1}{\bar{\delta}^2} \text{ and } E_k\left[\left|\left(\frac{1}{\eta_{k+1}} - \frac{1}{\eta_k}\right)\right|\right] \leq \epsilon_k \frac{\bar{\delta}+1}{\bar{\delta}^2} \text{ a.s., for any } k.$$

Proof is provided in [12].   ∎

Lemma 3. *The term $\alpha_k^m \to 0$ a.s., and, $\sum_k \epsilon_k |\alpha_k^m| < \infty$ a.s. for $m = \theta, \psi$.*

Proof: We will provide the proof for α_k^ψ and proof goes through in exactly similar line for α_k^θ. From equation (9), as in (10),

$$\alpha_k^\psi = E_k\left[L_{k+1}^\psi - \frac{\eta_{k+1}}{\eta_k} L_{k+1}^\psi\right], \text{ where } L_{k+1}^\psi = \frac{\mathbb{1}_{\{\eta_k > \delta\}}}{\eta_{k+1}} [G_{V,k+1} - (N_{k+1} - N_k)\psi_k].$$

By Lemma 2 and because $|G_{V,k+1} - (N_{k+1} - N_k)\psi_k| \leq 2$ a.s., we have:

$$\left|\alpha_k^\psi\right| \leq 2E_k\left[\left|\frac{1}{\eta_{k+1}} - \frac{1}{\eta_k}\right|\right] \leq 2\epsilon_k \frac{\bar{\delta}+1}{\bar{\delta}^2} \text{ a.s.} \tag{13}$$

Thus we have:

$$\sum_{k=1}^{\infty} \epsilon_k |\alpha_k^{\psi}| \le 2\frac{\bar{\delta}+1}{\bar{\delta}^2} \sum_{k=1}^{\infty} \epsilon_k^2 < \infty \quad a.s. \qquad \blacksquare$$

Lemma 4. $\sup_k E|L_k^m|^2 < \infty$ *for* $m = \theta, \psi, \eta$.

Proof: The result follows by Lemma 2 and (9) (for an appropriate C):

$$|L_k^{\theta}|^2 \le \frac{4}{\eta_k^2} \ a.s., \text{ and } |L_k^{\psi}|^2 \le \frac{4}{\eta_k^2} \ a.s., \text{ and } |L_k^{\eta}|^2 \le C < \infty, \ a.s. \qquad \blacksquare$$

Proof of Theorem 1: As in [10], we will show that the following sequence of piece-wise constant functions that start with Υ_k are equicontinuous in extended sense. Then the result follows from [10, Chapter 5, Theorem 2.2]. Define $(\Upsilon^k(t))_k := (\theta^k(t), \psi^k(t), \eta^k(t))_k$ where,

$$\theta^k(t) = \theta_k + \sum_{i=k}^{m(t_k+t)-1} \epsilon_k L_{k+1}^{\theta}, \quad \psi^k(t) = \psi_k + \sum_{i=k}^{m(t_k+t)-1} \epsilon_k L_{k+1}^{\psi}, \quad \eta^k(t))_k = \eta_k + \sum_{i=k}^{m(t_k+t)-1} \epsilon_k L_{k+1}^{\eta},$$

where $m(t) := \max\{k : t_k \le t\}$. This proof is exactly similar to that provided in the proof of [10, Chapter 5, Theorem 2.1] for the case with continuous g, except for the fact that $g(\cdot)$ in our case is not continuous. We will only provide differences in the proof steps towards $(\theta^k(t))_k$ sequence, and it can be proved analogously for others. Towards this we define $M_k^{\theta} = \sum_{i=0}^{k-1} \epsilon_i \delta M_i^{\theta}$ with $\delta M_k^{\theta} := L_{k+1}^{\theta} - g^{\theta}(\Upsilon_k) - \alpha_k^{\theta}$ as in [10] and show the required uniform continuity properties in view of Lemmas 3–4. Observe that $\eta_k \le 1 + N(0)/k$, $\theta_k \le 1$ for any k. Now the uniform continuity of integral terms like the following is achieved because our $g(\cdot)$ are bounded:

$$\left| \int_s^t g^{\theta}(\Upsilon^k(z))dz \right| \le \int_s^t |g^{\theta}(\Upsilon^k(z))|dz \le \int_s^t \frac{\lambda+r+b+2d_e}{\delta(d+b)} dz \le \bar{m}(t-s) \le \bar{m}\delta_1.$$

Such arguments lead to the required equicontinuity (details are in [12]). $\qquad \blacksquare$

Appendix B: ODE Attractors Related Proofs

Proof of Theorem 2: Let $\Upsilon := (\theta, \psi, \eta)$. Let $\hat{\Upsilon}$ represent the corresponding attractors from Table 1. Here $q(\theta, \psi) = \min\{\tilde{q}(\theta, \psi), 1\}$ with $\tilde{q}(\theta, \psi) = \beta\psi$.

We first consider the case where $\tilde{q}(\hat{\theta}, \hat{\psi}) < 1$. Further, note that one can re-write ODEs, $\dot{\Upsilon} = g(\Upsilon)$, as below:

$$\dot{\theta} = \frac{\mathbb{1}_{\{\eta>\delta\}} A\theta}{\eta\varrho}, \quad \dot{\psi} = \frac{\mathbb{1}_{\{\eta>\delta\}} B\psi}{\eta\varrho}, \text{ and } \dot{\eta} = \mathbb{1}_{\{\eta>\delta\}} C, \text{ where}$$

$A = A(\Upsilon) := (1 - \theta - \psi)\lambda - r - b$, $B = B(\Upsilon) := (1 - \theta - \psi)\beta\nu - b$ and $C = C(\Upsilon) := {(b-d)}/{\varrho} - \eta$. To this end, we define the following Lyapunov function based on the regimes of parameters:

$$
V(\Upsilon) := \begin{cases}
\left(\hat{A}(\theta)\right)^2 + \psi\left(\hat{B}(\psi)\right)^2 + C(\eta)(\hat{\eta} - \eta), & \text{if } \beta < \rho\mu, \rho > 1, \\
\theta\left(\hat{A}(\theta)\right)^2 + \left(\hat{B}(\psi)\right)^2 + C(\eta)(\hat{\eta} - \eta), & \text{if } \beta > \rho\mu, \text{ and } \rho > 1 \\
\theta\left(\hat{A}(\theta)\right)^2 + \psi\left(\hat{B}(\psi)\right)^2 + C(\eta)(\hat{\eta} - \eta), & \text{if } \rho < 1,
\end{cases}
$$

where $\hat{A}(\theta) := A(\theta, \hat{\psi}, \hat{\eta})$, $\hat{B}(\psi) := B(\hat{\theta}, \psi, \hat{\eta})$. We complete this proof using the above functions and the details are in [12]. ∎

Lemma 5. *Let $\hat{\theta}, \hat{\psi} > 0$. If $\tilde{q}(\hat{\theta}, \hat{\psi}) \neq 1$, there exists a Lyapunov function such that $(\hat{\theta}, \hat{\psi}, \hat{\eta})$ is locally asymptotically stable attractor for ODE (11) in the sense of Lyapunov.*

Proof: We use similar notations as in previous proof. Let us first consider the case where $\tilde{q}(\hat{\theta}, \hat{\psi}) < 1$, i.e., $q(\hat{\theta}, \hat{\psi}) = \tilde{q}(\hat{\theta}, \hat{\psi})$. Then, one can choose a neighborhood (further smaller, if required) such that $\hat{q} - \delta < q(\theta, \psi) < \hat{q} + \delta$, and $q(\theta, \psi) = \tilde{q}(\theta, \psi)$ for some $\delta > 0$. Define the following Lyapunov function (for some $w_1, w_2 > 0$, which would be chosen appropriately later):

$$
V(\Upsilon) := w_1(\hat{\theta} - \theta)\hat{A}(\theta) + w_2(\hat{\psi} - \psi)\hat{B}(\psi) + C(\hat{\eta} - \eta), \text{ where} \tag{14}
$$

$\hat{A}(\theta) := 1 - \theta - \hat{\psi} - \frac{1}{\rho}$, and $\hat{B}(\psi) := \hat{q}(1 - \hat{\theta} - \psi) - \mu\psi$ (recall $\hat{q} := q(\hat{\theta}, \hat{\psi})$). Call $\tilde{\theta} := \hat{\theta} - \theta$ and $\tilde{\psi} := \hat{\psi} - \psi$. The derivative of $V(\Upsilon(t))$ with respect to time is:

$$
\dot{V} = \langle \nabla V, g(\Upsilon) \rangle \tag{15}
$$

$$
= -\left(\hat{A}(\theta) + \tilde{\theta}\right)\frac{A\theta\lambda w_1}{\eta\varrho} - \left(\hat{B}(\theta) + \tilde{\psi}(\hat{q} + \mu)\right)\frac{B\nu w_2}{\eta\varrho} - (C(\eta) + \hat{\eta} - \eta)C.
$$

One can prove that the last component, i.e., $-(C + \hat{\eta} - \eta)C$ is strictly negative in an appropriate neighborhood of $\hat{\Upsilon}$ as in proof of Theorem 2. Now, we proceed to prove that other terms in $\dot{V}$ (see (15)) are also strictly negative in a neighborhood of $\hat{\Upsilon}$.

Consider the term[4] $\left(\hat{A}(\theta) + \tilde{\theta}\right)A$, call it A_1:

$$
A_1 = 2\tilde{\theta}A = 2\left(\tilde{\theta}^2 + \tilde{\theta}\tilde{\psi}\right) = 2\left(\left(\tilde{\theta}c_1 + \frac{1}{2c_1}\tilde{\psi}\right)^2 + (1 - c_1^2)\tilde{\theta}^2 - \frac{1}{4c_1^2}\tilde{\psi}^2\right), \tag{16}
$$

where c_1 will be chosen appropriately in later part of proof. Similarly the term corresponding to B is (details in [12]),

---

[4] Observe that $\hat{A}(\theta)A = (\hat{A}(\hat{\theta}) + \tilde{\theta})A$, and $\hat{A}(\hat{\theta}) = 0$.

$$B_1 = 2(\hat{q} + \mu) \left[\widetilde{\psi}^2 \left(\mu + p_1(\Upsilon) + \hat{q}(1 - c_2^2) \right) - \frac{1}{4c_2^2 \hat{q}} (\hat{q} - p_2(\Upsilon))^2 \widetilde{\theta}^2 \right]$$
$$+ 2(\hat{q} + \mu)\hat{q} \left(c_2 \widetilde{\psi} + \frac{1}{2c_2} \left(1 - \frac{p_2(\Upsilon)}{\hat{q}} \right) \widetilde{\theta} \right)^2 . \tag{17}$$

Thus, we get: $\dot{V} < -A_1 \frac{\theta \lambda w_1}{\eta \varrho} - B_1 \frac{\nu w_2}{\eta \varrho}$. Now, for $\dot{V}$ to be negative, we need (using terms, in (16) and (17), corresponding to $\widetilde{\theta}^2, \widetilde{\psi}^2$):

$$\nu(\hat{q} + \mu) \leq \frac{4c_2^2 \hat{q}}{w_2 (\hat{q} - p_2(\Upsilon))^2} \lambda w_1 (1 - c_1^2)\theta, \text{ and}$$

$$\frac{1}{2c_1^2} \theta \lambda w_1 \leq 2\nu w_2 (\hat{q} + \mu) \left(\mu + \hat{q}(1 - c_2^2) + p_1(\Upsilon) \right).$$

By appropriately choosing the constants (for various agents), we complete the proof (details in [12]). ∎

Appendix C: ESS Related Proofs

Lemma 6. *Let* $\rho > 1$. *Assume* $\tilde{q}(\Upsilon) \neq 1$ *where* $q(\Upsilon) = \min\{\tilde{q}(\Upsilon), 1\}$. *Consider a policy* $\pi(\hat{\beta})$ *where* $\pi \in \Pi$ *and* $\hat{\Upsilon}$ *is the attractor of the corresponding ODE* (11). *Let* $\hat{\Upsilon}_\epsilon$ *be attractor corresponding to* ϵ-*mutant of this policy,* $\pi_\epsilon(\hat{\beta}, p)$ *for some* $p \in [0, 1]$. *Then, i) there exists an* $\bar{\epsilon}(p) > 0$ *such that the attractor is unique and is a continuous function of* ϵ *for all* $\epsilon \leq \bar{\epsilon}$ *with* $\hat{\Upsilon}_0 = \hat{\Upsilon}$.
ii) Further $\bar{\epsilon}$ *could be chosen such that the sign of* $h(\hat{\Upsilon}_\epsilon)$ *remains the same as that of* $h(\hat{\Upsilon})$ *for all* $\epsilon \leq \bar{\epsilon}$, *when the latter is not zero.* ∎

Proof: We begin with an interior attractor. Such an attractor is a zero of a function like the following (e.g., for VFC1 it equals, see (11)):

$$\phi\lambda - r - b, \quad \min\left\{1, \psi\theta\hat{\beta}\right\} \phi\nu - b\psi, \text{ and } \frac{b - d}{\varrho} - \eta. \tag{18}$$

Under mutation policy, $\pi_\epsilon(\hat{\beta}, p)$, the function modifies to the following:

$$\phi\lambda - r - b, \quad \left((1 - \epsilon) \min\{1, \hat{\beta}\psi\theta\} + \epsilon p \right) \phi\nu - b\psi, \text{ and } \frac{b - d}{\varrho} - \eta. \tag{19}$$

By directly computing the zero of this function, it is clear that we again have unique zero and these are continuous[5] in ϵ (in some $\bar{\epsilon}$-neighbourhood) and that they coincide with $\hat{\Upsilon}$ at $\epsilon = 0$. Further using Lyapunov function as defined in the corresponding proofs (with obvious modifications) one can show that these zeros are also attractors in the neighborhood. The remaining part of the proof is completed in [12]. The last result follows by continuity of h function (2). ∎

---

[5] When $\tilde{q}(\Upsilon) > 1$, the zeros are $(\epsilon p + (1-\epsilon))/(\mu\rho)$, otherwise they are the zeros of a quadratic equation with varying parameters, we have real zeros in this regime.

References

1. Sahneh, F.D., Chowdhury, F.N., Scoglio, C.M.: On the existence of a threshold for preventive behavioral responses to suppress epidemic spreading. Sci. Rep. **2**(1), 1–8 (2012)
2. Piraveenan, M., et al.: Optimal governance and implementation of vaccination programs to contain the COVID-19 pandemic. arXiv preprint arXiv:2011.06455 (2020)
3. Hethcote, H.W.: Qualitative analyses of communicable disease models. Math. Biosci. **28**, no. 3–4 (1976)
4. Boguná, M., Pastor-Satorras, R.: Epidemic spreading in correlated complex networks. Phys. Rev. E **66**(4), 047104 (2002)
5. Iwamura, Y., Tanimoto, J.: Realistic decision-making processes in a vaccination game. Physica A: Statistical Mechanics and its Applications 494 (2018)
6. Li, Q., Li, M., Lv, L., Guo, C., Lu, K.: A new prediction model of infectious diseases with vaccination strategies based on evolutionary game theory. Chaos Solitons Fractals **104**, 51–60 (2017)
7. Bhattacharyya, S., Bauch, C.T.: "Wait and see" vaccinating behaviour during a pandemic: a game theoretic analysis. Vaccine 29, no. 33 (2011)
8. Armbruster, B., Beck, E.: Elementary proof of convergence to the mean-field model for the SIR process. J. Math. Biol. **75**(2), 327–339 (2017). https://doi.org/10.1007/s00285-016-1086-1
9. Cooke, K.L., Van Den Driessche, P.: Analysis of an SEIRS epidemic model with two delays. J. Math. Biol. **35**(2), 240–260 (1996)
10. Kushner, H., Yin, G.G.: Stochastic approximation and recursive algorithms and applications, vol. 35. Springer Science & Business Media (2003)
11. Webb, J.N.: Game theory: decisions, interaction and Evolution. Springer Science & Business Media (2007)
12. Singh, V., Agarwal, K., Shubham, Kavitha, V.: Evolutionary Vaccination Games with premature vaccines to combat ongoing deadly pandemic. arXiv preprint arXiv:2109.06008 (2021)

Quantitative Analysis of Attack Defense Trees

Nihal Pekergin$^{(\boxtimes)}$ and Sovanna Tan

LACL, Univ Paris Est Créteil, 94010 Créteil, France
{nihal.pekergin,sovanna.tan}@u-pec.fr

Abstract. The quantitative analysis of Attack Tree models brings insights on the underlying security-critical systems. Having information on temporal behaviours of such systems lets us check whether at a given time, the probability that the system is compromised is less than a critical threshold or not. Moreover the evaluation of the countermeasure efficiency and the determination of eventual reinforcements of security-critical systems are very important. In this paper, we extend the approach proposed in [11] for numerical analysis of the Attack Tree models to the Attack Defense Tree analysis. The completion times of attacks and countermeasures are defined by finite discrete random variables. The output distribution of the root of an Attack Defense Tree is computed by a bottom-up approach. However the size of the output distribution can become quickly very large. We prove that the method which consists in deriving bounding distributions of reduced sizes by means of the stochastic comparison method can be used in the presence of counter-measure gates.

Keywords: Attack defense tree · Discrete probability distribution · Stochastic bounds

1 Introduction

The graphical formalisms such as Fault Trees (FT) [14] and Attack Trees(AT) are commonly used models for safety and security analysis. In Attack Trees the leaves are the actions taken intentionally by attackers called basic attacks. A leaf turns out *True* when the underlying attack is successful. The complex attack scenarios may be specified by combining basic attacks with logical gates. The output of a gate is *True* if the subsystem having this gate as root is compromised. Thus the root of the AT turns out *True* when the whole system is compromised.

The Attack Trees were first proposed in [15]. Recently, several works have been done in the literature (see [8,17] and the references therein). These works can be classified as semantical approaches to give a rigorous, mathematical definition of AT trees and their extensions; generation approaches to study the construction of such models; quantitative approaches to propose efficient algorithms and techniques for the quantitative analysis.

© ICST Institute for Computer Sciences, Social Informatics and Telecommunications Engineering 2021
Published by Springer Nature Switzerland AG 2021. All Rights Reserved
Q. Zhao and L. Xia (Eds.): VALUETOOLS 2021, LNICST 404, pp. 207–220, 2021.
https://doi.org/10.1007/978-3-030-92511-6_13

In the first AT models the logical connectors were limited to the AND and OR gates, but they have been extended by including some dynamical gates to the Attack-Fault Trees [10], and by adding countermeasures [13,17] to the Attack-Defense Trees. For the static quantitative evaluation of such models, the leaves (basic events) are specified by the success probability of the underlying event. By applying the standard bottom-up algorithm, the success probability of attack scenarios specified by an AT can be computed [13]. Temporal behaviours of safety-security critical systems are primordial. For instance, the time for an attack to be successful is an important indicator of the safety properties. Thus, it is important to check whether at a given time the probability that the system is compromised is not greater than a critical threshold.

In [2], the authors propose to specify the time to success of an attack by *Acyclic Phase Type*(APH) distributions. The distributions associated with leaves may be any continuous distribution since fitting algorithms exist in the literature to approximate a continuous distribution by a APH distribution. By considering independence of basic attacks, the random variables of the gate outputs can be computed as the maximum (AND gate), the minimum (OR gate) and convolution of (SEQ gate) of the underlying input distributions. However the successive applications of these operators may lead to a state space explosion of the output distributions. To overcome this problem, the output distributions are *compressed* to construct approximate, smaller sizes APH distributions by algorithms of cubic computational complexity.

In [11] we have proposed an approach to quantitatively analyze AT based on ideas similar to those exposed in [2] but with radically different techniques. The temporal behaviors of basic events are specified by discrete probability distributions. We prevent the size explosion problem by replacing the output distribution with reduced-size bounding distributions in the sense of the strong stochastic ordering ($\leq_{st}$). Intuitively speaking, if two distributions are ordered in the sense of this order: $d_1 \leq_{st} d_2$, then the cumulative probability distribution of d_1 is always greater or equal to the cumulative probability distribution of d_2 (d_2 takes larger values than d_1). In other words given a time (t), in the upper bounding distribution the probability that the time to success is greater or equal to t is greater or equal to the probability computed by the original distribution.

The lower and upper reduced-size bounding distributions can be derived due to the monotonicity properties of gates (AND, OR, SEQ). Roughly speaking, monotonicity can be explained with the following property: if the input distributions are replaced by upper (resp. lower) bounding distributions the output distribution is also a upper (resp. lower) bounding distribution. Indeed these gates satisfy the monotonicity in the sense of the $\leq_{st}$ ordering since the related operations are non decreasing functions on the inputs.

The size reduction can be performed with several compressing algorithms having different computational complexities. The naive algorithm has linear complexity. It consists in merging the successive atoms and in putting the sum of corresponding probabilities to the largest value for the upper bound and to the smallest one for the lower bound. In this framework, it is possible to consider continuous distributions for the times of events and then to apply bounding discretization In [1], the algorithms to construct a bounding distribution in the sense of the $\leq_{st}$ ordering of size K for an input distribution of size N, with $K \leq N$ are given. The optimal algorithm with respect to a given non decreasing reward has $O(KN^2)$ computational complexity, but several greedy algorithms with lower computational complexities have been also presented.

In this paper, we propose to extend this approach to evaluate the counter-measures associated with AND gates as proposed in [7,13]. We first explain how the countermeasure nodes which are included in AT models to counter attacks can be considered in our framework. The times for the attack and the counter-measure to succeed have contrary impacts on the whole temporal behaviours of the underlying Attack Defense Tree. Roughly speaking if the time needed to complete a countermeasure decreases, the attack takes longer to succeed. Thus the underlying operation does not have non decreasing monotonicity property and its analysis needs some attention. We first show that adding a countermea-sure input to an AND gate does not increase the size of the output distribution. We propose not to compress the countermeasure node (leaf) and show that it is still possible to compress other distributions.

These last years, many works to enhance the design and analysis of attack-defense trees with formal methods such as model checking, automata theory, con-straint solving, Bayesian networks have been done. The stochastic game inter-pretation of Attack-Defense trees and its analysis with PRISM-Games model checker is given in [3]. The priced-timed automata modelling of ATs and analysis with UPPAAL model checker can be found in [5,9]. The extended asynchronous multi-agent systems formalism has been proposed in [12]. These works show the increasing interest of the community for the quantitative analysis of such models.

The paper is organized as follows. In Sect. 2 we present the analysis of ATs with the countermeasure nodes. We illustrate this approach with an example of the literature in Sect. 3. Finally we conclude and give perspectives.

2   Quantitative Evaluation of Attack Defense Trees

2.1   Bounding Distributions for Attack Trees

We have proposed in [11], the quantitative evaluation of Attack Trees with con-junction (AND), disjunction (OR) and sequential conjunction (noted as SEQ or $SAND$). A finite discrete probability distribution is associated with each Basic Attack (BA). This random variable is indeed the time at which the corre-sponding BA occurs. The input of a gate associated with this distribution turns out $(True)$ at this time. Similarly, the output distribution of a gate corresponds to the time at which the subsystem having this gate as root is compromised. Therefore the output distribution of the root of an Attack Tree denotes the time when the whole system is compromised. The model is observed during time inter-val $0 \leq t < MT$. The time value MT represents indeed the infinity such that events which occur at this date do not indeed happen. The output distribution of the formerly stated gates can be defined as a function of input distributions [2,11]. Let $I_1\ I_2$ be the input distributions. Under independent basic attacks and discrete distributions, the output distributions are numerically computed as:

- $\mathcal{P}(O_{OR}(I_1, I_2) = a)$
  $= \mathcal{P}(I_1 = a).\mathcal{P}(I_2 > a) + \mathcal{P}(I_2 = a).\mathcal{P}(I_1 > a) + \mathcal{P}(I_1 = a) \times \mathcal{P}(I_2 = a)$
- $\mathcal{P}(O_{AND}(I_1, I_2) = a)$
  $= \mathcal{P}(I_1 = a).\mathcal{P}(I_2 < a) + \mathcal{P}(I_2 = a).\mathcal{P}(I_1 < a) + \mathcal{P}(I_1 = a) \times \mathcal{P}(I_2 = a)$
- $\mathcal{P}(O_{SEQ}(I_1, I_2) = a) = \sum_k \mathcal{P}(I_1 = k) \times \mathcal{P}(I_2 = (a - k))$

Thanks to the tree structure of the model, the output distributions of the gates and finally the output distribution of the root (the time for the system to be compromised) is computed by the bottom-up approach.

The sizes of distributions (the number of atoms) can increase rapidly due to the successive applications of the operations associated with the gates. For AND, OR gates, the size of the output distribution may be the sum of the input distribution sizes, $O(l_1 + l_2)$. However for the convolution operation, the number of atoms may be the product of the input distribution sizes, $(O(l_1 \times l_2))$. The main drawback of this approach is the explosion of the number of atoms after successive computations. We have proposed to reduce the number of atoms and construct bounding distributions by compressing [11]. The computation complexity of the output distribution depends on the sizes of the input distributions. Let l_1, l_2 be two input distributions sizes and $l = \max(l_1, l_2)$. The output distributions of the above gates can be computed by a naive algorithm with complexity $\Theta(l_1 \times l_2)$. The computation complexities are indeed bounded by $\Theta(l \times \log l)$ (Discrete Fast Fourier transform for convolution and sorting for other gates). Obviously, the size reduction will be favoring for computation complexities and also for the numerical stability since we consider probability vectors.

The ability to control the distribution sizes during the bottom-up analysis of an AT is of great importance for the point of view of the algorithmic complexity. In [4,11], we have shown that the reduced-size bounding output distributions can be derived by considering bounding input distributions for AND, OR, SEQ gates. These bounding distributions are in the sense of the stochastic strong order ($\leq_{st}$) associated with non decreasing functions. Thus if two random variables are ordered in the $\leq_{st}$ order then their non decreasing rewards are also ordered:

$$A \leq_{st} A^u \quad \Leftrightarrow \quad \mathrm{E}[f(A)] \leq \mathrm{E}[f(A^u)] \tag{1}$$

for all non decreasing function f, when the expectations E exist. For instance, if $A \leq_{st} A^u$, then $\mathrm{E}[A] \leq \mathrm{E}[A^u]$.

The monotonicity properties of AND, OR, SEQ are proved in [4,11]. Let I_1, I_2 be the input distributions and $GATE(I_1, I_2)$ be the output distribution of a AND, OR, SEQ gate. The output of the the gate is upper (resp. lower) bounded if it is subjected to upper (resp. lower) bounded input distributions:

Property 1.

$$I_1 \leq_{st} I_1^{up}, I_2 \leq_{st} I_2^{up} \Rightarrow GATE(I_1, I_2) \leq_{st} GATE(I_1^{up}, I_2^{up})$$

The monotonicity results from the fact that the output of these gates are non decreasing functions of their inputs. This monotonicity property lets us construct the reduced-size distributions and diminish the algorithmic complexity.

The reduced-size bounding distributions in the sense of the $\leq_{st}$ ordering can be constructed in several manners. The naive approach to construct an upper bounding (resp. lower) distribution to divide the size by m is to take m successive values, to delete the smallest (resp. greatest) one and to include its probability to the greatest (resp. smallest) value [16]. In [1], an algorithm to construct for an arbitrary distribution of size N bounding distributions of size $K < N$ is proposed. This algorithm is optimal such that it is the smallest (resp. greatest) $\leq_{st}$ (Eq. (1)) upper (resp. lower) bound respect to a non decreasing reward (for example expectation). is given: and any positive non decreasing reward function , This optimal algorithm is based on a graph optimization problem and can be computed by a dynamical programming approach with $\Theta(KN^2)$. Greedy algorithms with less computational complexities have been also proposed. The sizes of the bounding distributions K can be reduced decrementally by taking care of the accuracy and the computational complexity trade-off.

2.2 *SI-AND* Gate with Countermeasure

Attack Trees are extended to Attack-Defense Trees (ADT) with two actors: attacker and defender. The goal of each actor is to counter the other one. We consider the semantic defined in [7,13] for the countermeasure nodes (defenders). Let A and D be respectively discrete distributions representing the time for an attack to be successful and the time to a countermeasure to be operational. The output of this gate turns out *True* if the input of the attack A is *True* and the input of the countermeasure is not yet operational thus *False*. Let O be the output distribution of a $SI\text{-}AND(A, D)$ gate having attack A and countermeasure D as inputs. The output distribution of this gate with mutually independent inputs is computed as:

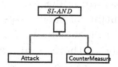

Fig. 1. $SI\text{-}AND$ gate with a single inverter on countermeasure input

$$P(O = i) = P(A = i).P(D > i), \quad i \neq MT \tag{2}$$

$$P(O = MT) = 1 - \sum_{j|j \neq MT} P(O = j) \tag{3}$$

Eq. (2) means that attack at time i is successful, if the countermeasure is not operational at time i. The probability of this event is the product of the probability to have an attack at time i and the probability that the countermeasure is not operational at time i which means that the countermeasure becomes operational at a time later than i. Equation (3) is derived since the probabilities are summed up to 1. This is the probability that the output does not turn out $True$ (the probability that the attack will not happen) during the observation time interval $[0 - MT[$.

The temporal properties of this gate with discrete random inputs A, D specified by their state space, V and the probability vectors, P are given in Table 1. The state space of the output, V_O will be the same as the state space of the attack, V_A. The probability that A is successful at time 2 is always 0.25, since the earliest time the countermeasure is operational is time 4. Thus the attack at time 2 can not be countered. The probability that A is successful at time 5 is the product of the probability that attack is successful at time 5 and the probability that the countermeasure is operational later than time 5, so it is $0.2.(1 - 0.15) = 0.17$.

Table 1. Input and output distributions for the gate given in Fig. 1

Attack Dist.				
V_A	2	5	8.5	10	15
P_A	0.25	0.2	0.3	0.15	0.1

Bound on Attack Dist.		
V_{A^u}	4	10	15
P_{A^u}	0.25	0.65	0.1

Defense Dist.		
V_D	4	7	15
P_D	0.15	0.4	0.45

Output Dist.				
V_O	2	5	8.5	10	15
P_O	0.25	0.17	0.135	0.0675	0.3775

Bound on Output Dist.		
V_{O^u}	4	10	15
P_{O^u}	0.2125	0.2925	0.495

2.3 Monotonicity Properties

Taking benefice of the monotonicity of a $SI\text{-}AND$ gate with countermeasure D is more subtle since the two inputs of this gate have contrary roles. We assume that the leaves which are countermeasure nodes are not compressed. Notice that we do not need to compress the output of such gates since the output distribution has the same state space as its attack input whatever the state space of the countermeasure input is. Therefore there is no explosion of the state space of the output due to the countermeasure input. However the attack input may be a bounding distribution of a subsystem. We must show that if the attack input is a bounding distribution, the output distribution of the $SI\text{-}AND$ gate will remain a bounding distribution.

We first illustrate this property with an example. First let us write $\leq_{st}$ inequalities (Eq. (1)) for $A \leq_{st} A^u$ where n_A denotes the size of the vectors and the indices of vectors are $1 \leq i \leq n_A$:

$$i \in \{n_A - 1, n_A - 2, \cdots, 2, 1\}, \quad \sum_{\{k|k>i\}} P_A[k] \leq \sum_{\{j|V_{A^u}[j]>V_A[i]\}} P_{A^u}[j].$$

Since the sum of probabilities is equal to 1, these inequalities can be written as follows:

$$i \in \{1, 2, \cdots, n_A - 2, n_A - 1\}, \quad \sum_{\{k|k\leq i\}} P_A[k] \geq \sum_{\{j|V_{A^u}[j]\leq V_A[i]\}} P_{A^u}[j].$$

The inequalities for A and the reduced size upper bound, A^u are

$$0.1 \leq 0.1; \quad 0.25 \leq 0.75; \quad 0.55 \leq 0.75; \quad 0.75 \leq 0.75$$

The output distributions can be computed by Eqs. (2), (3) with $MT = 15$. The inequalities for $O \leq_{st} O^u$ are

$$0.3775 \leq 0.495; \quad 0.0675 + 0.3775 \leq 0.2925 + 0.495;$$

$0.135 + 0.0675 + 0.3775 \leq 0.2925 + 0.495; \quad 0.17 + 0.135 + 0.0675 + 0.3775 \leq 0.2925 + 0.495$.

Therefore we conclude that $O \leq_{st} O^u$.

Proposition 1. *We consider the output distributions for a SI-AND gate with countermeasure D. We note the output of the gate with attack input A and A^u as follows:*

$$O = SI\text{-}AND(A, D), \quad O^u = SI\text{-}AND(A^u, D)$$

We have the following property:

$$\text{If } A \leq_{st} A^u \text{ then } O \leq_{st} O^u.$$

Proof. For the sake of simplicity, we assume that A and A^u take values in the same set which is the union set of their respective state spaces: $\mathcal{V}_A \cup \mathcal{V}_{A^u}$. This is indeed the case if we put null probabilities when the random variable does not take the corresponding value. Let N be the size of the union set. The vectors corresponding to A and A^u are indexed by $1 \leq i \leq N$.

$$\mathcal{V}_A[i] = \mathcal{V}_{A^u}[i], \quad 1 \leq i \leq N \text{ and } \mathcal{V}_A[N] = \mathcal{V}_{A^u}[N] = MT.$$

With these notations, the inequalities for $A \leq_{st} A^u$ as :

$$
\begin{aligned}
i = 1, \quad & \mathcal{P}_A[1] \geq \mathcal{P}_{A^u}[1] \\
i = 2, \quad & \mathcal{P}_A[1] + \mathcal{P}_A[2] \geq \mathcal{P}_{A^u}[1] + \mathcal{P}_{A^u}[2] \\
& \vdots \\
i = N - 1, \quad & \sum_{j=1}^{N-1} \mathcal{P}_A[j] \geq \sum_{j=1}^{N-1} \mathcal{P}_{A^u}[j] \\
i = N, \quad & \sum_{j=1}^{N} \mathcal{P}_A[j] = \sum_{j=1}^{N} \mathcal{P}_{A^u}[j] = 1
\end{aligned}
$$

The $O \leq_{st} O^u$ inequalities that must be satisfied for the $\leq_{st}$ comparison on the output distributions of the $SI\text{-}AND$ gate (Eq. (2) and (3)).

The first inequality, $i = 1$:

$$\mathcal{P}_A[1] \left(\sum_{\substack{k | \\ \mathcal{V}_D[k] > \mathcal{V}_A[1]}} \mathcal{P}_D[k] \right) \geq \mathcal{P}_{A^u}[1] \left(\sum_{\substack{k | \\ \mathcal{V}_D[k] > \mathcal{V}_A[1]}} \mathcal{P}_D[k] \right)$$

This is indeed the first inequality for $A \leq_{st} A^u$ but multiplied in both parts by the same positive value $(\sum_{k|V_D[k]>V_A[1]})$. Thus the first inequality is satisfied. The second inequality, $i = 2$:

$$\mathcal{P}_A[1]\left(\sum_{\substack{k| \\ V_D[k]>V_A[1]}} \mathcal{P}_D[k]\right) + \mathcal{P}_A[2]\left(\sum_{\substack{k| \\ V_D[k]>V_A[2]}} \mathcal{P}_D[k]\right)$$

$$\geq \mathcal{P}_{A^u}[1]\left(\sum_{\substack{k| \\ V_D[k]>V_A[1]}} \mathcal{P}_D[k]\right) + \mathcal{P}_{A^u}[2]\left(\sum_{\substack{k| \\ V_D[k]>V_A[2]}} \mathcal{P}_D[k]\right)$$

This can be written as follows:

$$(\mathcal{P}_A[1] + \mathcal{P}_A[2]])\left(\sum_{\substack{k| \\ V_D[k]>V_A[2]}} \mathcal{P}_D[k]\right) + \mathcal{P}_A[1]\left(\sum_{\substack{k| \\ V_D[k]\leq V_A[2] \\ V_D[k]>V_A[1]}} \mathcal{P}_D[k]\right)$$

$$\geq (\mathcal{P}_{A^u}[1] + \mathcal{P}_{A^u}[2]])\left(\sum_{\substack{k| \\ V_D[k]>V_A[2]}} \mathcal{P}_D[k]\right) + \mathcal{P}_{A^u}[1]\left(\sum_{\substack{k| \\ V_D[k]\leq V_A[2] \\ V_D[k]>V_A[1]}} \mathcal{P}_D[k]\right)$$

Similarly to the above case, this inequality can be derived from the two first inequalities for $A \leq_{st} A^u$ by multiplying both parts by the same values.

The inequality for $i = m$:

$$\sum_{j=1}^{m}\mathcal{P}_A[j]\left(\sum_{\substack{k| \\ V_D[k]>V_A[j]}} \mathcal{P}_D[k]\right) \geq \sum_{j=1}^{m}\mathcal{P}_{A^u}[j]\left(\sum_{\substack{k| \\ V_D[k]>V_A[j]}} \mathcal{P}_D[k]\right)$$

This inequality can be organized as follows:

$$\sum_{j=1}^{m}\mathcal{P}_A[j]\left(\sum_{\substack{k| \\ V_D[k]>V_A[m]}} \mathcal{P}_D[k]\right) + \sum_{j=1}^{m-1}\mathcal{P}_A[j]\left(\sum_{\substack{k| \\ V_D[k]\leq V_A[m] \\ V_D[k]>V_A[m-1]}} \mathcal{P}_D[k]\right) + \cdots$$

$$\cdots + \sum_{j=1}^{2} \mathcal{P}_A[j] \left(\sum_{\substack{k| \\ \mathcal{V}_D[k]\leq\mathcal{V}_A[3] \\ \mathcal{V}_D[k]>\mathcal{V}_A[2]}} \mathcal{P}_D[k] \right) + \mathcal{P}_A[1] \left(\sum_{\substack{k| \\ \mathcal{V}_D[k]\leq\mathcal{V}_A[2] \\ \mathcal{V}_D[k]>\mathcal{V}_A[1]}} \mathcal{P}_D[k] \right)$$

$$\geq \sum_{j=1}^{m} \mathcal{P}_{A^u}[j] \left(\sum_{\substack{k| \\ \mathcal{V}_D[k]>\mathcal{V}_A[m]}} \mathcal{P}_D[k] \right) + \sum_{j=1}^{m-1} \mathcal{P}_{A^u}[j] \left(\sum_{\substack{k| \\ \mathcal{V}_D[k]\leq\mathcal{V}_A[m] \\ \mathcal{V}_D[k]>\mathcal{V}_A[m-1]}} \mathcal{P}_D[k] \right) +$$

$$\cdots + \sum_{j=1}^{2} \mathcal{P}_{A^u}[j] \left(\sum_{\substack{k| \\ \mathcal{V}_D[k]\leq\mathcal{V}_A[3] \\ \mathcal{V}_D[k]>\mathcal{V}_A[2]}} \mathcal{P}_D[k] \right) + \mathcal{P}_{A^u}[1] \left(\sum_{\substack{k| \\ \mathcal{V}_D[k]\leq\mathcal{V}_A[2] \\ \mathcal{V}_D[k]>\mathcal{V}_A[1]}} \mathcal{P}_D[k] \right)$$

It can be seen that the inequality for $i = m$ can be derived from the m first inequalities for $A \leq_{st} A^u$ by multiplying both part by the same positive values. This completes the proof. □

This proposition shows that the monotonicity properties are satisfied to construct bounding output distributions, if the distributions at leaves associated with countermeasures are not compressed. Therefore it is possible to compress distributions to construct reduced-size output distributions of Attack-Defence models to overcome state space explosion problem. Remind that the countermeasures do not increase the size of the output distribution, thus there is no need apriori to compress discrete countermeasure distributions.

The presented gate specifications and the above algebraic proof are rather implementation oriented. Notice that these can be given with means of operations on input random variables, I_1, I_2 and the observation time MT:

$$AND(I_1, I_2) = \min(MT, \max(I_1, I_2))$$
$$OR((I_1, I_2) = \min(MT, \min(I_1, I_2))$$
$$SEQ(I_1, I_2) = \min(MT, I_1 + I_2)$$

As stated before, the monotonicity of these gates results from the generic definition of $\leq_{st}$ order (Eq. (1)) since the corresponding functions are non decreasing with respect to input parameters. The $SI\text{-}AND(A, D)$ gate with the observation time fixed to a constant value MT can be defined as:

$$SI\text{-}AND(A, D) = A \, \mathbb{1}_{A<\min(D,MT)} + MT \, \mathbb{1}_{A>=\min(D,MT)}$$

The left-side of the equation represents the case when the attack can take place while the right-side corresponds to the case when either the countermeasure (defense) prevents the attacks or the observation time is over. The above monotonicity proof of this gate can be also directly established since this gate is defined as a non decreasing function with respect to A. The opposite impacts of attack and defense inputs can be seen in the above equation, since the output is a non increasing function with respect to D while it is non decreasing with respect to A.

In the case where distributions are continuous, the proposed methodology can be started with constructing bounding discrete distributions. In such cases, since attacks and defenses have contrary roles, if one aims to construct upper (resp. lower) bounds on the overall Attack Defense trees, upper bounds (lower) must be considered for the attacks while lower (upper) bounds must be considered for the defenses.

3   Case Studies

In this section we consider a case study from [6] in which temporal behaviours of attack defense trees are studied through the underlying Markov chain. The attack tree is given in Fig. 2. This example has three sequential steps. An attacker wants to compromise a server by *running malicious scripts* on it. To do so, they must first identify the target network address by *scanning the network*. Then they have to gain root access to run the malicious scripts. Thus the root of the attack tree is a *SEQ* gate. To prevent network address discovery, a *mutable network* with a dynamic topology reconfiguration is used as a countermeasure. This is represented with a *SI-AND* gate (*target identified*). There is an alternative to perform the second step, the *privilege escalation* : either to *use a security*

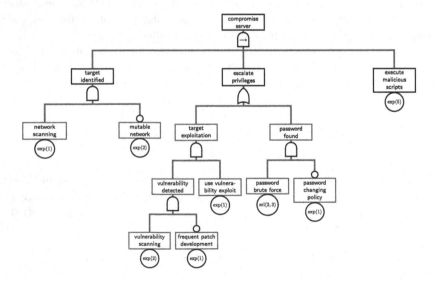

Fig. 2. Attack Tree *Compromise Server.*

vulnerability of the target or to *guess its root password*. It is represented with an *OR* gate. To do the *target exploitation*, the attacker must *scan the target vulnerabilities*. *Frequent security patch development* and frequent updates are used as countermeasure. This is materialized by a *SI-AND* gate. Once a vulnerability has been discovered, the ability to use it, called *use vulnerability exploit* on the AT, is needed. This is represented by *AND* gate. Another countermeasure, a *password changing policy* is used to prevent the *password brute force attack* represented with a third *SI-AND* gate. In [6], completion times of basic attacks and countermeasures have been taken as truncated exponential and Erlang distributions. The *network scanning* basic attack is an exponential of rate 1. The *mutable network* defence is an exponential of rate 2. The *vulnerability scanning* attack is an exponential of rate 2. The *frequent patch development* defence is an exponential of rate 1. The *use vulnerability exploit* is an exponential of rate 1. The *password brute force* attack is an Erlang distribution with shape 2 and rate 3. The *password changing policy* defence is an exponential of rate 1. The *execute malicious scripts* basic attack is an exponential of rate 5. We consider truncated exponential distributions taking values in time interval $[0 - 10/rate]$, and then discretize them. The discrete input distributions are constructed in this manner.

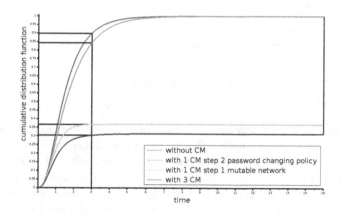

Fig. 3. Countermeasures effects

First we describe the behavior of the system according to the number of countermeasures applied. This corresponds to sub trees of the tree drawn on Fig. 2 with less *SI-AND* gates. On Fig. 3, one can see how the countermeasures modify the system behaviour during time interval $[0 - 16[$. Thus time 16 represents the infinity, and the events happening at time 16 indeed do not occur. The red curve shows the system behaviour when no countermeasure is applied. It corresponds to the attack tree with no *SI-AND* gate. At time $t = 3$ time unit, the probability that the attack succeeds is 0.89. If we consider the system, with only the *password changing policy* during the second step of the attack. The tree has now a single *SI-AND* gate. The computation results are shown on the

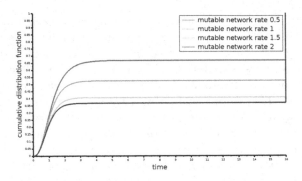

Fig. 4. Countermeasure distribution influence

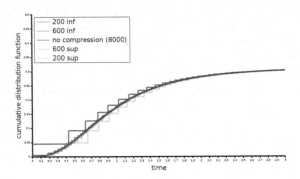

Fig. 5. Compression

orange curve. The probability at $t = 3$ time unit diminishes slightly to 0.84. On the green curve for the tree with only one countermeasure *mutable network*, the probability is only 0.37. This countermeasure is very efficient. The blue curve represents the system behaviour of the whole tree with three countermeasures. The probability at $t = 3$ time unit that the attack succeed is then only 0.30. Notice that for the green and the blue curves the cumulative distributions sum up to 1 at time 16 (∞). Therefore the probability that an attack occurs in the green curve is 0.37 while it is 0.30 in the blue curve. There is no jump at time 16 for the orange curve since the considered countermeasure is an input of an *OR* gate. Thus, the effect of the countermeasure is neutralized when the target exploitation happens in the other input.

Secondly we study the influence of a countermeasure distribution. We consider the tree with only one countermeasure *mutable network*. The other countermeasures are deleted (the operational time is defined as ∞). The Fig. 4 shows that the probability p that the attack succeeds increases when the rate λ of the countermeasure exponential distribution decreases. For $\lambda = 0.5$, $p = 0.67$, for $\lambda = 1$, $p = 0.52$, for $\lambda = 1.5$, $p = 0.41$ and for $\lambda = 2$, $p = 0.37$. Notice that the discrete input distributions are constructed as explained above.

Finally, we illustrate how compression influences the calculations on the full tree of Fig. 2. We apply the naive approach with linear complexity to construct bounding distributions. In the SEQ gate computations, the size of the results are bounded with different values. In each case a lower bound and an upper bound of the cumulative function are given. All the input distributions are discretized with 100 bins. If we do not bound the size of the results, the final result has about 8000 values. The results are shown in Fig. 5. The blue curve represents the cumulative function without any compression. We see that if we bound the result size to 600, the lower bound given by the red curve and the upper bound given by the orange curve still give usable values even for the small time values. The computations are less accurate when we bound the result size to 200. That is illustrated by the green curves.

4 Conclusion

In this paper, we extend the framework to analyze Attack Trees given in [11] to Attack Defense Trees with countermeasures. The completion times of attacks and countermeasures are defined by finite discrete random variables. The output distributions of the gates and finally the root of the tree are computed by a bottom-up approach. However the sizes of the output distributions can become quickly very large. We propose to derive bounding distributions with reduced sizes by means of the stochastic comparison method. In practise, we obtain fairly accurate bounds by compressing the output distribution even with linear computational complexity algorithms. Therefore the quantitative evaluation of Attack Trees with the presence of countermeasure can be efficiently provided by the proposed methodology.

We consider to extend this approach to the other extensions of Attack Trees and to the case when the basic events are not mutually independent.

References

1. Aït-Salaht, F., Castel-Taleb, H., Fourneau, J.-M., Pekergin, N.: Stochastic bounds and histograms for network performance analysis. In: Balsamo, M.S., Knottenbelt, W.J., Marin, A. (eds.) EPEW 2013. LNCS, vol. 8168, pp. 13–27. Springer, Heidelberg (2013). https://doi.org/10.1007/978-3-642-40725-3_3
2. Arnold, F., Hermanns, H., Pulungan, R., Stoelinga, M.: Time-dependent analysis of attacks. In: Abadi, M., Kremer, S. (eds.) POST 2014. LNCS, vol. 8414, pp. 285–305. Springer, Heidelberg (2014). https://doi.org/10.1007/978-3-642-54792-8_16
3. Aslanyan, Z., Nielson, F., Parker, D.: Quantitative verification and synthesis of attack-defence scenarios. In: IEEE CSF 2016, pp. 105–119 (2016)
4. Fourneau, J.M., Pekergin, N.: A numerical analysis of dynamic fault trees based on stochastic bounds. In: Campos, J., Haverkort, B.R. (eds.) QEST 2015. LNCS, vol. 9259, pp. 176–191. Springer, Cham (2015). https://doi.org/10.1007/978-3-319-22264-6_12

5. Gadyatskaya, O., Hansen, R.R., Larsen, K.G., Legay, A., Olesen, M.C., Poulsen, D.B.: Modelling attack-defense trees using timed automata. In: Fränzle, M., Markey, N. (eds.) FORMATS 2016. LNCS, vol. 9884, pp. 35–50. Springer, Cham (2016). https://doi.org/10.1007/978-3-319-44878-7_3

6. Jhawar, R., Lounis, K., Mauw, S.: A stochastic framework for quantitative analysis of attack-defense trees. In: Barthe, G., Markatos, E., Samarati, P. (eds.) STM 2016. LNCS, vol. 9871, pp. 138–153. Springer, Cham (2016). https://doi.org/10.1007/978-3-319-46598-2_10

7. Kordy, B., Mauw, S., Radomirovic, S., Schweitzer, P.: Attack-defense trees. J. Log. Comput. **24**(1), 55–87 (2014)

8. Kordy, B., Piètre-Cambacédès, L., Schweitzer, P.: Dag-based attack and defense modeling: Don't miss the forest for the attack trees. Comput. Sci. Rev. **13–14**, 1–38 (2014)

9. Kumar, R., Ruijters, E., Stoelinga, M.: Quantitative attack tree analysis via priced timed automata. In: Sankaranarayanan, S., Vicario, E. (eds.) FORMATS 2015. LNCS, vol. 9268, pp. 156–171. Springer, Cham (2015). https://doi.org/10.1007/978-3-319-22975-1_11

10. Kumar, R., Stoelinga, M.: Quantitative security and safety analysis with attack-fault trees. In: HASE 2017, pp. 25–32 (2017)

11. Pekergin, N., Tan, S., Fourneau, J.-M.: Quantitative attack tree analysis: stochastic bounds and numerical analysis. In: Kordy, B., Ekstedt, M., Kim, D.S. (eds.) GraMSec 2016. LNCS, vol. 9987, pp. 119–133. Springer, Cham (2016). https://doi.org/10.1007/978-3-319-46263-9_8

12. Petrucci, L., Knapik, M., Penczek, W., Sidoruk, T.: Squeezing state spaces of (attack-defence) trees. In: Pang, J., Sun, J. (eds.) ICECCS 2019, pp. 71–80 (2019)

13. Roy, A., Seong, D., Kim, K.T.: Act:towards unifying the constructs of attack and defense trees. Securtiy Commun. Netw. **3**, 1–15 (2011)

14. Ruijters, E., Stoelinga, M.: Fault tree analysis: a survey of the state-of-the-art in modeling, analysis and tools. Comput. Sci. Rev. **15**, 29–62 (2015)

15. Schneier, B.: Attack trees: modeling security threats. Dr. Dobb's J. Softw. Tools **24**(12), 21–29 (1999)

16. Tancrez, J.S., Semal, P., Chevalier, P.: Histogram based bounds and approximations for production lines. Eur. J. of Oper. Res. **197**(3), 1133–1141 (2009)

17. Widel, W., Audinot, M., Fila, B., Pinchinat, S.: Beyond 2014: formal methods for attack tree-based security modeling. ACM Comput. Surv. **52**(4), 75:1–75:36 (2019)

Modeling and Cycle Analysis of Traffic Signal Systems Based on Timed Networks

Yingxuan Yin[1], Cailu Wang[2], Haiyong Chen[1], and Yuegang Tao[1(✉)]

[1] School of Artificial Intelligence, Hebei University of Technology, Tianjin 300130, People's Republic of China
yuegangtao@hebut.edu.cn
[2] School of Automation, Beijing Institute of Technology, Beijing 100081, People's Republic of China

Abstract. The traffic signal control systems offer an effective settlement for the city traffic management. In this paper, the operation process of traffic signal systems for one-way street and crossroad is described by using timed Petri nets. Furthermore, the linear dynamic equations of the systems are established by the max-plus algebra. Based on these, the cycle time of traffic signal systems is calculated. It is found out that the cycle time of traffic signal systems for one-way street is independent of the travel time between two traffic lights.

Keywords: Traffic signal control systems · Timed Petri nets · Max-plus algebra · Cycle time

1 Introduction

The discrete-event systems with synchronization but no concurrency or choice occur can be described by models that are 'linear' in the max-plus algebra with maximisation and addition as basic operations, although they are nonlinear in the conventional algebra. Such kind of systems are the max-plus linear systems [2,3,7,8,15], which have a variety of applications in modeling and control of manufacturing systems, traffic networks, screening systems, robot systems, etc. (see e.g. [1,5,9,11,14,18–20,24,25]).

Petri net is a graphical description tool for max-plus linear systems modeling (see e.g., [2]). By establishing the Petri net model of a max-plus linear system, some important information about the structure and dynamic behaviors of the system can be got and used to evaluate and optimize the performance of max-plus linear systems.

The traffic light is a kind of signal device located at road intersection or pedestrian crossing, which is used to indicate when it is safe to drive, ride or

Supported by National Natural Science Foundation of China, Grant/Award Numbers: 61741307, 61903037, 61976242, 62073117; China Postdoctoral Science Foundation: 2020M670164.

Q. Zhao and L. Xia (Eds.): VALUETOOLS 2021, LNICST 404, pp. 221–232, 2021.
https://doi.org/10.1007/978-3-030-92511-6_14

walk with common color code. With the increase of the number of vehicles, traffic congestion, traffic accidents and transportation delays are increasing in urban arteries around the world. Therefore, the research of urban traffic control systems becomes more and more valuable.

In urban roads, traffic light control is a cheaper and more effective method, and now it is widely used in almost every city. In fact, the traffic light control systems of urban vehicles can be regarded as a complex discrete-event system with synchronization and no concurrency or choice occur, and has been modeled by Petri nets. For example, Maia et al. [22] used Petri nets to model traffic lights and designs a feedback controller. Huang et al. [16] proposed a new method to design and analyze the urban traffic signal system using synchronous timed Petri nets. Qi et al. [23] developed the normal and emergency traffic light control strategies to prevent large-scale traffic congestion caused by accidents, and introduced their control logic in detail through timed Petri nets. Luo et al. [21] used timed Petri nets to model the H-strategy traffic signal system, verified the real-time and reversibility of the proposed timed Petri nets model, as well as the traffic congestion problem and avoidance method in the model.

As mentioned above, the traffic light control system is a complex discrete-event system with strong nonlinearity. It is an effective way to study nonlinear systems by representing them as linear forms. Max-plus algebra can transform the traffic light control system into a 'linear' system in the sense of algebra. For the traffic light control system of one-way street and crossroad, this paper will use timed Petri nets to describe the operation process of the system, and use the max-plus algebra method to establish its linear dynamic equation. On this basis, the cycle time of the traffic light control system is analyzed.

The remainder of this paper is organized as follows. Section 2 presents some preliminaries about the max-plus algebra. Section 3 and Sect. 4 model and analysis the operation process of the traffic light control system for one-way street and crossroad, respectively. Conclusions and future works are discussed in Sect. 5.

2   Preliminaries

Let us begin our discussion with basic concepts and results from Petri nets and the max-plus algebra, which can be consulted for more details in [2–4, 7, 15, 17] unless noted otherwise.

2.1   Petri Nets

A continuous Petri net is defined by

$$\mathcal{N} = (\mathcal{P}, \mathcal{Q}, M, \rho, m, \tau),$$

where

1. $\mathcal{P}$ is a finite set whose elements are called places;
2. $\mathcal{Q}$ is a finite set whose elements are called transitions;

3. $M \in (\mathbb{R}^+)^{\mathcal{P} \times \mathcal{Q} \cup \mathcal{P} \times \mathcal{Q}}$ are the arc multipliers that is M_{pq} (resp. M_{qp}) denotes the number of arcs from transition q to place p (resp. from place p to transition q), where $\mathbb{R}^+$ is the set of positive real numbers;
4. $\rho : \mathcal{Q} \times \mathcal{P} \to \mathbb{R}^+$ verifying:

$$\sum_{q \in p^{\mathrm{out}}} \rho_{qp} = 1, \ \forall p \in \mathcal{P}$$

is the routing policy which gives the imposed proportion of fluid going from place p to transition q with respect to quantity of fluid entering place p;
5. $m \in (\mathbb{R}^+)^{\mathcal{P}}$ is the initial marking, that is, m_p is the amount of fluid available at place p at starting time;
6. $\tau \in (\mathbb{R}^+)^{\mathcal{P}}$ is the holding time which is the time that a molecule of fluid has to stay in place p before leaving.

Figure 1 is an example of Petri net.

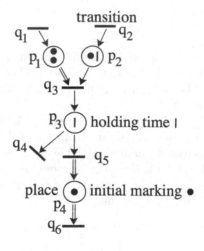

Fig. 1. A Petri net

2.2   Max-Plus Algebra

Let $\mathbb{R}$ be the set of real numbers, $\mathbb{N}$ be the set of natural numbers and $\mathbb{N}^+$ be the set of positive integers. For $n \in \mathbb{N}^+$, denote by $\mathbb{N}_n$ the set $\{1, 2, \cdots, n\}$.

For $a, b \in \mathbb{R} \cup \{-\infty\}$, let

$$a \oplus b = \max\{a, b\} \text{ and } a \otimes b = a + b,$$

where $\max\{a, -\infty\} = a$ and $a + (-\infty) = -\infty$. $(\mathbb{R} \cup \{-\infty\}, \oplus, \otimes)$ is called the max-plus algebra and simply denoted by $\mathbb{R}_{\mathrm{max}}$. In $\mathbb{R}_{\mathrm{max}}$, $-\infty$ is the zero element denoted by ε, and 0 is the identity element denoted by e.

Let $\mathbb{R}_{\max}^{m \times n}$ be the set of $m \times n$ matrices with entries in $\mathbb{R}_{\max}$. The basic operations of max-plus algebra can be extended to max-plus matrices as follows:

For $A = (a_{ij})$, $B = (b_{ij}) \in \mathbb{R}_{\max}^{m \times n}$, $(A \oplus B)_{ij} = a_{ij} \oplus b_{ij}$;

For $A = (a_{ij}) \in \mathbb{R}_{\max}^{m \times r}$ and $B = (b_{ij}) \in \mathbb{R}_{\max}^{r \times n}$, $(A \otimes B)_{ij} = \bigoplus_{u=1}^{r} a_{iu} \otimes b_{uj}$;

For $A = (a_{ij}) \in \mathbb{R}_{\max}^{m \times n}$ and $d \in \mathbb{R}_{\max}$, $(d \circ A)_{ij} = d \otimes a_{ij}$.

For $A \in \mathbb{R}_{\max}^{n \times n}$ and $l \in \mathbb{N}^{+}$, the lth power of A is defined by

$$A^{\otimes l} = \underbrace{A \otimes A \otimes \cdots \otimes A}_{l}.$$

For $A = (a_{ij})$, $B = (b_{ij}) \in \mathbb{R}_{\max}^{m \times n}$, $A \leqslant B$ if $a_{ij} \leqslant b_{ij}$ for any $i \in \mathbb{N}_m$ and $j \in \mathbb{N}_n$. For $A, B \in \mathbb{R}_{\max}^{m \times n}$ and $\alpha, \beta \in \mathbb{R}_{\max}^{n}$, if $A \leqslant B$ and $\alpha \leqslant \beta$, then $A \otimes \alpha \leqslant B \otimes \beta$.

The zero matrix and identity matrix in $\mathbb{R}_{\max}^{n \times n}$ are

$$\mathcal{E} = \begin{pmatrix} \varepsilon & \varepsilon & \cdots & \varepsilon \\ \varepsilon & \varepsilon & \cdots & \varepsilon \\ \vdots & \vdots & & \vdots \\ \varepsilon & \varepsilon & \cdots & \varepsilon \end{pmatrix} \text{ and } I = \begin{pmatrix} e & \varepsilon & \cdots & \varepsilon \\ \varepsilon & e & \cdots & \varepsilon \\ \vdots & \vdots & & \vdots \\ \varepsilon & \varepsilon & \cdots & e \end{pmatrix},$$

respectively.

A permutation matrix is a square binary matrix which has exactly one entry of e in each row and each column and εs elsewhere. If P is a permutation matrix, then $P^{\mathsf{T}} P = I$, where P^{T} is the transpose of P. Multiplying matrix A by P from left will permute the rows of A.

For $A = (a_{ij}) \in \mathbb{R}_{\max}^{n \times n}$, the *precedence graph* of A, denoted by $\mathcal{G}(A)$, is a weighted digraph with n nodes and an arc from j to i if $a_{ij} \neq \varepsilon$, in which case the weight of this arc receives the numerical value of a_{ij} (see, e.g., [12]). A square matrix A is said to be irreducible if $\mathcal{G}(A)$ is strongly connected, otherwise, A is said to be reducible. Harary [13] proved that for any max-plus matrix A, there exists a permutation matrix P, such that

$$P^{\mathsf{T}} A P = \begin{pmatrix} A^{(1)} & \mathcal{E} & \cdots & \mathcal{E} \\ * & A^{(2)} & \cdots & \mathcal{E} \\ \vdots & \vdots & & \vdots \\ * & * & \cdots & A^{(l)} \end{pmatrix},$$

where $A^{(i)}$ ($i \in \mathbb{N}_l$) is irreducible. Obviously, A is irreducible if and only if $l = 1$. In other words, any reducible matrix can be changed into a lower triangular block matrix by renumbering rows and columns.

3 Traffic Light Control System of One-Way Street

Consider a one-way street as shown in Fig. 2. There exists a traffic light and a sidewalk in both c_0 and c_1. The traffic light control system of this one-way street is recorded as $\mathcal{T}_1$. In real life, there is a phenomenon that drivers wait for

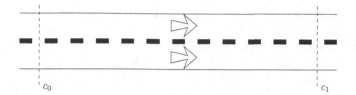

Fig. 2. A one-way street

a red light at c_0, and then wait for a red light at T_1 not far away. This kind of design makes drivers and passengers especially prone to anxiety. Analysing the operation process of traffic light control systems can reduce their waiting times.

When the vehicles travel on the road, they must observe the following rules: During the time when the green light at c_0 is on continuously, the vehicle platoon passing through c_0 can pass through c_1 without being stopped by the red light at c_1. The red light at c_1 is to ensure the safety of pedestrians. Then, the operation process of system T_1 is as follows. 1) The green light is on at c_0, passing through a certain length of vehicle platoon. When this length of vehicle platoon is over, the red light at c_0 is on. During the red light is on, pedestrians cross the road. After pedestrians cross the road, the green light is on. 2) When the vehicle platoon reaches c_1, the green light at c_1 is on. When the vehicle platoon of this length is over, the red light is on, then the pedestrians cross the road. Then repeat the above process.

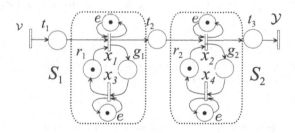

Fig. 3. Timed Petri nets of system T_1

Suppose that the length of the vehicle platoon through the signal is proportional to the duration of the green light. The traffic light control system T_1 can be represented by a timed Petri nets shown in Fig. 3, in which the vehicle platoon is represented by tokens crossing the graph from transition v to transition y. Each of these tokens is assumed to represent a virtual platoon of vehicles (see for instance [10, 22] for a presentation of this concept). Virtual means that a token represents the presence of vehicles even when the vehicles are not there, that is, in this model the traffic light system runs independently of the virtual platoon since we do not assume sensors of virtual presence on the road. In other words, input v is not a constraint and assumed to be fired an infinite number of time, which is equivalent to $v = \varepsilon$.

In Fig. 3, S_1 and S_2 represent the traffic lights at c_0 and c_1, respectively. x_1 and x_3 represent the green light and red light at c_0, respectively. x_2 and x_4 represent the green light and red light at c_1, respectively. g_1 is the duration of the green light at c_0, which is also the time required for a certain length of vehicle platoon to fully pass c_0. r_1 is the duration of the red light at c_0, and also the time required for a certain platoon of people to completely pass the sidewalk at c_0. g_2 is the duration of the green light at c_1, which is also the time required for a certain length of vehicle platoon to fully pass c_1. r_2 is the duration of the red light at c_1, and also the time required for a certain platoon of people to completely pass the sidewalk at c_1. The duration t_i is the shortest time allowed for the vehicle platoon from c_0 to c_1, for example, t_2 is the shortest time for a vehicle platoon to leave the first traffic light and arrive at the second.

The occurrence of the event follows the following constraints:

1) The red light can only be on after the green light is off, and the green light can only be on after red light is off.
2) The traffic lights change in a flash.
3) The red light is on long enough for pedestrians to pass through the sidewalk, and when the green light is on, there is no pedestrian on the sidewalk.

Let $x_1(k)$ be the kth time when the green light of S_1 turns on, $x_2(k)$ be the kth time when the green light of S_2 turns on, $x_3(k)$ be the kth time when the red light of S_1 turns on, and $x_4(k)$ be the kth time when the red light of S_2 turns on. Then the operation process of system T_1 can be modeled as system of equations

$$\begin{cases} x_1(k) = \max\{x_1(k-1), x_3(k-1) + r_1, v + t_1\}, \\ x_2(k) = \max\{x_1(k) + t_2, x_4(k-1) + r_2, x_2(k-1)\}, \\ x_3(k) = \max\{x_1(k) + g_1, x_3(k-1)\}, \\ x_4(k) = \max\{x_2(k) + g_2, x_4(k-1)\}. \end{cases}$$

The above system is nonlinear in conventional algebra. Since $v = \varepsilon$, by replacing operations max and $+$ by $\oplus$ and $\otimes$ in the max-plus algebra, respectively, it can be formulated as the max-plus equations

$$\begin{cases} x_1(k) = x_1(k-1) \oplus x_3(k-1) \otimes r_1, \\ x_2(k) = x_1(k) \otimes t_2 \oplus x_4(k-1) \otimes r_2 \oplus x_2(k-1), \\ x_3(k) = x_1(k) \otimes g_1 \oplus x_3(k-1), \\ x_4(k) = x_2(k) \otimes g_2 \oplus x_4(k-1). \end{cases}$$

Let

$$x(k) = \begin{pmatrix} x_1(k) \\ x_2(k) \\ x_3(k) \\ x_4(k) \end{pmatrix}, \quad A_0 = \begin{pmatrix} \varepsilon & \varepsilon & \varepsilon & \varepsilon \\ t_2 & \varepsilon & \varepsilon & \varepsilon \\ g_1 & \varepsilon & \varepsilon & \varepsilon \\ \varepsilon & g_2 & \varepsilon & \varepsilon \end{pmatrix} \quad \text{and} \quad A_1 = \begin{pmatrix} e & \varepsilon & r_1 & \varepsilon \\ \varepsilon & e & \varepsilon & r_2 \\ \varepsilon & \varepsilon & e & \varepsilon \\ \varepsilon & \varepsilon & \varepsilon & e \end{pmatrix}.$$

Then, one can obtain the following max-plus linear system

$$x(k) = A_0 x(k) \oplus A_1 x(k-1), \quad k = 1, 2, \ldots \tag{1}$$

By iterating Eq. (1), one can obtain

$$x(k) = A_0^3 x(k) \oplus \left(A_0^2 \oplus A_0 \oplus I \right) A_1 x(k-1).$$

By a direct calculation, one has $A_0^3 = \mathcal{E}$. Hence,

$$x(k) = \left(A_0^2 \oplus A_0 \oplus I \right) A_1 x(k-1) := A x(k-1), \tag{2}$$

where

$$A = \begin{pmatrix} e & \varepsilon & r_1 & \varepsilon \\ t_2 & e & t_2 r_1 & r_2 \\ g_1 & \varepsilon & g_1 r_1 & \varepsilon \\ t_2 g_2 & g_2 & t_2 g_2 r_1 & g_2 r_2 \end{pmatrix}.$$

Obviously, A is reducible, which can become a lower triangular matrix denoted by A_l. Let

$$A_l = P^\mathsf{T} A P = \begin{pmatrix} e & r_1 & \varepsilon & \varepsilon \\ g_1 & g_1 r_1 & \varepsilon & \varepsilon \\ t_2 & t_2 r_1 & e & r_2 \\ t_2 g_2 & t_2 g_2 r_1 & g_2 & g_2 r_2 \end{pmatrix} = \begin{pmatrix} A^{(1)} & \mathcal{E} \\ * & A^{(2)} \end{pmatrix},$$

where

$$A^{(1)} = \begin{pmatrix} e & r_1 \\ g_1 & g_1 r_1 \end{pmatrix}, A^{(2)} = \begin{pmatrix} e & r_2 \\ g_2 & g_2 r_2 \end{pmatrix}.$$

Let $\lambda(A^{(1)})$, $\lambda(A^{(2)})$ and $\lambda(A)$ be the maximum eigenvalue of $A^{(1)}$, $A^{(2)}$ and A, respectively. The eigenvalue of a matrix is equal to the maximum cycle mean of the precedence graph of the matrix (see e.g., [15]). By a direct calculation, one has

$$\lambda(A^{(1)}) = \frac{g_1 r_1}{2} \oplus g_1 r_1 = g_1 r_1,$$

$$\lambda(A^{(2)}) = \frac{g_2 r_2}{2} \oplus g_2 r_2 = g_2 r_2,$$

$$\lambda(A) = \frac{g_1 r_1}{2} \oplus \frac{g_2 r_2}{2} \oplus g_1 r_1 \oplus g_2 r_2 = g_1 r_1 \oplus g_2 r_2.$$

In order to ensure the stability of the system, it is required to let $\lambda(A^{(1)}) = \lambda(A^{(2)})$ (see e.g., [6] Theorem 5), i.e., $g_1 r_1 = g_2 r_2$. Then, one has $\lambda(A) = g_2 r_2$. The cycle time of a stable max-plus linear system is the maximum eigenvalue of the matrix (see e.g., [15]). Hence, the cycle time of system $\mathcal{T}_1$ is $\lambda(A) = g_2 r_2$. It can be seen that the cycle time of the system is independent of t_2, in other words, the travel time of the vehicle between c_0 and c_1 has no affect on the cycle time of the system. In fact, the traffic light changes continuously during the time when the vehicle is driving between c_0 and c_1. Therefore, no matter how long the driving time is, the cycle time of the system has nothing to do with the driving time. This model ensures that the vehicle platoon passing through c_0 can pass through c_1 without being stopped by the red light at c_1.

4   Traffic Light Control System of crossroad

Consider a crossroad shown in Fig. 4. There exists a group of traffic lights in the North-South direction and the East-West direction. The two traffic lights in the same group operate in the same way, which are represented by NS and WE, respectively. The traffic light control system of this crossroad is represented by T_2. The driving direction of the vehicle is shown in Fig. 5, that is, when the green light is on, the vehicles can only go straight in the current direction, turn left or right. Considering that in many city crossroads, whether red or green, vehicles can turn right, this paper does not consider the case of vehicles turning right at the crossroad.

The operation process of system T_2 is as follows: The green light in NS is on, the vehicles of North-South direction can go straight or turn left. After the green light lasts for a period of time, the red light in NS is on. Then the green light in WE is on, the vehicles of East-West direction can go straight or turn left. After the green light lasts for a period of time, the red light in WE is on. Then repeat the above process.

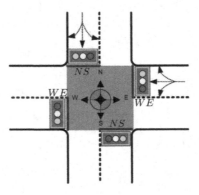

Fig. 4. An elementary crossroad (Color figure online)

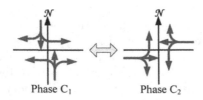

Fig. 5. The direction of the vehicles (Color figure online)

The traffic light control system T_2 can be represented by timed Petri nets as shown in the Fig. 6, in which NS and WE represent a group of traffic lights in North-South and East-West directions, respectively; x_1 and x_3 represent the green light and red light in NS, respectively; x_2 and x_4 represent the green light and red light in WE, respectively; g_1 and g_2 represent time of duration of the green light in NS and WE, respectively; r_1 and r_2 represent time of duration of the red light in NS and WE, respectively.

To ensure the safety of the driver, let us make the following constraints:

1) The red light can only be on after the green light is off, and the green light can only be on after red light is off.
2) The traffic lights change in a flash.
3) The red light is on long enough for pedestrians to pass through the sidewalk, and when the green light is on, there is no pedestrian on the sidewalk.
4) When there are no vehicles going straight or turning left in the North-South (resp. East-West) direction in the gray part as shown in the Fig. 4, the green light in the East-West (resp. North-South) direction can only be on.

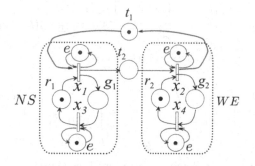

Fig. 6. Timed Petri nets of system T_2

With the above constraints, t_1 (resp. t_2) represents the shortest time from entering the gray part shown in Fig. 4 to leaving the gray part completely for the whole vehicle platoon in the East-West (resp. North-South) direction.

Similar to Sect. 3, this section uses vehicle platoon to study the system. Assume that the moving speed of the vehicle platoon in each direction is the same, and the length of the vehicle platoon passing through the signal light is proportional to the duration of the green light. Taking the South-North moving vehicle platoon as an example, the moving process of the vehicle platoon is as follows: 1) The green light in S is on, and the head of the vehicle platoon enters the gray part of Fig. 4. Since the duration of the green light of NS is directly proportional to the length of the vehicle platoon passing the traffic signal, the time required for the whole platoon to enter the gray part is g_1; 2) After the tail of the vehicle platoon just enters the gray part, the vehicle platoon will move in the gray part; 3) When the head of the platoon reaches the traffic signal light in N, the vehicle platoon starts to pass through the traffic signal light, and leaves the gray part. Therefore, the time from the head of the platoon to the tail of the platoon to leave the gray part is g_1; 4) After the whole platoon leaves the gray part, the green light is on in WE. The movement process of the vehicle platoon in other directions is similar to the above. From the above moving process of the vehicle platoon, it can be seen that $t_2 > g_1{}^2$ and $t_1 > g_2{}^2$.

Let $x_1(k)$ be the kth time when the green light in NS turns on, $x_2(k)$ be the kth time when the green light in WE turns on, $x_3(k)$ be the kth time when the red light of NS turns on, and $x_4(k)$ be the kth time when the red light of WE turns on, then the operation process of system T_2 can be modeled as the nonlinear system of equations

$$\begin{cases} x_1(k) = \max\{x_1(k-1), x_2(k-1) + t_1, x_3(k-1) + r_1,\}, \\ x_2(k) = \max\{x_1(k) + t_2, x_4(k-1) + r_2, x_2(k-1)\}, \\ x_3(k) = \max\{x_1(k) + g_1, x_3(k-1)\}, \\ x_4(k) = \max\{x_2(k) + g_2, x_4(k-1)\}. \end{cases}$$

By replacing operations max and $+$ by $\oplus$ and $\otimes$ in the max-plus algebra, respectively, it can be formulated as the max-plus equations

$$\begin{cases} x_1(k) = x_1(k-1) \oplus x_2(k-1) \otimes t_1 \oplus x_3(k-1) + r_1, \\ x_2(k) = x_1(k) \otimes t_2 \oplus x_4(k-1) \otimes r_2 \oplus x_2(k-1), \\ x_3(k) = x_1(k) \otimes g_1 \oplus x_3(k-1), \\ x_4(k) = x_2(k) \otimes g_2 \oplus x_4(k-1). \end{cases}$$

Let

$$x(k) = \begin{pmatrix} x_1(k) \\ x_2(k) \\ x_3(k) \\ x_4(k) \end{pmatrix}, \quad A_0 = \begin{pmatrix} \varepsilon & \varepsilon & \varepsilon & \varepsilon \\ t_2 & \varepsilon & \varepsilon & \varepsilon \\ g_1 & \varepsilon & \varepsilon & \varepsilon \\ \varepsilon & g_2 & \varepsilon & \varepsilon \end{pmatrix} \quad \text{and} \quad A_1 = \begin{pmatrix} e & t_1 & r_1 & \varepsilon \\ \varepsilon & e & \varepsilon & r_2 \\ \varepsilon & \varepsilon & e & \varepsilon \\ \varepsilon & \varepsilon & \varepsilon & e \end{pmatrix}.$$

Then, one can obtain the following max-plus linear system

$$x(k) = A_0 x(k) \oplus A_1 x(k-1), \quad k = 1, 2, \dots \tag{3}$$

By iterating Eq. (3), one can obtain

$$x(k) = A_0^3 x(k) \oplus \left(A_0^2 \oplus A_0 \oplus I \right) A_1 x(k-1).$$

By a direct calculation, one has $A_0^3 = \mathcal{E}$. Hence,

$$x(k) = \left(A_0^2 \oplus A_0 \oplus I \right) A_1 x(k-1) := Ax(k-1), \tag{4}$$

where

$$A = \begin{pmatrix} e & t_1 & r_1 & \varepsilon \\ t_2 & t_1 t_2 & t_2 r_1 & r_2 \\ g_1 & g_1 t_1 & g_1 r_1 & \varepsilon \\ t_2 g_2 & t_1 t_2 g_2 & t_2 g_2 r_1 & g_2 r_2 \end{pmatrix}.$$

Obviously, A is irreducible. Let $\lambda(A)$ be the maximum eigenvalue of A. By a direct calculation, one has

$$\lambda(A) = \frac{t_1 t_2 g_1 r_1}{2} \oplus \frac{t_1 t_2 g_2 r_2}{2} \oplus \frac{r_1 r_2}{2} \oplus \frac{t_1 t_2 r_1 r_2 g_1 g_2}{3} \oplus g_1 r_1 \oplus g_2 r_2,$$

that is, the cycle time of system T_2 is $\lambda(A)$.

5   Conclusion

This paper models the one-way street and crossroad traffic signal system as a max-plus linear system. Then, the cycle time of traffic signal system is calculated to improve the operating efficiency of the vehicle. It is found out that the cycle time of the one-way street traffic signal system is independent of the travel time between two traffic lights. The future work will focus on the modeling and performance analysis of traffic signal systems at multiple-direction street and multiple crossroads. With the increase of directions, the number of traffic lights and the dimension of the system will inevitably be increased. The future work is concerned with setting the appropriate constraints and reduce the dimension of systems. In addition, how to optimize and control such complex traffic signal systems based on the Petri net model is also of interest.

References

1. Adzkiya, D., De Schutter, B., Abate, A.: Computational techniques for reachability analysis of max-plus-linear systems. Automatica **53**, 293–302 (2015)
2. Baccelli, F., Cohen, G., Olsder, G.J., Quadrat, J.P.: Synchronization and Linearity: An Algebra for Discrete Event Systems. Wiley, New York (1992)
3. Butkovič, P.: Max-linear Systems: Theory and Algorithms. Springer, London (2010). https://doi.org/10.1007/978-1-84996-299-5
4. Cohen, G., Gaubert, S., Quadrat, J.P.: Asymptotic throughput of continuous Petri nets. In: Proceedings of the IEEE-CDC, New Orleans (1995)
5. Cohen, G., Dubois, D., Quadrat, J.P., Viot, M.: A linear-system-theoretic view of discrete-event processes and its use for performance evaluation in manufacturing. IEEE Trans. Autom. Control **30**(3), 210–220 (1985)
6. Cohen, G., Moller, P., Quadrat, J.P., Viot, M.: Linear system theory for discrete event systems. In: Proceedings of the 23rd IEEE Conference on Decision and Control, pp. 539–544 (1984)
7. Cuninghame-Green, R.: Minimax Algebra. Springer, Heidelberg (1979). https://doi.org/10.1007/978-3-642-48708-8
8. De Schutter, B., van den Boom, T., Xu, J., Farahani, S.S.: Analysis and control of max-plus linear discrete-event systems: an introduction. Discrete Event Dyn. Syst. **30**(1), 25–54 (2019). https://doi.org/10.1007/s10626-019-00294-w
9. Esmaeil Zadeh Soudjani, S., Adzkiya, D., Abate, A.: Formal verification of stochastic max-plus-linear systems. IEEE Trans. Autom. Contr. **61**(10), 2861–2876 (2016)
10. Garcia, T., Cury, J.E.E., Kraus, W., Demongodin, I.: Traffic light coordination of urban corridors using max-plus algebra, Buenos Aires-Argentina, pp. 103–108 (2007)
11. Gonçalves, V.M., Maia, C.A., Hardouin, L.: On max-plus linear dynamical system theory: the observation problem. Automatica **107**, 103–111 (2020)
12. Gondran, M., Minoux, M.: Graphs and Algorithms. Wiley, New York (1984)
13. Harary, F.: A graph theoretic approach to matrix inversion by partitioning. Numer. Math. **4**(1), 128–135 (1962)
14. Hardouin, L., Shang, Y., Maia, C.A., Cottenceau, B.: Observer-based controllers for max-plus linear systems. IEEE Trans. Autom. Control **62**(5), 2153–2165 (2017)

15. Heidergott, B., Olsder, G.J., van der Woude, J.: Max-Plus at Work: Modeling and Analysis of Synchronized Systems. Princeton University Press, New Jersey (2006)
16. Huang, Y., Weng, Y., Zhou, M.: Modular design of urban traffic-light control systems based on synchronized timed Petri nets. IEEE Trans. Intell. Transp. Syst. **15**(2), 530–539 (2014)
17. Peterson, J.L.: Petri Nets Theory and the Modeling of Systems, vol. 25 (2010)
18. Kersbergen, B., Rudan, J., van den Boom, T., De Schutter, B.: Towards railway traffic management using switching Max-plus-linear systems. Discrete Event Dyn. Syst. **26**(2), 183–223 (2016). https://doi.org/10.1007/s10626-014-0205-7
19. Lai, A., Lahaye, S., Giua, A.: State estimation of max-plus automata with unobservable events. Automatica **105**, 36–42 (2019)
20. Lopes, G.A.D., Kersbergen, B., De Schutter, B., van den Boom, T., Babuška, R.: Synchronization of a class of cyclic discrete-event systems describing legged locomotion. Discrete Event Dyn. Syst. **26**(2), 225–261 (2015). https://doi.org/10.1007/s10626-014-0206-6
21. Luo, J., Huang, Y., Weng, Y.: Design of variable traffic light control systems for preventing two-way grid network traffic jams using timed Petri nets. IEEE Trans. Intell. Transp. Syst. **21**, 3117–3127 (2019)
22. Maia, C.A., Hardouin, L., Cury, J.E.R.: Some results on the feedback control of Max-plus linear systems under state constrains. In: 52nd IEEE Conference on Decision and Control, pp. 6992–6997 (2013)
23. Qi, L., Zhou, M., Luan, W.: A two-level traffic light control strategy for preventing incident-based urban traffic congestion. IEEE Trans. Intell. Transp. Syst. **19**(1), 13–24 (2018)
24. Shang, Y., Hardouin, L., Lhommeau, M., Maia, C.A.: An integrated control strategy to solve the disturbance decoupling problem for Max-plus linear systems with applications to a high throughput screening system. Automatica **63**, 338–348 (2016)
25. van den Boom, T.J.J., Muijsenberg, M., De Schutter, B.: Model predictive scheduling of semi-cyclic discrete-event systems using switching max-plus linear models and dynamic graphs. Discrete Event Dyn. Syst. **30**(4), 635–669 (2020). https://doi.org/10.1007/s10626-020-00318-w

NS3 Based Simulation Framework for 5G-IoV Networks

Yu He, Jiang Wu$^{(\boxtimes)}$, Jiaxin Li, and Zhanbo Xu

Xi'an Jiaotong University, Xi'an 710049, China
{hy1997023,leejiaxin}@stu.xjtu.edu.cn,
{jwu,zbxu}@sei.xjtu.edu.cn

Abstract. With the continuous advancement of Fifth Generation (5G) cellular network, the application scenarios of mobile communications have been transformed into three application scenarios for eMBB (enhanced mobile broadband), mMTC (massive Machine Type of Communication), and uRLLC (ultra-high-reliability and ultra-low-latency communications). The Internet of Vehicles (IoV) uses the network infrastructure to allow cars to be connected to new radio technologies, and can be supported by 5G networks. Therefore, the society's applications for IOV are also increasing. At the same time, due to the mobility of vehicles, the network performance of IOV is facing great challenges. The performance of the network is not only affected by the network equipment, but also depends on the network architecture and various protocols of the network. Effective performance evaluation of communication network can guide the further development of IOV technology. As an industry-recognized and iteratively updated network simulator, NS3 can effectively simulate the network and output key network evaluation indicators to evaluate network performance. Based on NS3, this article simulates the scenarios where a large amount of traffic transfers in the Internet of Vehicles, counts the key indicators of network evaluation, shows how to simulate and evaluate the network through NS3, and makes recommendations for network performance improvement.

Keywords: 5G · Internet of vehicles · NS3

1 Introduction

The 5G network puts forward the goal of "all things interconnection" and three application scenarios of Enhanced Mobile Broadband (eMBB), Massive Internet of Things (mMTC), and High Reliability and Low Latency (uRLLC) [1]. Compared with the 4G network rate, eMBB can provide higher speed, mobility and spectrum efficiency, and can meet the needs of 4K/8K ultra-high-definition video, VR/AR and other high-traffic applications, and provide users with a better experience. mMTC and uRLLC are brand-new scenarios launched for vertical industries. They are designed in terms of traffic density, connection density, end-to-end delay, and reliability to meet the needs of massive IoT connections, car

© ICST Institute for Computer Sciences, Social Informatics and Telecommunications Engineering 2021
Published by Springer Nature Switzerland AG 2021. All Rights Reserved
Q. Zhao and L. Xia (Eds.): VALUETOOLS 2021, LNICST 404, pp. 233–242, 2021.
https://doi.org/10.1007/978-3-030-92511-6_15

networking, industrial control, and smart factories. And other applications, to promote the transformation of 5G from the mobile Internet era to the Internet of Everything era [2].

With the development of 5G, the Internet of Vehicles (IoV), which uses the network infrastructure to allow cars to be connected to new radio technologies, has been greatly developed but it also faces many challenges [3]. Due to the high-speed mobility of the vehicle itself, on the one hand, the vehicle communication channel becomes worse, the vehicle is frequently accessed and switched between networks and cells, the reliability of vehicle communication is reduced, and the communication is frequently interrupted; on the other hand, The physical distance between the vehicle and the core network is relatively long, and the same service requested by the vehicle in high-speed movement is forwarded through several roadside units and other infrastructures, which leads to the repeated configuration of network resources, causes serious problems such as network congestion, and causes communication delays to increase [4]. Therefore, it is of great significance to select and research core technologies to build and ensure the Internet of Vehicles communication. Simulating the scenes in the communication network and evaluating the performance of the communication network can guide us how to choose and tackle key technologies.

The NS3 project extensively draws on the technical experience of mainstream emulators NS2, YANS and GTNets, implemented in C++ language, compatible with the popular Python nowadays, NS2 has been ported to NS3, so that it has good performance in terms of function implementation, version updates, user experience, etc. [5]. Its core and various functional modules are completed by C++ code, and provide a wealth of extended interfaces to the outside world, and modules can be changed or added or removed as needed. NS3 includes a network component simulation interface and an event scheduler, which can simulate the real communication "behavior" by executing related events. At the same time, it also has a perfect tracking mechanism, which is convenient for users to analyze data, transmission process and analysis results [6]. It is very meaningful to use NS3 to simulate and evaluate the communication network [7]. Therefore, we use NS3 as the basis to simulate the IOV communication network scenario and analyze the communication network performance.

Based on the NS3 network simulator, this paper builds a network simulation environment, simulates the scene where a large amount of traffic occurs node transfer in IOV caused by the mobility of vehicles, outputs key network indicators, evaluates network performance, and give some instructive suggestions to guides the selection and research of further key technologies, the framework as shown in the Fig. 1. This article makes the following contributions: (1) Realize the evaluation of communication network performance through NS3 simulation; (2) Simulate scenarios in the IOV communication network and analyze the challenges that the IOV network may face; (3) Guide the selection and further research of key technologies.

The rest of the paper is organized as follows: Sect. 2 give an introduction to NS3 and describe how to simulate NS3. Section 3 simulates the scenario where a

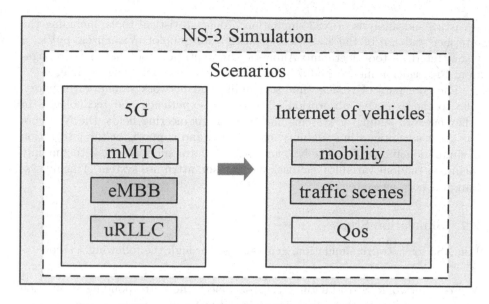

Fig. 1. NS3 based simulation framework for 5G-IoV networks.

large amount of traffic occurs node transfer in IOV, and analyzes the simulation results. Finally, the conclusion of our work is given in Sect. 4.

2 NS3 Simulator

2.1 NS3 Overview

NS3 is a discrete event simulation software, NS3 is a discrete event simulation software, its core and modules are implemented in C++ language. In essence, NS3 can be considered as a dynamic or static library. These library functions (API) can be called by simulation programs written in C++. Similarly, NS3 also supports Python language library functions. Python simulation programs can call the API provided by NS3 in the same way that C++ language calls library functions [6].

NS3 is composed of a series of distinct functional modules. Common modules are: Core: NS3 kernel module, which implements the basic mechanisms of NS3, such as smart pointer (Ptr), attribute, callback, random ariable, logging, tracing, and event scheduler. Network: network data packet module, generally used in simulation. Internet: it implements related protocol suites on TCP/IPv4 and IPv6, including IPv4, IPv6, ARP, UPP, TCP, neighbor discovery and other related protocols. At present, most networks are based on the Internet protocol stack. Applications: concentrate the commonly used application layer protocol. Mobility: mobile model module. Topolopy-read: take the specified trajectory file data and generate the corresponding network topology according to the specified format. Status: statistical framework module to facilitate data collection, statistics and analysis of NS3 simulation. Tools: statistical tools, including the interface and use of the statistical drawing tool gnuplot. Visualizer: pyViz, a visual interface tool. Netanim: Animation presentation tool Netnim. Propagation: Propagation model module. Flow-monitor: flow monitoring module.

The core part of NS3 is time scheduling and network simulation support. Due to the extensive absorption of successful experience and technologies in other network simulator systems and modern engineering fields, the NS3 core has great advantages in scalability, modularity, and support for the integration of simulation and reality. The NS3 network simulator supports the kernel mainly including random variables, callback mechanism, attribute system, Tracing system and Logging system.

2.2 Simulation Steps

Use NS3 for network simulation, generally go through the following 4 steps.

Write Network Simulation Scripts. Select the corresponding simulation module according to the actual simulation object and simulation scenario: such as wired local area network (CSMA) or wireless local area network (Wi-Fi): whether the node needs to be mobile (mobility); what application to use; whether it needs energy (Energy) management; what routing protocol to use (internet, aodv, etc.); whether a visual interface such as animation presentation (visualizer, netanim) is needed, etc. If the network to be built is a relatively new network, such as a delay tolerant network (DTN), or the reader wants to develop a protocol of his own design, such as a routing protocol, a mobile model, an energy management model of his own design, use NS3 When testing, if there is no corresponding module support at present, then you need to design and develop your own network simulation module. For specific methods, refer to the subsequent section on adding modules.

Select or Develop Corresponding Modules. With the corresponding modules, you can build a network simulation environment. The general process of writing NS3 simulation script is as follows. 1. Generating node: The node in NS3 is equivalent to an empty computer shell. Next, the computer needs to install the software and hardware required by the network, such as network cards, applications, protocol stacks; 2. Install network equipment: different Network types have different network devices, which provide different channels, physical layers and MAC layers, such as CSMA, Wi-Fi, WIMAX and point-to-point; 3. Install the protocol stack: TCP is generally used in NS3 networks /IP protocol stack, select specific protocols based on the network, such as UDP or TCP, which routing protocol to choose (OLSR, AODV, Global, etc.) and configure the corresponding IP address for it, NS3 supports both IPV4 and IPv6; 4. Install the application layer protocol: select the corresponding application layer protocol according to the selected transport layer protocol, but sometimes you need to write your own code to generate network data traffic at the application layer;

5. Other configurations: such as whether the node is moving, whether energy management is required; 6. Start the simulation: After the entire network scene is configured, start the simulation.

Analysis of Simulation Results. There are generally two kinds of simulation results: one is network scenario, and the other is network data. Network scenes, such as node topology, mobile models, can generally be visually observed through the visual interface (PyViz or NetAnim); network data can also have simple statistics under the visual interface, in addition, through a special statistical framework (status) or Collect, count and analyze the corresponding network data through the tracing system provided by NS3, such as data packet delay, network traffic, packet loss rate, and node message buffer queue.

Adjust the Network Configuration Parameters or Modify the Source Code According to the Simulation Results. Sometimes the actual results are far from the expectations. At this time, we need to analyze the reasons, whether there is a problem with the network parameters or the protocol itself, NS3., and then redesign, resimulate, and repeat until a satisfactory result is achieved.

3   Simulation Scenario and Results

3.1   Simulation Scenario Description

Considering that with the development of 5G, more and more vehicles are connected to the network for data interaction during the driving process, such as in-vehicle online conferences, in-vehicle online surgery, in-vehicle movie viewing, and autonomous driving applications, which will lead to vehicles The flow of access to the network has greatly increased, and these applications have extremely high requirements for the low latency and stability of the network [8]. For the vehicle flow in an area, when the vehicle flow is in the city, the connected base stations are constantly switching, so that a large amount of traffic data is constantly being switched in the base station, which has a great impact on the performance of the network. Therefore, we simulate and analyze the scene where the flow of vehicles is driving in the city, and a large amount of traffic is constantly switched as the base station of the vehicle is switched [9].

Specifically, the considered simulation scenario is the following: We assume a network with 14 nodes, node 0, node 1, node 2, node 11, node 9 and node 13 represent 6 different base stations, node 6 represents the remote host, and the vehicle generates information through the base station and the remote host. Therefore, there is already traffic between each base station and the remote host node (0–6, 2–6, 1–6, 11–6, 9–6, 13–6), and then part of the vehicle flow initially connected to node 2 is switched to be connected to node 0 due to driving, so that part of the traffic is transferred from node 2 to node 1. In this way, the scene of a large number of traffic transfers due to the handover of vehicle flow base stations is simulated.

The network topology used in the simulation is the topology provided by the National Science Foundation as Fig. 2, and the simulated traffic data comes from D. Raca, etc. [10] and we aggregated and scaled this traffic, D. Raca present a 5G trace dataset collected from a major Irish mobile operator. The dataset is generated from two mobility patterns (static and car), and across two application patterns(video streaming and file download). The dataset is composed of client-side cellular key performance indicators (KPIs) comprised of channel-related metrics, context-related metrics, cell- related metrics and throughput information. These metrics are generated from a well-known non-rooted Android network monitoring application, G-NetTrack Pro. The simulation parameters are shown in Table 1.

Based on the above, we conducted two simulations, one simulation occurred traffic transfer from node 2 to node 0, and the other did not.

Table 1. Simulation paramaeters

Paramate	Example
Simulation program	NS3.29
Operating system	Ubuntu 18.04
Routing protocol	OSPF
Number of node	14
Protocal transport packet size	User datagram protocol OnOff application

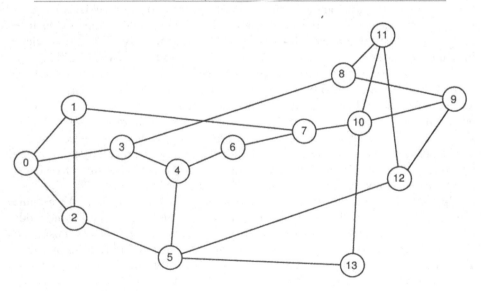

Fig. 2. Network topology.

3.2   Results

We call the situation where the traffic is transferred as scenario 1, and the situation where it does not happen is called scenario 2.

Throughput: the total number of data bytes successfully accepted in units of time and describe network data rate condition [11]. Figure 3 shows the throughput of node 0 to node 6 in different scenarios, Fig. 4 shows the throughput of node 2 to node 6 in different scenarios. We can see that in scenario 1 the traffic has shifted at 30 s.

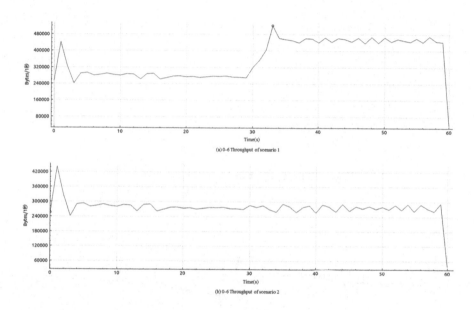

Fig. 3. Throughput of node 0 to node 6.

RTT (Round-Trip Time): Round trip delay [12]. It is an important performance indicator in a computer network. It represents the total time delay experienced from the start of the sender sending data until the sender receives the confirmation from the receiver (the receiver sends the confirmation immediately after receiving the data). Figure 5 shows RTT of node 0 to node 6 in different scenarios, Fig. 6 shows RTT of node 2 to node 6 in different scenarios. We can see that when the traffic shifts, the RTT curve of the network fluctuates greatly, which means that when the traffic shifts, the delay of some applications fluctuates, which is intolerable for some scenarios, such as automatic Driving, in-vehicle surgery. Therefore, for IOV, how to reduce the delay fluctuation and network shock caused by the large-scale transfer of traffic is a key problem.

3.3   Analysis and Suggestion

The simulation results show that traditional IOV network technology often optimizes the current network based on the information known by the network. When

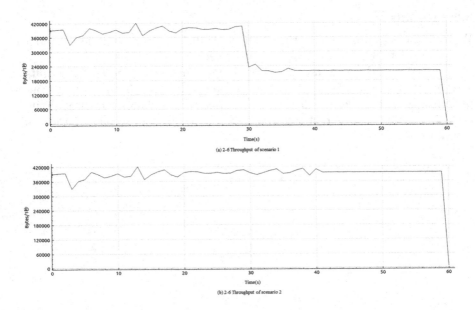

Fig. 4. Throughput of node 2 to node 6.

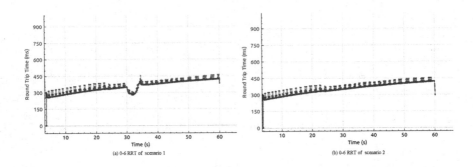

Fig. 5. RTT of node 0 to node 6.

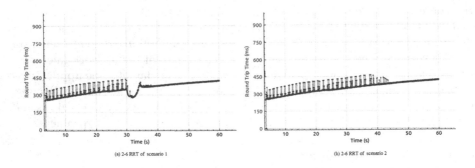

Fig. 6. RTT of node 2 to node 6.

the network changes, it will often cause performance degradation and network shock. The method of optimizing the network based on the existing information is not enough to meet the increasing requirements, so you can consider using the large amount of data in the network and the city to make a prediction on the vehicle and the network, and provide a priori information for network optimization. The improvement of network performance will be of great significance. For the Internet of Vehicles, vehicle behavior includes not only vehicle movement behavior, but also vehicle communication behavior. The two influence each other and are inseparable. On the one hand, the movement behavior of vehicles affects the temporal and spatial distribution characteristics of vehicle communication behavior, and communication behaviors usually increase in places with many vehicles; on the other hand, vehicle communication behavior can change the movement behavior of vehicles, such as a traffic accident or road congestion ahead, After vehicles exchange information through the entire network, subsequent vehicles change their trajectory. Therefore, considering vehicle movement behavior and communication behavior at the same time, and then accurately characterize the time and space characteristics of vehicle behavior, predict the trajectory and flow of vehicles, and pre-allocate resources and routes, which will effectively improve the performance of IOV.

4   Conclusion

This article first introduces the impact of the development of 5G technology on the communication network, focusing on the challenges that the increase in traffic and the movement of vehicles in the IOV network bring to the IOV network. In order to guide how to face the challenges, we propose the application of NS3 The network simulator carries on the simulation analysis to the concrete scene. Afterwards, we gave an overview of NS3 and introduced the simulation steps. Finally, we simulated specific scenarios and got the conclusion that the large-scale traffic transfer caused by the flow of vehicles will bring great challenges to the communication network, and put forward corresponding suggestions to solve these challenges.

Acknowledgements. The authors gratefully acknowledge the financial support received from the National Natural Science Foundation of China (Project Nos. U1766205).

References

1. Gpp, T.S.: 3rd generation partnership project; technical specification group services and system aspects; generic authentication architecture (gaa); generic bootstrapping architecture (release 6)
2. Shafique, K., Khawaja, B.A., Sabir, F., Qazi, S., Mustaqim, M.: Internet of things (IoT) for next-generation smart systems: a review of current challenges, future trends and prospects for emerging 5g-IoT scenarios. IEEE Access **8**, 23022–23040 (2020)

3. Ge, X., Li, Z., Li, S.: 5g software defined vehicular networks. IEEE Commun. Mag. **55**(7), 87–93 (2017)
4. Zhou, H., Wenchao, X., Chen, J., Wang, W.: Evolutionary v2x technologies toward the internet of vehicles: challenges and opportunities. Proc. IEEE **108**(2), 308–323 (2020)
5. Siraj, M.S., Ajay, M., Gupta, K., Rinku-Badgujar, M.: Network simulation tools survey
6. Carneiro, G.: Ns-3: Network simulator (March 2010)
7. Saleh, S.A.: A comparative performance analysis of manet routing protocols in various propagation loss models using ns3 simulator (2019)
8. Wenchao, X., Zhou, H., Cheng, N., Lyu, F., Shi, W., Chen, J., Shen, X.: Internet of vehicles in big data era. IEEE/CAA J. Automatica Sinica **5**(1), 19–35 (2018)
9. Storck, C.R., Duarte-Figueiredo, F.: A 5g v2x ecosystem providing internet of vehicles. Sensors **19**(3), 550 (2019)
10. Raca, D., Leahy, D., Sreenan, C.J., Quinlan, J.J.: Beyond throughput, the next generation: a 5g dataset with channel and context metrics. In: Proceedings of the 11th ACM Multimedia Systems Conference, MMSys 2020, pp. 303–308, New York, NY, USA, Association for Computing Machinery (2020)
11. Xiao, Y., Rosdahl, J.: Throughput and delay limits of IEEE 802.11. IEEE Commun. Lett. **6**(8), 355–357 (2002)
12. Brown, P.: Resource sharing of TCP connections with different round trip times. In: Proceedings IEEE INFOCOM 2000. Conference on Computer Communications. Nineteenth Annual Joint Conference of the IEEE Computer and Communications Societies (Cat. No.00CH37064), vol. 3, pp. 1734–1741 (2000)

Variance Reduction for Matrix Computations with Applications to Gaussian Processes

Anant Mathur[1(✉)], Sarat Moka[2], and Zdravko Botev[1]

[1] University of New South Wales,
High Street, Kensington Sydney, NSW 2052, Australia
anant.mathur@unsw.edu.au
[2] Macquarie University, 192 Balaclava Rd, Macquarie Park, NSW 2113, Australia

Abstract. In addition to recent developments in computing speed and memory, methodological advances have contributed to significant gains in the performance of stochastic simulation. In this paper we focus on variance reduction for matrix computations via matrix factorization. We provide insights into existing variance reduction methods for estimating the entries of large matrices. Popular methods do not exploit the reduction in variance that is possible when the matrix is factorized. We show how computing the square root factorization of the matrix can achieve in some important cases arbitrarily better stochastic performance. In addition, we detail a factorized estimator for the trace of a product of matrices and numerically demonstrate that the estimator can be up to 1,000 times more efficient on certain problems of estimating the log-likelihood of a Gaussian process. Additionally, we provide a new estimator of the log-determinant of a positive semi-definite matrix where the log-determinant is treated as a normalizing constant of a probability density.

Keywords: stochastic simulation · Variance reduction · Gaussian processes

1 Introduction

In many scientific computing and machine learning problems we are required to estimate the entries of a positive semi-definite (PSD) matrix $\mathbf{A} = (a_{i,j}) \in \mathbb{R}^{n \times n}$, where $\mathbf{A}$ is not explicitly available, but is rather only available as a black box function. That is to say, we assume we only have access to an oracle function that for any $\mathbf{x} \in \mathbb{R}^n$ can compute $\mathbf{A}\mathbf{x}$. Current methods [3] employ Monte Carlo simulations to produce unbiased estimates for $a_{i,j}$ via operations of the kind $\mathbf{A}\mathbf{Z}$ where $\mathbf{Z} \in \mathbb{R}^n$ is a random vector drawn from a pre-defined distribution. Hutchinson [12] was the first to use this methodology to approximate matrix traces, $\mathrm{tr}(\mathbf{A}) = \sum_i a_{i,i}$ and Bekas et al. [3] extended this method to estimate the actual diagonal values $a_{i,i}$. Significant innovations [1,2,8,16,18] in this area

Q. Zhao and L. Xia (Eds.): VALUETOOLS 2021, LNICST 404, pp. 243–261, 2021.
https://doi.org/10.1007/978-3-030-92511-6_16

of research have sought to improve the performance of these estimators through variance reduction and other means, effectively improving their scalability to larger matrices.

One major application of this kind of estimation is fitting Gaussian processes where one is required to evaluate the functions of the black box matrices: $\text{tr}(\mathbf{K}^{-1}\frac{\partial \mathbf{K}}{\partial \theta})$ and $\ln|\mathbf{K}| = \text{tr}(\ln(\mathbf{K}))$, where $\mathbf{K} \in \mathbb{R}^{n \times n}$ is a covariance matrix parameterized by θ, and is of dimension n - the size of the data-set. In these cases, explicitly computing the matrix functions roughly requires $O(n^3)$ operations, making their evaluation infeasible for any matrix of reasonable size whereas, $\mathbf{K}^{-1}\boldsymbol{x}$ and $\log(\mathbf{K})\boldsymbol{x}$ can be estimated in m iterations via a Krylov Subspace method that takes $O(mn^2)$ time. Generally, $m \ll n$ iterations are required with each iteration only requiring a matrix-vector-multiplication operation with $\mathbf{K}$.

The scope of black box matrix estimation, however, is not limited to Gaussian processes. For example, one can estimate statistical leverages for a sparse linear model by approximating the diagonals of the projection matrix $\mathbf{A} = \mathbf{X}(\mathbf{X}^\top\mathbf{X})^{-1}\mathbf{X}^\top$ as in [7] or one can precondition neural networks by estimating the diagonals of a Hessian matrix [6]. Furthermore, black box trace estimation has many applications including but not limited to evaluating the Cramer-Rao lower bound for the variance of an unbiased estimator [4] and approximating spectral densities [14].

Martens et al. [15] suggest that the variance for the Monte Carlo estimate of $a_{i,j}$ can depend on the matrix factorization of $\mathbf{A}$, and our work extends their ideas further. Our paper explores how taking the square root factorization of $\mathbf{A} = \mathbf{A}^{1/2}\mathbf{A}^{1/2}$ can reduce variance with little to no additional cost to CPU time.

In particular, our paper makes the following contributions:

1. We illustrate the advantages of black box matrix estimation with factorization over estimation without factorization. In particular, we consider the estimator of Bekas et al. [3] for the diagonal elements that uses no factorization and compare it with the factorized square root estimator to show that the former can exhibit an arbitrarily larger variance compared to the latter. No such analysis has been given in [15]. We further provide analysis on how the distribution of $\boldsymbol{Z}$ effects the variance of the factorized estimator. This includes variance results for when $\boldsymbol{Z}$ are simulated to be uniformly distributed on the surface of the unit radius n-dimensional sphere centered at the origin.

2. We introduce an estimator for the trace of a product of matrices, $\text{tr}(\mathbf{AW})$ where $\mathbf{A} \in \mathbb{R}^{n \times n}$ is a PSD matrix and $\mathbf{W} \in \mathbb{R}^{n \times n}$ is a symmetric matrix, see also [19] and [10]. This estimator exploits square root factorization to achieve variance reduction. We provide numerical examples for the problem of estimating the log-likelihood of a Gaussian process to suggest that the estimator can be up to 1,000 times more efficient in performance, compared to existing methods, for commonly used covariance matrices with dimension n up to 20,000.

3. We propose a novel Monte Carlo estimator for the log-determinant, $\ln|\mathbf{A}|$ where $\mathbf{A} \in \mathbb{R}^{n \times n}$ is assumed to be a PSD matrix. Existing Monte Carlo

methods for approximating the log-determinant require evaluating expressions such as $\ln(\mathbf{K})\mathbf{Z}$ via the Lanczos algorithm [11, 21]. Our proposed method instead relies on the more memory efficient operation of solving the linear-system $\mathbf{K}^{-1}\mathbf{Z}$ via the conjugate gradient method.

This paper is organized as follows. In Sect. 2 we review how one can estimate the entries of the PSD matrix $\mathbf{A}$ via matrix vector multiplications with the random vector $\mathbf{Z}$. We then explain how factorization can attain variance reduction and we provide three examples to illustrate when factorization provides greater performance. In this section, we also discuss how one can efficiently compute the factorized estimator. In Sect. 3, we introduce a Monte Carlo estimator for $\mathrm{tr}(\mathbf{AW})$ and we illustrate the performance improvement attained when the estimator is applied to the problem of estimating the log-likelihood of a Gaussian process. In Sect. 4 we discuss the existing Monte Carlo method for estimating $\ln|\mathbf{A}|$ via the Lanczos algorithm [21] and we introduce a new Monte Carlo estimator which treats the log-determinant as the normalizing constant of a probability density. Finally, we discuss implications of our research and areas of future work. All proofs are available in the Appendix unless stated otherwise.

2 Estimating Matrix Elements

Suppose we wish to estimate the entries of the PSD matrix $\mathbf{A} = (a_{i,j}) \in \mathbb{R}^{n \times n}$. Let $\mathbf{BC}^{\top}$ be an arbitrary matrix factorization of the matrix $\mathbf{A}$, with $\mathbf{B}, \mathbf{C} \in \mathbb{R}^{n \times p}$. Let $\mathbf{Z} \in \mathbb{R}^p$ be a random vector such that $\mathbb{E}\left[\mathbf{Z}\mathbf{Z}^{\top}\right] = \mathbf{I}_p$. Then, a rank-1 unbiased estimate of $\mathbf{A}$ is given in [15]:

$$\hat{\mathbf{A}} = \mathbf{B}\mathbf{Z}\mathbf{Z}^{\top}\mathbf{C}^{\top} = (\mathbf{B}\mathbf{Z})(\mathbf{C}\mathbf{Z})^{\top}. \tag{1}$$

To approximate $a_{i,j}$ we evaluate $[\mathbf{B}\mathbf{Z}]_i \times [\mathbf{C}\mathbf{Z}]_j$, and to approximate the diagonal of $\mathbf{A}$ we evaluate $(\mathbf{B}\mathbf{Z}) \odot (\mathbf{C}\mathbf{Z})$, where $\odot$ is defined as the element-wise matrix product. If we select the trivial factorization, $\mathbf{A} = \mathbf{A}\mathbf{I}_n$, that is when $\mathbf{B} = \mathbf{A}$ and $\mathbf{C} = \mathbf{I}_n$, we obtain the well known Bekas et al. [3] estimator $\hat{\ell}_{\text{Bekas}}$ for the diagonal,

$$\hat{\ell}_{\text{Bekas}} = \mathbf{Z} \odot \mathbf{A}\mathbf{Z}. \tag{2}$$

Various sampling distributions for $\mathbf{Z}$ have been proposed, including the Rademacher distribution, whose entries are either -1 or 1 each with probability 0.5. One can also sample $\mathbf{Z}$ from the standard Gaussian, uniformly from the set of standard basis vectors or uniformly from the surface of the unit radius n-dimensional sphere centered at the origin.

We note that Bekas et al. [3] suggest scaling the diagonal estimate $\hat{\ell}_{\text{Bekas}}$ element-wise by the vector $\mathbf{Z} \odot \mathbf{Z}$. This scaling has no effect when $\mathbf{Z}$ is Rademacher as $\mathbf{Z} \odot \mathbf{Z}$ is equal to the vector of ones. When $\mathbf{Z}$ is Gaussian, the performance with scaling is improved, as proven by Kaperick [13]. However, even with scaling the variance of the Gaussian estimator is still marginally higher than that of the Rademacher estimator.

The estimator $\hat{\ell}_{\text{Bekas}}$ does not guarantee non-negative diagonal entries when $\mathbf{A}$ is a PSD matrix. To mitigate this, we can use a decomposition of the form $\mathbf{A} = \mathbf{B}\mathbf{B}^\top$, that is, when $\mathbf{C} = \mathbf{B}$. The entries of the ensuing estimate $\mathbf{B}\mathbf{Z} \odot \mathbf{B}\mathbf{Z}$ are now guaranteed to be non-negative. We note that this decomposition is not unique and can for example be the Cholesky decomposition, or the symmetric square root factorization $\mathbf{A} = \mathbf{A}^{1/2}\mathbf{A}^{1/2}$. Moreover, if the functional form of $\mathbf{A}$ is known, it may be exploitable and the factorization $\mathbf{A} = \mathbf{B}\mathbf{B}^\top$ can be constructed without computing the symmetric square root or Cholesky factorization.

Example 1 (Diagonals of Inverse Matrices). Suppose we wish to estimate the variances for the OLS estimator $\hat{\boldsymbol{\beta}} = (\mathbf{X}^\top\mathbf{X})^{-1}\mathbf{X}^\top\boldsymbol{y}$, where $\mathbf{X} \in \mathbb{R}^{n \times p}$, and $\boldsymbol{y} \in \mathbb{R}^n$. Assume our observations are independent and $\mathbb{V}\text{ar}(y_i) = \sigma^2$ for all $i = 1, \ldots, n$. Then, the covariance matrix $\mathbf{A} = \mathbb{V}\text{ar}\left[\hat{\boldsymbol{\beta}}\right]$ is,

$$\mathbf{A} = \sigma^2(\mathbf{X}^\top\mathbf{X})^{-1}\mathbf{X}^\top\mathbf{I}_n\mathbf{X}(\mathbf{X}^\top\mathbf{X})^{-1} = \sigma^2(\mathbf{X}^\top\mathbf{X})^{-1}.$$

Our goal is to estimate the diagonal entries of $\mathbf{A}$. In this example, we do not need to compute the square root of $\mathbf{A}$ directly (e.g., $\mathbf{B} = \mathbf{A}^{1/2}$). Instead, we can use conjugate gradient methods, since $\mathbf{A} = \mathbf{B}\mathbf{B}^\top$, where,

$$\mathbf{B} = \sigma(\mathbf{X}^\top\mathbf{X})^{-1}\mathbf{X}^\top.$$

Thus, the factorized diagonal estimator is $\hat{\ell} = \mathbf{B}\mathbf{Z} \odot \mathbf{B}\mathbf{Z}$ and the computation $(\mathbf{X}^\top\mathbf{X})^{-1}\boldsymbol{c}$, where $\boldsymbol{c} = \mathbf{X}^\top\mathbf{Z}$ can be solved via the conjugate gradient method.

2.1   Variance Analysis

Martens et al. [15] provides variance formulas for estimator (1) when $\mathbf{Z}$ is drawn from the standard Gaussian and Rademacher distribution. In this section we state these formulas along with the variance formula for when $\mathbf{Z}$ is drawn uniformly on the surface of the unit radius n-dimensional sphere centered at the origin. For the following analysis we assume that $\hat{\mathbf{A}} = (\hat{a}_{i,j}) = (\mathbf{B}\mathbf{Z})(\mathbf{C}\mathbf{Z})^\top$ and $\mathbf{B}, \mathbf{C} \in \mathbb{R}^{n \times n}$. Furthermore, we denote $\boldsymbol{b}_{i,:}$ and $\boldsymbol{c}_{i,:}$ as the column vectors containing the i-th rows of the matrices $\mathbf{B}$ and $\mathbf{C}$.

Gaussian: When $\mathbf{Z} \sim \mathcal{N}(\mathbf{0}, \mathbf{I}_n)$, the variance for $\hat{a}_{i,j}$ is,

$$\mathbb{V}\text{ar}_G[\hat{a}_{i,j}] = \|\boldsymbol{b}_{i,:}\|^2\|\boldsymbol{c}_{j,:}\|^2 + a_{i,j}^2, \tag{3}$$

and when $\mathbf{B} = \mathbf{A}$, $\mathbf{C} = \mathbf{I}_n$, and $i = j$, we obtain the variance for the diagonal estimator of Bekas et al. [3], that is,

$$\mathbb{V}\text{ar}_G^{\text{AI}}[\hat{a}_{i,i}] = 2a_{i,i}^2 + \sum_{j \neq i} a_{i,j}^2. \tag{4}$$

Theorem 1 (Theorem 4.1 in Martens et al. [15]). *For the factorization* $\mathbf{A} = \mathbf{B}\mathbf{C}^\top$, $\mathrm{Var}_G[\hat{a}_{i,i}]$ *is minimized when* $\mathbf{C} = \mathbf{B}$ *with resulting variance,*

$$\mathrm{Var}_G^{\mathbf{B}\mathbf{B}^\top}[\hat{a}_{i,i}] = 2a_{i,i}^2. \tag{5}$$

Our next result contributes to our understanding of using a spherical probing vector.

Uniform Spherical: Suppose $\mathbf{Y} \sim \mathcal{N}(\mathbf{0}, \mathbf{I}_n)$ then $\mathbf{Z} = \mathbf{Y}/\|\mathbf{Y}\|$ is uniformly distributed on the surface of the unit radius n-dimensional sphere centered at the origin.

Theorem 2 (Spherical Estimator for Matrix Elements). *Suppose $\mathbf{Z}$ is uniformly distributed on the surface of the unit radius n-dimensional sphere centered at the origin and $\mathbf{A} = \mathbf{B}\mathbf{C}^\top$. Then,*

$$\mathbb{E}\left[n\,(\mathbf{B}\mathbf{Z})\,(\mathbf{C}\mathbf{Z})^\top\right] = \mathbf{A},$$

and $\hat{a}_{i,j} = n\,[\mathbf{B}\mathbf{Z}]_i \times [\mathbf{C}\mathbf{Z}]_j$ *has variance,*

$$\mathrm{Var}_S[\hat{a}_{i,j}] = \frac{n}{n+2}\|\boldsymbol{b}_{i,:}\|^2\|\boldsymbol{c}_{j,:}\|^2 + \frac{n-2}{n+2}a_{i,j}^2.$$

As we can see, when n is large, $\mathrm{Var}_S[\hat{a}_{i,j}] \approx \mathrm{Var}_G[\hat{a}_{i,j}]$.

Rademacher: Martens et al. [15] give the following variance for $\hat{a}_{i,j}$ when $\mathbf{Z} \in \mathbb{R}^n$ is Rademacher,

$$\mathrm{Var}_R[\hat{a}_{i,j}] = \mathrm{Var}_G[\hat{a}_{i,j}] - 2\sum_k c_{i,k}^2 b_{j,k}^2.$$

$$= \|\boldsymbol{b}_{i,:}\|^2\|\boldsymbol{c}_{j,:}\|^2 + a_{i,j}^2 - 2\sum_k c_{i,k}^2 b_{j,k}^2.$$

For the two cases where 1) $\mathbf{B} = \mathbf{A}$, $\mathbf{C} = \mathbf{I}$ (unfactorized) and 2) $\mathbf{C} = \mathbf{B}$ (factorized), we have the variances for the diagonal entry estimates,

$$\mathrm{Var}_R^{\mathbf{A}\mathbf{I}}[\hat{a}_{i,i}] = \sum_{j \neq i} a_{i,j}^2. \tag{6}$$

$$\mathrm{Var}_R^{\mathbf{B}\mathbf{B}^\top}[\hat{a}_{i,i}] = 2a_{i,i}^2 - 2\sum_k b_{i,k}^4. \tag{7}$$

These formulas indicate $\mathrm{Var}_R^{\mathbf{B}\mathbf{B}^\top} \leq \mathrm{Var}_G^{\mathbf{B}\mathbf{B}^\top}$ and $\mathrm{Var}_R^{\mathbf{A}\mathbf{I}}[\hat{a}_{i,i}] \leq \mathrm{Var}_G^{\mathbf{A}\mathbf{I}}[\hat{a}_{i,i}]$; therefore we recommend the Rademacher as the best choice for $\mathbf{Z}$.

Extending Results on Rademacher: Theorem 1 shows that $\mathbf{A} = \mathbf{B}\mathbf{B}^\top$ is the most optimal factorization when $\mathbf{Z}$ is Gaussian. This is not guaranteed for the Rademacher case and our contribution is to analyze the effect of the square root factorization in the Rademacher case. For example, for PSD matrices with growing off-diagonals with increasing size, $\mathrm{Var}_R^{\mathbf{AI}}[\hat{a}_{i,i}]$ can be made arbitrarily large, while $\mathrm{Var}_R^{\mathbf{BB}^\top}[\hat{a}_{i,i}]$ remains bounded above by $2a_{i,i}^2$. We now show an example matrix that behaves this way.

Example 2 (Stochastic matrices). Suppose we have x, y such that $x > y \geq 0$. Consider the symmetric matrix $\mathbf{A}$ of size $n \times n$ with

$$
a_{i,j} = \begin{cases} x, & \text{if } i = j, \\ y, & \text{if } i \neq j. \end{cases}
$$

For this matrix $\mathbf{A}$, we can show that the eigenvalues (or singular values) are $x+py$ with multiplicity 1 and $x - y$ with multiplicity $n - 1$. Since all the eigenvalues are positive, $\mathbf{A}$ is a positive definite matrix. In particular,

$$
\mathrm{Var}_R^{\mathbf{AI}}[\hat{a}_{i,i}] - \mathrm{Var}_R^{\mathbf{BB}^\top}[\hat{a}_{i,i}] \geq \sum_{j \neq i} a_{i,j}^2 - 2a_{i,i}^2 = (n-1)y^2 - 2x^2.
$$

Therefore, $\mathrm{Var}_R^{\mathbf{AI}}[\hat{a}_{i,i}] \geq \mathrm{Var}_R^{\mathbf{BB}^\top}[\hat{a}_{i,i}]$ for all $i = 1, \ldots, n$ if $y \geq x\sqrt{2/(n-1)}$. Furthermore, the factorized estimator is substantially better than the unfactorized estimator when the ratio $\sqrt{(n-1)}\,y/x \to \infty$ as $n \to \infty$.

For instance, let $\varepsilon_n = \frac{1}{(n-1)^\gamma}$ for some constant $1/2 < \gamma < 1$. Then, with $x = \varepsilon_n$ and $y = \frac{1-\varepsilon_n}{n-1}$, the matrix $\mathbf{A}$ is a doubly stochastic matrix for all n, because each row and each column sum to $x + \cdot(n-1)y = 1$. Note that $x > y$ for sufficiently large values of n. Furthermore,

$$
\frac{\sqrt{(n-1)}\,y}{x} = \frac{\sqrt{n-1}}{(n-1)^{-\gamma}} \frac{1-\varepsilon_n}{n-1} = (1-\varepsilon_n)(n-1)^{\gamma-1/2},
$$

which goes to ∞ as $n \to \infty$. In other words, the square root estimator can be arbitrarily better than the naive one.

We now show an example where the unfactorized estimator exhibits an arbitrary larger variance compared to the factorized estimator when estimating an arbitrary matrix element.

Example 3 (Toeplitz matrices with exponential decay). Consider the Gram matrix $\mathbf{A} = \{k(x_i, x_j)\}_{i,j=1,\ldots,n}$, where k is the Squared Exponential kernel parameterized by $\boldsymbol{\theta} = (\theta_1, \theta_2)^\top$,

$$
k(x_i, x_j) = \theta_1 \exp\left[\frac{-(x_i - x_j)^2}{2\theta_2^2}\right],
$$

for $x_1, \ldots, x_n \in \mathbb{R}$. For simplicity let $\theta_1 = 1$, forcing the diagonal entries to be unitary. Suppose $\{x_i\}_{i=1,\ldots,n}$ are equally separated with spacing $\frac{1}{n-1}$, then $\mathbf{A}$ is

Toeplitz. The entries of $\mathbf{A}$ can be written as,

$$a_{i,j} = \exp\left[\frac{-(i-j)^2}{2(n-1)^2\theta_2^2}\right].$$

For the estimate $\hat{a}_{i,j}$, the variance of the unfactorized estimator has the following lower bound,

$$\operatorname{Var}_R^{\mathbf{AI}}[\hat{a}_{i,j}] = \sum_{k=1}^{n} \exp\left[\frac{-(i-k)^2}{(n-1)^2\theta_2^2}\right] - \exp\left[\frac{-(i-j)^2}{(n-1)^2\theta_2^2}\right]$$

$$\geq n\exp(-\theta_2^{-2}) - 1.$$

If we let $\mathbf{B} = (b_{i,j}) \in \mathbb{R}^{n\times n}$ be the square root matrix of $\mathbf{A}$, then we obtain the upper bound for the variance of the factorized estimator,

$$\operatorname{Var}_R^{\mathbf{BB}^\top}[\hat{a}_{i,j}] = 2 - 2\sum_m b_{i,m}^2 b_{j,m}^2 \leq 2.$$

In this example, $\operatorname{Var}_R^{\mathbf{AI}}[\hat{a}_{i,j}]$ can be made arbitrary larger than the variance for the factorized estimator. That is to say, for all $\delta > 0$ there exists n' such that $\operatorname{Var}_R^{\mathbf{AI}}[\hat{a}_{i,j}] - \operatorname{Var}_R^{\mathbf{BB}^\top}[\hat{a}_{i,j}] > \delta$ for all $n > n' = \lceil(\delta+3)\exp(\theta_2^{-2})\rceil$.

We now show an example of the matrix $\mathbf{A}$ where the unfactorized Rademacher estimator exhibits lower variance than the factorized estimator.

Example 4 (Tridiagonal matrices). It is clear that a PSD tridiagonal matrix should be an example where we expect the unfactorized estimator to perform on par or better than the symmetrized estimator due to its minimal off-diagonal entries. While it is safe to assume that this case is rare in application, it is still a worthwhile example to show when factorizing $\mathbf{A}$ may not be advantageous. Let $\mathbf{A} \in \mathbb{R}^{n\times n}$ be the symmetric tridiagonal matrix,

$$\mathbf{A} = \begin{bmatrix} 1 & c & & & \\ c & 1 & c & & \\ & c & \ddots & \ddots & \\ & & \ddots & \ddots & c \\ & & & c & 1 \end{bmatrix}.$$

Then, the Cholesky decomposition of $\mathbf{A}$ is $\mathbf{A} = \mathbf{BB}^\top$, where $\mathbf{B} \in \mathbb{R}^{n\times n}$ is a lower bi-diagonal matrix with entries given by the following recursive formulas,

$$b_{i,i} = \sqrt{1 - b_{i,i-1}^2}, \quad \text{and} \quad b_{i,i-1} = \frac{c}{b_{i-1,i-1}}, \tag{8}$$

where $b_{1,1} = 1$ and $b_{2,1} = c$. By using (6), (7) and (8) we can show,

$$\operatorname{Var}_R^{\mathbf{BB}^\top}[\hat{a}_{i,i}] = 4c^2\left[\frac{1}{b_{i-1,i-1}^2} - \frac{c^2}{b_{i-1,i-1}^4}\right] = 4c^2\frac{b_{i,i}^2}{b_{i-1,i-1}^2}, \quad \text{and} \tag{9}$$

$$\mathbb{V}\mathrm{ar}_R^{\mathbf{AI}^\top}[\hat{a}_{i,i}] = \begin{cases} c^2, & \text{if } i = 1 \text{ or } n, \\ 2c^2, & \text{otherwise.} \end{cases}$$

By numerically evaluating (8) we observe that $\frac{b_{i,i}^2}{b_{i-1,i-1}^2}$ is non-decreasing in i, hence $\mathbb{V}\mathrm{ar}_R^{\mathbf{BB}^\top}[\hat{a}_{i,i}]$ is non-decreasing in i. We further observe that as the limit $n \to \infty$,

$$\mathbb{V}\mathrm{ar}_R^{\mathbf{BB}^\top}[\hat{a}_{n,n}] \uparrow 4c^2.$$

Based on these numerical observations,

$$\frac{\mathbb{V}\mathrm{ar}_R^{\mathbf{BB}^\top}[\hat{a}_{i,i}]}{\mathbb{V}\mathrm{ar}_R^{\mathbf{AI}}[\hat{a}_{i,i}]} \leq \frac{\mathbb{V}\mathrm{ar}_R^{\mathbf{BB}^\top}[\hat{a}_{n,n}]}{\mathbb{V}\mathrm{ar}_R^{\mathbf{AI}}[\hat{a}_{n,n}]},$$

and as $n \to \infty$,

$$\frac{\mathbb{V}\mathrm{ar}_R^{\mathbf{BB}^\top}[\hat{a}_{n,n}]}{\mathbb{V}\mathrm{ar}_R^{\mathbf{AI}}[\hat{a}_{n,n}]} \uparrow 4.$$

Therefore, $\mathbb{V}\mathrm{ar}_R^{\mathbf{BB}^\top}[\hat{a}_{i,i}]$ will be at most $4 \times \mathbb{V}\mathrm{ar}_R^{\mathbf{AI}}[\hat{a}_{i,i}]$ for large values of n.

Numerical simulations with the square root factorization $\mathbf{B} = \mathbf{A}^{1/2}$ indicate a similar conclusion. In particular, for certain values of c, $\mathbb{V}\mathrm{ar}_R^{\mathbf{AI}}[\hat{a}_{i,i}] > \mathbb{V}\mathrm{ar}_R^{\mathbf{BB}^\top}[\hat{a}_{i,i}]$, with a relative difference of at most 0.4. This relative difference remains bounded above by 0.4 as $n \to \infty$, whereas in Examples 2 and 3 the relative difference, when the factorized estimator outperforms the unfactorized estimator, goes to ∞ as $n \to \infty$. In plain English, while the downside of using the factorized estimator is limited, the upside can be unlimited as the size of the matrix grows.

2.2   Computing $\mathbf{A}^{1/2}\mathbf{Z}$

We briefly review well-known methods for efficiently computing with the square root factorization. As discussed in the introduction, black box estimation relies on the ability to efficiently evaluate matrix-vector operations. Thus, when using the factorization $\mathbf{A} = \mathbf{BB}^\top$ we must consider how to evaluate $\mathbf{BZ}$. If the functional form of $\mathbf{A}$ is not exploitable or not known, the only practical choice for $\mathbf{B}$ is $\mathbf{B} = \mathbf{A}^{1/2}$, the unique symmetric square root of $\mathbf{A}$.

Chow and Saad [5] outline how to compute $\mathbf{A}^{1/2}\mathbf{Z}$ via the well-known Lanczos tridiagonalization algorithm. The Lanczos algorithm takes the black box matrix $\mathbf{A}$ and the initial vector $\mathbf{Z}$ and after $m \leq n$ iterations outputs the matrices $\mathbf{T}_m$ and $\mathbf{V}_m$ that satisfy,

$$\mathbf{T}_m = \mathbf{V}_m^\top \mathbf{A}\mathbf{V}_m,$$

where $\mathbf{V}_m \in \mathbb{R}^{n \times m}$ contains orthonormal columns, and $\mathbf{T}_m \in \mathbb{R}^{m \times m}$ is a symmetric tridiagonal matrix. We can then use the approximation $\mathbf{V}_m^\top f(\mathbf{A})\mathbf{V}_m \approx f\left(\mathbf{V}_m^\top \mathbf{A}\mathbf{V}_m\right) = f(\mathbf{T}_m)$ and take $f(\mathbf{A}) = \mathbf{A}^{1/2}$ to obtain the approximation,

$$\boldsymbol{x}_m^* = \|\mathbf{Z}\|\mathbf{V}_m\mathbf{T}_m^{1/2}\boldsymbol{e}_1. \tag{10}$$

Generally $\|x_m^* - A^{1/2}Z\|$ can be made small when $m \ll n$, thus allowing efficient computation and storing of $T_m^{1/2}$. A discussion on the convergence rates for the Lanczos approximation can be found in Chow and Saad [5]. In general, this method has $O(mn^2)$ time complexity.

Alternatively, we can evaluate $A^{1/2}Z$ using the contour integral quadrature (CIQ) of Pleiss et al. [17]. This method also offers $O(mn^2)$ cost where m, the number of iterations is often less than 100 even for matrices of size up to 50,000. This method can be implemented via the GPyTorch framework [9].

3   Estimating the Trace of a Product of Matrices

In this section we introduce an estimator for the trace of a product of matrices, $\mathrm{tr}\,(AW)$, where $A \in \mathbb{R}^{n \times n}$ is a PSD matrix and $W \in \mathbb{R}^{n \times n}$ is a symmetric matrix. Applications of such traces abound and an example with Gaussian processes is given later on. If the goal is to estimate $\mathrm{tr}(A)$, we can take the sum of the estimator $\hat{\ell}_{\text{Bekas}}$ to obtain the well known Hutchinson estimator,

$$\mathrm{tr}(\hat{A}) = Z^\top A Z, \tag{11}$$

where $\mathbb{E}[ZZ^\top] = I_n$. We note that there is no difference in the estimator when using the factorization $A = BB^\top$ as,

$$\sum_i [BZ]_i \times [BZ]_i = Z^\top BB^\top Z = Z^\top A Z.$$

However, suppose we want to estimate $\mathrm{tr}(AW)$, where $A \in \mathbb{R}^{n \times n}$ is a PSD matrix and $W \in \mathbb{R}^{n \times n}$ is a symmetric matrix (not necessarily PSD). We can then use the following estimator,

$$\hat{\psi}_{\text{sqrt}} = Z^\top B^\top WBZ, \tag{12}$$

where $B = A^{1/2}$, see also [19] and [10]. To see this is unbiased, observe that due to the invariant cyclic nature of the trace,

$$\mathbb{E}\left[Z^\top B^\top WBZ\right] = \mathbb{E}\left[\mathrm{tr}(ZZ^\top B^\top WB)\right] = \mathrm{tr}(AW).$$

3.1   Variance Analysis

We now provide comparative variance analysis of the square root estimator (12) when Z is drawn from the standard Gaussian, from the Rademacher distribution, and uniformly on the surface of the unit radius n-dimensional sphere centered at the origin. For the following analysis we refer to the square root estimator as,

$$\hat{\psi}_{\text{sqrt}} = Z^\top B^\top WBZ,$$

where $B = A^{1/2}$, and the standard Bekas et al. [3] estimator as,

$$\hat{\psi} = Z^\top AWZ.$$

Gaussian

Theorem 3 (Lemma 9 in Avron and Toledo [2]). *Suppose* $\boldsymbol{\Sigma} \in \mathbb{R}^{n \times n}$, *and* $\boldsymbol{Z} \sim \mathcal{N}(\mathbf{0}, \mathbf{I}_n)$. *Let* $\boldsymbol{\Sigma}' = (\boldsymbol{\Sigma} + \boldsymbol{\Sigma}^\top)/2$. *Then,*

$$\mathrm{Var}\left[\boldsymbol{Z}^\top \boldsymbol{\Sigma} \boldsymbol{Z}\right] = 2\|\boldsymbol{\Sigma}'\|_{\mathrm{F}}^2, \tag{13}$$

where $\|\mathbf{X}\|_{\mathrm{F}} := \sqrt{\mathrm{tr}(\mathbf{X}^\top \mathbf{X})} = \sqrt{\sum_{i,j} (\mathbf{X}_{i,j})^2}$.

Following from this theorem we have the following.

Lemma 1 (Comparing the Variance of the Gaussian Trace Estimator). *If* $\boldsymbol{Z} \sim \mathcal{N}(\mathbf{0}, \mathbf{I}_n)$, *then* $\mathbb{E}[\hat{\psi}_{\mathrm{sqrt}}] = \mathbb{E}[\hat{\psi}] = \mathrm{tr}(\mathbf{AW})$,

$$\mathrm{Var}_G[\hat{\psi}] = \mathrm{tr}\left((\mathbf{AW})^\top (\mathbf{AW})\right) + \mathrm{tr}\left((\mathbf{WA})^\top (\mathbf{AW})\right), \quad and \tag{14}$$

$$\mathrm{Var}_G[\hat{\psi}_{\mathrm{sqrt}}] = 2\,\mathrm{tr}\left((\mathbf{WA})^\top (\mathbf{AW})\right). \tag{15}$$

Furthermore,

$$\mathrm{Var}_G[\hat{\psi}_{\mathrm{sqrt}}] \leq \mathrm{Var}_G[\hat{\psi}].$$

Proof Using Theorem 3, we obtain the variance formulas (14) and (15). To prove $\mathrm{Var}[\ell_{\mathrm{sqrt}}] \leq \mathrm{Var}[\hat{\psi}]$ we need to show,

$$\mathrm{tr}\left((\mathbf{WA})^\top (\mathbf{AW})\right) \leq \mathrm{tr}\left((\mathbf{AW})^\top (\mathbf{AW})\right).$$

As $\langle \mathbf{X}, \mathbf{Y} \rangle_{\mathrm{F}} := \mathrm{tr}(\mathbf{X}^\top \mathbf{Y})$ defines the Frobenius inner-product, the inequality holds true from the Cauchy-Schwarz inequality.

Uniform Spherical

Theorem 4 (Variance of the Spherical Quadratic Form). *Suppose* $\boldsymbol{\Sigma} \in \mathbb{R}^{n \times n}$ *and* $\boldsymbol{Z}$ *is uniformly distributed on the surface of the unit radius n-dimensional sphere centered at the origin. Let* $\boldsymbol{\Sigma}' = (\boldsymbol{\Sigma} + \boldsymbol{\Sigma}^\top)/2$. *Then,*

$$\mathrm{Var}\left[n\boldsymbol{Z}^\top \boldsymbol{\Sigma}' \boldsymbol{Z}\right] = \frac{n}{n+2} \times 2\left(\|\boldsymbol{\Sigma}'\|_{\mathrm{F}}^2 - \frac{(\sum_i \boldsymbol{\Sigma}'_{i,i})^2}{n}\right). \tag{16}$$

Following from this theorem we have the following.

Lemma 2 (Variance of the Spherical Trace Estimator). *If* $\boldsymbol{Z}$ *is uniformly distributed on the surface of the unit radius n-dimensional sphere centered at the origin, then* $\mathbb{E}[n\hat{\psi}_{\mathrm{sqrt}}] = \mathbb{E}[n\hat{\psi}] = \mathrm{tr}(\mathbf{AW})$,

$$\mathrm{Var}_S\left[n\hat{\psi}\right] = \frac{n}{n+2}\mathrm{Var}_G\left[\hat{\psi}\right] - \frac{2}{n+2}\left(\sum_i \left[\frac{\mathbf{AW} + \mathbf{WA}}{2}\right]_{i,i}\right)^2, \quad and \tag{17}$$

$$\mathrm{Var}_S\left[n\hat{\psi}_{\mathrm{sqrt}}\right] = \frac{n}{n+2}\mathrm{Var}_G\left[\hat{\psi}_{\mathrm{sqrt}}\right] - \frac{2}{n+2}\left(\sum_i [\mathbf{B}^\top \mathbf{WB}]_{i,i}\right)^2. \tag{18}$$

As we can see, $\mathrm{Var}_S \approx \mathrm{Var}_G$ for large values of n.

Rademacher

Theorem 5 (Lemma 1 in Avron and Toledo [2]). *Suppose* $\Sigma \in \mathbb{R}^{n \times n}$, *and* $Z \in \mathbb{R}^n$ *follows a Rademacher distribution. Let* $\Sigma' = (\Sigma + \Sigma^\top)/2$. *Then,*

$$\mathrm{Var}\left[Z^\top \Sigma Z\right] = 2\left[\|\Sigma'\|_F^2 - \sum_i \Sigma'^2_{i,i}\right]. \tag{19}$$

Following from this theorem we have the following.

Lemma 3 (Variance of the Rademacher Trace Estimator). *If* Z *is a* n-*dimensional Rademacher random vector, then* $\mathbb{E}[\hat{\psi}_{\mathrm{sqrt}}] = \mathbb{E}[\hat{\psi}] = \mathrm{tr}(\mathbf{AW})$,

$$\mathrm{Var}_R\left[\hat{\psi}\right] = \mathrm{Var}_G\left[\hat{\psi}\right] - 2\sum_i \left[\frac{\mathbf{AW} + \mathbf{WA}}{2}\right]^2_{i,i}, \quad \text{and} \tag{20}$$

$$\mathrm{Var}_R\left[\hat{\psi}_{\mathrm{sqrt}}\right] = \mathrm{Var}_G\left[\hat{\psi}_{\mathrm{sqrt}}\right] - 2\sum_i [\mathbf{B}^\top \mathbf{WB}]^2_{i,i}. \tag{21}$$

Analogous to our conclusions for estimating arbitrary matrix elements, the above results indicate that the Rademacher estimator is guaranteed to be better than Gaussian estimator for estimating the trace of $\mathbf{AW}$. Moreover, when Z is Gaussian, we have proven $\hat{\psi}_{\mathrm{sqrt}}$ always performs better than $\hat{\psi}$. However, this is not always true when Z is Rademacher.

Let $\mathbf{\Lambda} = (\mathbf{AW} + \mathbf{WA})/2$, and $\mathbf{\Omega} = \mathbf{B}^\top \mathbf{WB}$. Then $\mathrm{Var}_R\left[\hat{\psi}_{\mathrm{sqrt}}\right] > \mathrm{Var}_R\left[\hat{\psi}\right]$ if and only if,

$$2\sum_i \left[(\mathbf{\Lambda}_{i,i})^2 - (\mathbf{\Omega}_{i,i})^2\right] > \sum_{i,j}\left[(\mathbf{\Lambda}_{i,j})^2 - (\mathbf{\Omega}_{i,j})^2\right].$$

Numerical experiments indicate this inequality can hold when the contribution of $\|\mathbf{\Lambda}\|_F$ and $\|\mathbf{\Omega}\|_F$ is concentrated on the diagonal. In these cases both $\hat{\psi}_{\mathrm{sqrt}}$ and $\hat{\psi}$ are very accurate when Z is Rademacher. However, when most of the contribution of the Frobenius norm stems from the off-diagonals we observe cases where $\|\mathbf{\Lambda}\|_2^2 \gg \|\mathbf{\Omega}\|_2^2$ and $\mathrm{Var}_R\left[\hat{\psi}\right] \gg \mathrm{Var}_R\left[\hat{\psi}_{\mathrm{sqrt}}\right]$. We now show such an example in the context of maximum likelihood estimation for Gaussian processes.

3.2   Gaussian Processes

Gaussian processes (GP's) are used in a variety of modelling applications mainly through the following regression setting. Let $X = \{x_1, \ldots, x_n\}$ be a set of n data points and let $y = (y_1, \ldots, y_n)^\top$ be the vector of regression values under the model,

$$y_i = f(x_i) + \epsilon_i, \quad \text{where } \epsilon_i \sim \mathcal{N}(0, \sigma^2).$$

In GP regression, we assume the following prior on the function values $f = (f(x_1), \ldots, f(x_n))^\top$,

$$p(f|x_1, \ldots, x_n) = \mathcal{N}(\mathbf{0}, \mathbf{K}).$$

The covariance matrix $\mathbf{K}$ has entries $\mathbf{K}_{i,j} = \{k(\boldsymbol{x}_i, \boldsymbol{x}_j | \boldsymbol{\theta})\}$, where k is a covariance function with hyper-parameters $\boldsymbol{\theta}$. A popular choice for k is the Squared Exponential covariance function,

$$k(\boldsymbol{x}_i, \boldsymbol{x}_j) = \theta_1 \exp\left[\frac{-\|\boldsymbol{x}_j - \boldsymbol{x}_i\|^2}{2\theta_2^2}\right].$$

To fit a GP we must select the hyper-parameters $\boldsymbol{\theta} = (\theta_1, \theta_2)^\top$ by maximizing the likelihood. Therefore, we need to evaluate the log-likelihood function,

$$\ln[\mathcal{L}(\boldsymbol{\theta}|\boldsymbol{y})] = -\frac{1}{2}\ln|\mathbf{K}_y| - \frac{1}{2}\boldsymbol{y}^\top \mathbf{K}_y^{-1}\boldsymbol{y} + \text{constant}, \tag{22}$$

and the score function,

$$\frac{\partial \ln[\mathcal{L}(\boldsymbol{\theta}|\boldsymbol{y})]}{\partial \theta_i} = -\frac{1}{2}\text{Tr}\left(\mathbf{K}_y^{-1}\frac{\partial \mathbf{K}_y}{\partial \theta_i}\right) + \frac{1}{2}\boldsymbol{y}^\top \mathbf{K}_y^{-1}\frac{\partial \mathbf{K}}{\partial \theta_i}\mathbf{K}_y^{-1}\boldsymbol{y}, \tag{23}$$

where $\mathbf{K}_y = \mathbf{K} + \sigma^2 \mathbf{I}_n$. For large n explicitly evaluating $\text{tr}\left(\mathbf{K}_y^{-1}\frac{\partial \mathbf{K}_y}{\partial \theta_i}\right)$ is infeasible. Instead, we must approximate the trace via simulation. Suppose we simulate M draws of the n-dimensional Rademacher random variable, $\{\boldsymbol{Z}_i\}_{i=1,\dots,M}$. Then, the Hutchinson estimator for $\text{tr}\left(\mathbf{K}_y^{-1}\frac{\partial \mathbf{K}_y}{\partial \theta_2}\right)$ is,

$$\hat{\psi} = \frac{1}{M}\sum_{i=1}^{M}\left(\mathbf{K}_y^{-1}\boldsymbol{Z}_i\right)^\top \left(\frac{\partial \mathbf{K}_y}{\partial \theta_i}\boldsymbol{Z}_i\right).$$

Instead, we can use the square root estimator,

$$\hat{\psi}_{\text{sqrt}} = \frac{1}{M}\sum_{i=1}^{M}\boldsymbol{c}_i^\top \frac{\partial \mathbf{K}_y}{\partial \theta_i}\boldsymbol{c}_i, \quad \text{where } \boldsymbol{c}_i = \mathbf{K}_y^{-1/2}\boldsymbol{Z}_i.$$

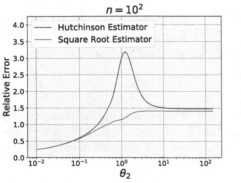
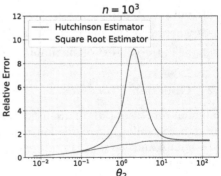

Fig. 1. True relative error as a function of θ_2 for $M = 1$ (a single replication), when $n = 10^2$ and $n = 10^3$. Relative Error $= \sqrt{\text{Var}(\hat{\psi})}/|\psi|$, where $\psi = \text{tr}\left(\mathbf{K}_y^{-1}\frac{\partial \mathbf{K}_y}{\partial \theta_2}\right)$.

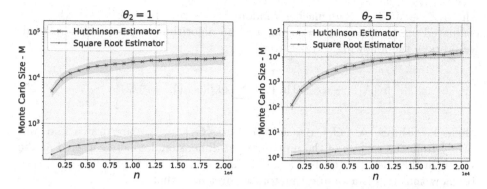

Fig. 2. Mean Monte Carlo size M required in simulation to obtain a mean squared error of 0.25 as n increases, when $\theta_2 = 1$ and $\theta_2 = 5$.

We can evaluate $\mathbf{K}_y^{-1/2} \mathbf{Z}_i$ with the Pleiss et al. [17] CIQ method. We observe that on average $\mathbf{K}_y^{-1/2} \mathbf{Z}_i$ takes double the CPU time to evaluate as compared to $\mathbf{K}_y^{-1} \mathbf{Z}_i$ via the conjugate gradient method; the latter being the required evaluation for the Hutchinson estimator. Importantly, both operations have $O(mn^2)$ complexity, where $m \ll n$.

To compare the performance of the estimators numerically, we fix $\sigma^2 = 0.1$ and $\theta_1 = 1$ and we let $\mathbf{X}$ be the n equidistant points spanning the interval $[0, 1]$. In Fig. 1 we evaluate the theoretical relative error of the two estimators when $M = 1$ and observe that the square root estimator always performs better than or equal to the Hutchinson estimator. For $\theta_2 \in [10^{-1}, 10^1]$ the relative error for Hutchinson estimator significantly deviates from the relative error of the square root estimator. In Fig. 2 we illustrate the differing performance as the size of the matrix increases in size. We observe that when $\theta_2 = 5$ the Hutchinson estimator can require up-to 1,000 times more Monte Carlo replications to obtain the same accuracy (mean squared error) than the square root estimator.

4   Estimating the Log-Determinant

In this section we propose a conditional Monte Carlo estimator for the log-determinant of a positive semi-definite matrix by treating the log-determinant as a normalizing constant of a probability density. As seen from (22), training GP's require the evaluation of the log-determinant of $\mathbf{K}_y$. Suppose $\mathbf{A} \in \mathbb{R}^{n \times n}$ is a PSD definite matrix. Then, $\ln |\mathbf{A}|$ can be evaluated in $O(n^3)$ time with the Cholesky decomposition. However, as previously stated this is infeasible for matrices with dimension greater than a few thousand. Existing log-determinant estimation techniques [9, 21], utilize the Hutchinson estimator via the following approximation,

$$\ln |\mathbf{A}| = \mathrm{tr}\left(\ln(\mathbf{A})\right) = \mathbb{E}[\mathbf{Z}^\top \ln(\mathbf{A})\mathbf{Z}] \approx \frac{1}{M} \sum_{i=1}^{M} \mathbf{Z}_i^\top \ln(\mathbf{A})\mathbf{Z}_i, \qquad (24)$$

where $Z_i \in \mathbb{R}^n$ is a Rademacher random vector and $\ln(\mathbf{A})Z_i$, like $\mathbf{A}^{1/2}Z_i$ is evaluated via the Lanczos tridiagonalization algorithm (see Sect. 2.2). Instead, we propose the following estimator,

$$\widehat{\ln|\mathbf{A}|} = -nc + 2\ln\left[\frac{1}{M}\sum_{i=1}^{M}\exp(-n(Y_i - c)/2)\right], \tag{25}$$

where $Y_i = \ln(\boldsymbol{\Theta}_i^\top \mathbf{A}^{-1}\boldsymbol{\Theta}_i)$, $c = \min_i Y_i$ and $\{\boldsymbol{\Theta}_i\}_{i=1,\dots,M}$ are drawn i.i.d uniformly from the surface of the unit radius n-dimensional sphere centered at the origin. In this case, $\mathbf{A}^{-1}\boldsymbol{\Theta}_i$ is evaluated via the conjugate gradient method [11]. To show that this is a valid estimator, we first note that,

$$\ln|\mathbf{A}| = 2\ln\int(2\pi)^{-\frac{n}{2}}\exp\left(-z^\top\mathbf{A}^{-1}z/2\right)dz.$$

This equation can be derived by noting the integral of the multivariate normal density is equal to one. Thus, our goal is to produce an unbiased estimate of the integral on the RHS.

To estimate $\int\exp(-z^\top\mathbf{A}^{-1}z/2)dz$ we switch to spherical coordinates. Let $Z \sim \mathcal{N}(\mathbf{0}, \mathbf{I}_n)$, then, Z can be represented as $Z = R\boldsymbol{\Theta}$ with $\boldsymbol{\Theta}$ being uniformly distributed on the surface of the unit radius n-dimensional sphere centered at the origin and $R \sim \chi_n$, independently. Then, we have the unbiased estimator,

$$\frac{1}{(2\pi)^{n/2}}\int\exp(-z^\top\mathbf{A}^{-1}z/2)dz = \mathbb{E}[(\boldsymbol{\Theta}^\top\mathbf{A}^{-1}\boldsymbol{\Theta})^{-n/2}] \approx \frac{1}{M}\sum_{i=1}^{M}(\boldsymbol{\Theta}_i^\top\mathbf{A}^{-1}\boldsymbol{\Theta}_i)^{-n/2}.$$

The details of this derivation are provided in the Appendix. We therefore obtain the conditional Monte Carlo estimate,

$$\widehat{\ln|\mathbf{A}|} = 2\ln\left[\frac{1}{M}\sum_{i=1}^{M}(\boldsymbol{\Theta}_i^\top\mathbf{A}^{-1}\boldsymbol{\Theta}_i)^{-n/2}\right].$$

However, for large n, accurate evaluations of $(\boldsymbol{\Theta}_i^\top\mathbf{A}^{-1}\boldsymbol{\Theta}_i)^{-n/2}$ are unlikely due to round-off errors. Instead, we let,

$$Y_i = \ln(\boldsymbol{\Theta}_i^\top\mathbf{A}^{-1}\boldsymbol{\Theta}_i).$$

We then evaluate the numerically stable approximation,

$$\widehat{\ln|\mathbf{A}|} = 2\ln\frac{1}{M}\sum_{i=1}^{M}\exp(-nY_i/2)$$

$$= -nc + 2\ln\left[\frac{1}{M}\sum_{i=1}^{M}\exp(-n(Y_i - c)/2)\right],$$

where $c = \min_i Y_i$. By evaluating the terms $\exp(-n(Y_i - c)/2)$ we can alleviate further round-off errors.

We note that as $\ln|\mathbf{A}| = -\ln|\mathbf{A}^{-1}|$, one can go about estimating $\ln|\mathbf{A}^{-1}|$ and therefore evaluate the terms $Y_i = \ln(\boldsymbol{\Theta}_i^\top \mathbf{A} \boldsymbol{\Theta}_i)$, which only require matrix vector multiplications with $\mathbf{A}$. As a numerical example, consider Fig. 3, where we estimate the log-determinant of the matrix $\mathbf{K}$, which is a squared exponential Gram matrix with $\theta_1 = 1$ and $\theta_2 = 1$.

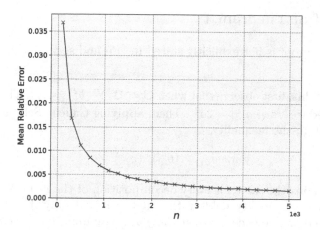

Fig. 3. Mean relative error of $\widehat{\ln|\mathbf{K}|}$, as n increases, where $\mathbf{K}$ is the squared exponential Gram matrix with $\theta_1 = 1$, $\theta_2 = 1$. Monte Carlo size: $M = 10$.

5   Conclusion

In this paper, we discuss how factorizing the PSD matrix $\mathbf{A}$ can reduce variance when trying to estimate its entries using only matrix vector operations of the form $\mathbf{A}\mathbf{Z}$ where $\mathbf{Z}$ is a random vector generated from a pre-defined distribution. We show in some examples that the unfactorized estimator can exhibit an arbitrary larger variance when compared to the factorized square root estimator. However, we also observe examples where the unfactorized estimator performs better than the factorized estimator, but the underperformance is relatively small and remains bounded from above as the size of the matrix grows. Whether this behavior of the underperformance holds true in general for all matrices $\mathbf{A}$ is an open question.

Furthermore, we show how factorization can achieve variance reduction when estimating $\mathrm{tr}(\mathbf{A}\mathbf{W})$, where $\mathbf{A} \in \mathbb{R}^{n \times n}$ is a PSD matrix and $\mathbf{W} \in \mathbb{R}^{n \times n}$ is a symmetric matrix. Similar to arbitrary matrix entry estimation, we observe numerical examples where the factorized estimator significantly outperforms the unfactorized estimator. However, when $\mathbf{Z}$ is Rademacher, this outperformance is not theoretically guaranteed for all choices of $\mathbf{A}$ and $\mathbf{W}$. Ideally, we'd like a theoretical framework for identifying when the factorized estimator will work most in our favor.

Finally, we provide a conditional Monte Carlo estimate for the log-determinant of a PSD matrix $\mathbf{A}$ by treating it as the normalizing constant of a probability density. We note that it is possible to implement Markov chain Monte Carlo (MCMC) techniques such as particle filters to provide better estimates for the quantity $\mathbb{E}\left[(\boldsymbol{\Theta}^\top \mathbf{A}^{-1} \boldsymbol{\Theta})^{-n/2}\right]$.

A   Proof of Theorem 1

The following proof is not explicitly stated in [15] and so we include it here for completeness.

Proof. To see this, first observe that when $\mathbf{C} = \mathbf{B}$, $\sum_j b_{i,j}^2 = a_{i,i}$. Therefore, from (3) we can deduce $\mathrm{Var}_G[a_{i,i}] = 2a_{i,i}^2$. Then, applying Cauchy-Schwarz inequality to (3), we obtain the inequality,

$$\mathrm{Var}_G[a_{i,i}] \geq (\boldsymbol{b}_{i,:})^\top (\boldsymbol{c}_{i,:}) + a_{i,i}^2.$$

Since $a_{i,i}^2$ remains the same for every decomposition of the form $\mathbf{A} = \mathbf{B}\mathbf{C}^\top$ and Cauchy-Schwarz inequality holds with equality if and only if $\boldsymbol{b}_{i,:}$ and $\boldsymbol{c}_{i,:}$ are linearly dependent, we conclude that $\mathrm{Var}_G[a_{i,i}]$ is minimized when $\mathbf{C} = \mathbf{B}$.

B   Uniform Spherical Estimator

B.1   Proof of Theorem 2

Proof. Suppose $\boldsymbol{Z} = (Z_1, \ldots, Z_n)^\top$ is uniformly distributed on the surface of the unit radius n-dimensional sphere centered at the origin. Using spherical coordinates, it is not difficult to show that the vector $\boldsymbol{X} = \boldsymbol{Z} \odot \boldsymbol{Z}$ follows a Dirichlet distribution over the simplex:

$$f(\boldsymbol{x}) = \frac{\pi^{n/2}}{\Gamma(n/2)} \prod_{i=1}^n x_i^{-1/2}, \quad x_i \in [0,1], \quad \sum_{i=1}^n x_i = 1.$$

Therefore, using the well-known mean and variance formula for the Dirichlet distribution we have,

$$\mathbb{E}[\boldsymbol{X}] = 1/n, \quad \mathrm{Var}[\boldsymbol{X}] = \frac{1}{n(n/2+1)} \left[\mathbf{I}_n - \frac{1}{n}\mathbf{1}\mathbf{1}^\top\right]. \tag{26}$$

Using spherical coordinates, we also know that $\mathbb{E}[Z_i Z_j] = 0$ for $i \neq j$. Therefore,

$$\mathbb{E}[\boldsymbol{Z}\boldsymbol{Z}^\top] = \frac{1}{n}\mathbf{I}_n,$$

and,

$$\mathbb{E}\left[n\,(\mathbf{B}\boldsymbol{Z})\,(\mathbf{C}\boldsymbol{Z})^\top\right] = n\mathbb{E}\left[\mathbf{B}\boldsymbol{Z}\boldsymbol{Z}^\top\mathbf{C}^\top\right] = \mathbf{A}.$$

To prove the variance, we first note that,

$$\mathbb{E}[(\hat{a}_{i,j}/n)^2] = \mathbb{E}[([\mathbf{B}\mathbf{Z}]_i \times [\mathbf{C}\mathbf{Z}]_j)^2] = \sum_{k,l,m,n} b_{i,k} c_{j,l} b_{i,m} c_{j,n} \mathbb{E}[Z_k Z_l Z_m Z_n].$$

Representing $\mathbf{Z}$ in spherical coordinates, we obtain the formula,

$$\mathbb{E}[Z_k Z_l Z_m Z_n] = c_1 [\delta_{kl}\delta_{mn} + \delta_{km}\delta_{ln} + \delta_{kn}\delta_{lm}] + (c_2 - 3c_1)[\delta_{kl}\delta_{lm}\delta_{mn}].$$

The constants c_1 and c_2 are given by,

$$c_1 = \mathrm{Var}[\mathbf{X}]_{p,q} + \mathbb{E}[\mathbf{X}]_p^2 = \frac{1}{n(n+2)}, \qquad p \neq q, \qquad \text{and}$$

$$c_2 = \mathrm{Var}[\mathbf{X}]_{p,p} + \mathbb{E}[\mathbf{X}]_p^2 = \frac{3}{n(n+2)}.$$

Thus $c_2 - 3c_1 = 0$, and,

$$\mathbb{E}[(a_{i,j}/n)^2] = c_1 \sum_{k,l,m,n} b_{ik} c_{jl} b_{im} c_{jn} [\delta_{kl}\delta_{mn} + \delta_{km}\delta_{ln} + \delta_{kn}\delta_{lm}]$$

$$= c_1 \left[(\mathbf{b}_{i,:}^\top \mathbf{c}_{j,:})^2 + (\mathbf{b}_{i,:}^\top \mathbf{b}_{i,:})(\mathbf{c}_{j,:}^\top \mathbf{c}_{j,:}) + (\mathbf{b}_{i,:}^\top \mathbf{c}_{j,:})^2 \right]$$

$$= c_1 \left[2a_{i,j}^2 + \|\mathbf{b}_{i,:}\|^2 \|\mathbf{c}_{i,:}\|^2 \right].$$

Therefore,

$$\mathrm{Var}_S[(\hat{a}_{i,j}/n)] = \mathbb{E}\left[(\hat{a}_{i,j}/n)^2\right] - a_{i,j}^2/n^2$$

$$= (2c_1 - 1/n^2)a_{i,j}^2 + c_1 \|\mathbf{b}_{i,:}\|^2 \|\mathbf{c}_{j,:}\|^2.$$

Hence, the variance for $\hat{a}_{i,j}$ is,

$$\mathrm{Var}_S[(\hat{a}_{i,j})] = n^2 \mathrm{Var}_S[(\hat{a}_{i,j}/n)]$$

$$= \frac{n-2}{n+2} a_{i,j}^2 + \frac{n}{n+2} \|\mathbf{b}_{i,:}\|^2 \|\mathbf{c}_{j,:}\|^2.$$

B.2   Proof of Theorem 4

Proof. Suppose $\mathbf{\Sigma}'$ has eigenvalues $\{\lambda_i\}$, and orthonormal eigendecomposition $\mathbf{Q}\mathbf{\Lambda}\mathbf{Q}^\top$. Then, as the random variable that is distributed uniformly on the surface of the n-dimensional sphere centered at the origin is invariant to orthogonal rotations, we obtain,

$$\mathrm{Var}[\mathbf{Z}^\top \mathbf{\Sigma}' \mathbf{Z}] = \mathrm{Var}[\mathbf{Z}^\top \mathbf{\Lambda} \mathbf{Z}] = \mathrm{Var}[\boldsymbol{\lambda}^\top \mathbf{X}] = \boldsymbol{\lambda}^\top \mathrm{Var}[\mathbf{X}]\boldsymbol{\lambda}.$$

We notice that $\mathrm{tr}(\mathbf{\Sigma}^\top \mathbf{\Sigma}) = \boldsymbol{\lambda}^\top \boldsymbol{\lambda}$. Therefore,

$$\mathrm{Var}[n\mathbf{Z}^\top \mathbf{\Sigma}' \mathbf{Z}] = \frac{n}{n+2} \times 2 \left(\mathrm{tr}(\mathbf{\Sigma}'^\top \mathbf{\Sigma}') - \frac{(\sum_i \mathbf{\Sigma}'_{i,i})^2}{n} \right).$$

To see that this gives the variance for $n\boldsymbol{Z}^\top \boldsymbol{\Sigma} \boldsymbol{Z}$, we note,

$$\boldsymbol{Z}^\top \boldsymbol{\Sigma}' \boldsymbol{Z} = \boldsymbol{Z}^\top \left[(\boldsymbol{\Sigma} + \boldsymbol{\Sigma}^\top)/2 \right] \boldsymbol{Z} = \boldsymbol{Z}^\top \boldsymbol{\Sigma} \boldsymbol{Z}.$$

We note that when $\boldsymbol{Z}$ is distributed uniformly on the unit radius complex sphere instead, the variance formula is given in Tropp [20].

C   Log-Determinant

Suppose $\boldsymbol{\Sigma} \in \mathbb{R}^{n \times n}$ is a PSD matrix. Let $\boldsymbol{Z} \sim \mathcal{N}(\mathbf{0}, \mathbf{I}_n)$. Then, $\boldsymbol{Z}$ can be represented as $\boldsymbol{Z} == R\boldsymbol{\Theta}$ with $\boldsymbol{\Theta}$ being uniformly distributed on the surface of the unit radius n-dimensional sphere centered at the origin and $R \sim \chi_n$, independently. Then using the standard normal as a change of measure we get the following,

$$\frac{1}{(2\pi)^{n/2}} \int \exp(-\boldsymbol{z}^\top \boldsymbol{\Sigma}^{-1} \boldsymbol{z}/2) \mathrm{d}\boldsymbol{Z} = \mathbb{E}\left[\exp(-\boldsymbol{Z}^\top \boldsymbol{\Sigma}^{-1} \boldsymbol{Z}/2 + \|\boldsymbol{Z}\|^2/2) \right]$$

$$= \mathbb{E}\left[\exp(-R^2 \boldsymbol{\Theta}^\top \boldsymbol{\Sigma}^{-1} \boldsymbol{\Theta}/2 + R^2/2) \right]$$

$$= \frac{1}{2^{n/2} \Gamma(n/2)} \mathbb{E}\left[\int_0^\infty r^{n/2-1} \exp(-r\boldsymbol{\Theta}^\top \boldsymbol{\Sigma}^{-1} \boldsymbol{\Theta}/2) \mathrm{d}r \right]$$

$$= \mathbb{E}\left[(\boldsymbol{\Theta}^\top \boldsymbol{\Sigma}^{-1} \boldsymbol{\Theta})^{-n/2} \right]$$

$$\approx \frac{1}{M} \sum_{i=1}^M (\boldsymbol{\Theta}_i^\top \boldsymbol{\Sigma}^{-1} \boldsymbol{\Theta}_i)^{-n/2}.$$

We use the following integral formula to evaluate the integral on line 3,

$$\int_0^\infty r^{n/2-1} \exp(-r\alpha/2) \mathrm{d}r = \alpha^{-n/2} 2^{n/2} \Gamma(n/2).$$

Thus a conditional Monte Carlo estimate for the log-determinant is,

$$\widehat{\ln |\boldsymbol{\Sigma}|} = 2 \ln \left[\frac{1}{M} \sum_{i=1}^M (\boldsymbol{\Theta}_i^\top \boldsymbol{\Sigma}^{-1} \boldsymbol{\Theta}_i)^{-n/2} \right].$$

References

1. Adams, R.P., et al.: Estimating the spectral density of large implicit matrices. arXiv preprint arXiv:1802.03451 (2018)
2. Avron, H., Toledo, S.: Randomized algorithms for estimating the trace of an implicit symmetric positive semi-definite matrix. J. ACM (JACM) 58(2), 1–34 (2011)
3. Bekas, C., Kokiopoulou, E., Saad, Y.: An estimator for the diagonal of a matrix. Appl. Numer. Math. 57(11–12), 1214–1229 (2007)
4. Casella, G., Berger, R.L.: Statistical inference. Cengage Learning (2021)
5. Chow, E., Saad, Y.: Preconditioned krylov subspace methods for sampling multivariate gaussian distributions. SIAM J. Sci. Comput. 36(2), A588–A608 (2014)

6. Dauphin, Y.N., De Vries, H., Bengio, Y.: Equilibrated adaptive learning rates for non-convex optimization. arXiv preprint arXiv:1502.04390 (2015)
7. Drineas, P., Magdon-Ismail, M., Mahoney, M., Woodruff, D.P.: Fast approximation of matrix coherence and statistical leverage. J. Mach. Learn. Res. **13**(1), 3475–3506 (2012)
8. Fitzsimons, J.K., Osborne, M.A., Roberts, S.J., Fitzsimons, J.F.: Improved stochastic trace estimation using mutually unbiased bases. arXiv preprint arXiv:1608.00117 (2016)
9. Gardner, J.R., Pleiss, G., Bindel, D., Weinberger, K.Q., Wilson, A.G.: Gpytorch: blackbox matrix-matrix gaussian process inference with gpu acceleration. arXiv preprint arXiv:1809.11165 (2018)
10. Geoga, C.J., Anitescu, M., Stein, M.L.: Scalable gaussian process computations using hierarchical matrices. J. Comput. Graph. Stat. **29**(2), 227–237 (2020)
11. Golub, G.H., Van Loan, C.F.: Matrix Computations, vol. 3. JHU Press (2013)
12. Hutchinson, M.F.: A stochastic estimator of the trace of the influence matrix for laplacian smoothing splines. Commun. Stat. Simul. Comput. **18**(3), 1059–1076 (1989)
13. Kaperick, B.J.: Diagonal Estimation with Probing Methods. Ph.D. thesis, Virginia Tech (2019)
14. Lin, L., Saad, Y., Yang, C.: Approximating spectral densities of large matrices. SIAM Rev. **58**(1), 34–65 (2016)
15. Martens, J., Sutskever, I., Swersky, K.: Estimating the hessian by back-propagating curvature. arXiv preprint arXiv:1206.6464 (2012)
16. Meyer, R.A., Musco, C., Musco, C., Woodruff, D.P.: Hutch++: optimal stochastic trace estimation. In: Symposium on Simplicity in Algorithms (SOSA), pp. 142–155. SIAM (2021)
17. Pleiss, G., Jankowiak, M., Eriksson, D., Damle, A., Gardner, J.R.: Fast matrix square roots with applications to gaussian processes and bayesian optimization. arXiv preprint arXiv:2006.11267 (2020)
18. Stathopoulos, A., Laeuchli, J., Orginos, K.: Hierarchical probing for estimating the trace of the matrix inverse on toroidal lattices. SIAM J. Sci. Comput. **35**(5), S299–S322 (2013)
19. Stein, M.L., Chen, J., Anitescu, M.: Stochastic approximation of score functions for gaussian processes. In: The Annals of Applied Statistics, pp. 1162–1191 (2013)
20. Tropp, J.A.: Randomized algorithms for matrix computations (2020)
21. Ubaru, S., Chen, J., Saad, Y.: Fast estimation of tr(f(a)) via stochastic lanczos quadrature. SIAM J. Matrix Anal. Appl. **38**(4), 1075–1099 (2017)

Author Index

Printed in the United States
by Baker & Taylor Publisher Services